Writing Windows WDM Device Drivers

Covers NT 4, Win 98, and Win 2000

Chris Cant

R&D Books
Lawrence, Kansas 66046

R&D Books
Miller Freeman, Inc.
1601 W. 23rd Street, Suite 200
Lawrence, KS 66046
USA

Cover art created by Robert Ward.

Distributed in the U.S. and Canada by:
Publishers Group West
P.O. Box 8843
Emeryville, CA 94662
ISBN: 0-87930-565-7

Miller Freeman
A United News & Media publication

Table of Contents

Chapter 6   Testing and Debugging **115**

Chapter 7   Dispatch Routines................. **131**

Preface

First, I suggest that you check the book's web site at www.phdcc.com/wdmbook/. These pages will list any errata, etc., and include a feedback page. This book was written mainly using Windows 98 and the Beta 2 version of Windows 2000, with some updates for the Beta 3 version. I will try to ensure that I post any necessary updates on the book's web site.

Although this book and the Microsoft documentation can give you a good start when writing device drivers, it is only when you write one yourself that you really learn about device drivers. Often it is only when you make mistakes that you find out how to write code correctly. I have tried to include some anecdotes of when I have erred and the techniques I have used to track down the problem and sort it out. Trying to explain the example code in detail in this book has helped iron out some wrinkles, so I recommend that you describe your code line by line to someone else.

Who Are You?

Read this book if you want to learn how to write device drivers for Windows 98 and Windows 2000. It covers the core Windows Driver Model (WDM) fully and looks at two types of system class driver in detail, for the Universal Serial Bus (USB) and Human Input Devices (HID).

The book also covers Window NT 4 and NT 3.51 drivers. These "NT style" drivers will also run in Windows 2000 and sometimes in Windows 98.

I suggest that you run all the examples and study their source code to get the full benefit. The book software includes some software called DebugPrint that lets you view the trace debug statements produced by the example drivers.

If you use a proprietary device driver toolkit, you should still find this book useful. First, the core WDM chapters give a background to the whole subject and might help explain how your toolkit works. Some features of WDM may not be covered by your toolkit and so will need to be implemented "by hand". More importantly, the class driver chapters give you the information you need to write drivers for various device categories.

Firmware or hardware engineers will also find it useful to read some sections of this book. When you design your external device, a software engineer will need to write a device driver to talk to it. Reading the first chapters of this book should let you speak "their language". Help them by providing clear specifications of your hardware at the appropriate level.

If you are currently a VxD driver writer, then many of the general concepts presented in this book will be familiar. However, you will find that the implementation techniques are quite different.

If you have written drivers for NT 4 and NT 3.51, some of the core WDM chapters will be familiar territory. However, Plug and Play and Power Management are crucial new areas of functionality. Supporting Windows Management Instrumentation (WMI) is desirable. The system class drivers will be new to you as well.

NT drivers should run unchanged in Windows 2000, and can sometimes run in Windows 98. However, you should seriously consider migrating to the Plug and Play if it's appropriate. Supporting Power Management will help to reduce power on and shutdown delays, making PCs more pleasant to use. The Direct Memory Access (DMA) system is used in a slightly different way by WDM device drivers.

Terminology

Whenever a new and important piece of terminology is introduced, it is highlighted in *italic text*. Kernel function names (e.g., `IoCompleteRequest`) and driver function names (e.g., **Wdm1Create**) are in **bold monospaced text**. Code examples are in `monospaced text`.

Acronyms are spelled out when they are first used. Please refer to the Glossary for a full list.

Globally Unique Identifiers (GUIDs) are long obscure strings used to identify Component Object Model (COM) objects and device interfaces. A full GUID such as `{C0CF0640-5F6E-11d2-B677-00C0DFE4C1F3}` is often abbreviated as `{C0CF0640...}`.

When I refer to "Windows", you can take it that I mean both Windows 98 and Windows 2000. "Windows" does not refer to NT 3.51 and NT 4 unless the text specifically says so. In case it is not obvious, W98 refers to Windows 98 and W2000 refers to Windows 2000. Windows 2000 was originally called NT 5.

Coding Style

You will have to endure the coding style that I have used for the book and its example source code. You will find that actual driver code on disk often contains more comments than the main text listings.

Most of the code examples are written using C++ although I do not make great use of C++ features. For example, there are no C++ classes used.

Boolean true and false values are represented by two different types. Kernel calls must use the `BOOLEAN` type, which is really an `unsigned char`, where zero means `FALSE` and non-zero means `TRUE`. For any other Boolean values, I use the intrinsic C++ `bool` type that has `false` and `true` values.

The main text often refers to Win32 applications. I have not yet come to grips with the implications of Win64, the 64-bit version of Windows 2000. However, you can assume that whatever calls a Win32 application can make, a Win64 application will also be able to make.

Of more importance, I have not yet found out what happens to device drivers in Win64 systems.

Thanks

Several people have helped directly and indirectly in making this book. First, thanks to Alec Erskine for drawing the cartoons and Ian Cuthbert for scanning them. Caz, Jenny, and Viv have stoically endured months of me "'putering too much". Berney Williams at R&D provided the initial impetus for the project and gave crucial support as the book was written. My brother John and my friend Robin Sillem have helped considerably with ideas and proofreading. Thanks to Vireo Software, Inc., for the review copies of Driver::Works and Driver::Agent, and for a useful web site.

Chapter 1

Introduction

In this book, I will tell you how to write some types of device driver for Windows. I will primarily describe the Windows Driver Model for Windows 98 and Windows 2000. Additionally, I will cover device drivers that also run in Windows NT 3.51 and NT 4, which I call "NT style" drivers.

A device driver provides a software interface to hardware connected to a computer. It is a trusted part of the operating system. User application programs can access hardware in a well-defined manner, without having to worry about how the hardware must be controlled. For example, a disk driver might hide the fact that data must be written in 512-byte chunks.

A *driver* is a piece of software that becomes part of the operating system kernel when it is loaded. A driver makes one or more *devices* available to the user mode programmer, each representing a physical or logical piece of hardware. For example, one physical hard disk may be viewed as two logical disks called C: and D:.

In Windows, a driver always makes a device look like a file. A handle to the device can be opened. An application program can then issue read and write requests to the driver, before the device handle is finally closed.

Clearly, there are many pieces of hardware that are essentially alike, because they share a bus or do similar tasks. Microsoft provides several generic drivers that perform these common tasks. Device drivers can use the facilities of these standard drivers. This approach makes it easier to share a common bus, and makes it simpler to write new drivers.

The task of writing a new driver, therefore, often starts by identifying which generic drivers can be used. A *stack of drivers*, layered one on top of each other, processes user requests in stages. A low-level *bus* driver might be used to handle all the basic communication with hardware. An intermediate *class* driver might provide the facilities that are common to a whole category of devices.

In Windows 98 and Windows 2000, device drivers must be designed according to the Windows Driver Model (WDM), which I describe in the following section. WDM is based on the device driver model used in Windows NT 4 and NT 3.51.

The Windows Driver Model

The Windows Driver Model has two separate but equally important aspects. First, the core model describes the standard structure for device drivers. Second, Microsoft provides a series of bus and class drivers for common types of devices.

The core WDM model describes how device drivers are installed and started, and how they should service user requests and interact with hardware. A WDM device driver must fit into the Plug and Play (PnP) system that lets users plug in devices that can be configured in software.

Microsoft provides a series of system drivers that have all the basic functionality needed to service many standard types of device. The first type of system driver supports different types of bus, such as the Universal Serial Bus (USB), IEEE 1394 (Firewire) and Audio port devices. Other class drivers implement standard Windows facilities such as Human Input Devices (HID) and kernel streaming. Finally, the Still Image Architecture (STI) provides a framework for handling still images, scanners, etc.

These system class drivers can make it significantly easier to write some types of device driver. For example, the USB system drivers handle all the low-level communications across this bus. A well defined interface is made available to other drivers. This makes it fairly straightforward to issue requests to the USB bus.

Source and Binary Compatibility

Originally Microsoft stated that WDM drivers would be binary compatible between Windows 98 and Windows 2000 x86, and source code compatible to Windows 2000 Alpha platforms. However, it now seems as though binary compatibility is not assured, even though the DDKs are unclear on the subject.

I have erred on the safe side, only installing drivers that have been built for the right operating system. That is, the Windows 98 Driver Development Kit (DDK) is used when building drivers for Windows 98, and the W2000 DDK for W2000.

If you use some WDM facilities that only appear in Windows 2000, then you may not achieve source code compatibility. For example, the W2000 USB system drivers support some features that are not available to W98 drivers.

I will look first at the core WDM functionality as a simple device driver is developed. Next, I will deal with drivers that have to use hardware resources such as accessing memory and handling interrupts. Finally, I will cover the USB and HID system drivers. Use the kernel routine IoIsWdmVersionAvailable to determine whether the required version WDM version is available. The DDK header files define two constants, WDM_MAJORVERSION and WDM_MINOR-VERSION. For Windows 98, these constants are 1 and 0. For Windows 2000, these are 1 and 0x10.

WDM vs. NT Style Drivers

Figure 1.1 gives a rough indication of the differences between WDM and NT style drivers. The rest of this book explains all the features mentioned in the figure.

The overlap between the two types of driver is considerable. Indeed, writing WDM and NT style drivers is essentially the same job. The main difference in the driver code is how devices are created.

In a WDM driver, the Plug and Play Manager tells you when a device is added or removed from the system. The PnP Manager uses installation INF files to find the correct driver for a new device. In contrast, an NT style driver has to find its own devices, usually in its initialisation routine. NT style drivers are usually installed using a special installation program.

The new bus and class drivers are only available to WDM device drivers. New WDM and NT style drivers should support the Power Management and the Windows Management Instrumentation features.

Figure 1.1   WDM and NT style device drivers

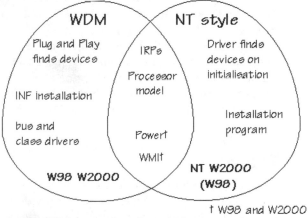

@ 1999 PHD Computer Consultants Ltd

Ready-to-Use Drivers

If you start writing a driver from scratch, it seems as though most of your code has nothing to do with talking to your device. There is much "infrastructure" to set up before you can perform some real Input and Output (I/O). This book should help you get started as I have tried to use some real, useful drivers as examples. Some drivers can be used directly, while others can form the basis of your own drivers.

A virtual device driver is used to explain the core WDM functionality. A virtual device does not use any real hardware. Three drivers, called Wdm1, Wdm2, and Wdm3, gradually implement more features. From the start, they provide a shared memory buffer so they could form the basis for other useful drivers. Indeed, the Wdm2 driver was used as the base for several other drivers in the book, including drivers that use the system drivers.

The DebugPrint software is used throughout the book to provide trace debugging output from drivers. The DebugPrint driver is examined in full in Chapter 14. You can use Debug-Print in your own device drivers.

WdmIo and PHDIo are general purpose drivers that can be used straightaway to provide access to simple hardware devices. A controlling Win32 program can use a simple but powerful command set to talk to the hardware. These drivers support interrupt-driven I/O. The example applications show how these generic drivers are used to talk to the parallel port.

The UsbKbd and HidKbd drivers both talk to a keyboard attached to the USB bus. The drivers illustrate the techniques required to use the USB and HID class drivers. Finally, the Win32 application HidKbdUser shows how a user mode application can find and talk to a HID device.

Book CD-ROM

The accompanying CD contains all the drivers mentioned previously. The full source of the drivers is included, along with the compiled executables. Each driver has at least one test Win32 program that puts it through its paces. In addition, the book has some other useful Win32 utility programs that aid device driver development. Chapter 4 has full instructions for installing the book software. Most of the test Win32 programs are console applications to make them easy to write and understand; there is no reason why the code should not be in fully fledged windowing or MFC applications.

To obtain the full benefit of the book, you should install all the example drivers and run the Win32 test program. All the drivers include trace debugging statements. If the DebugPrint software is installed, you can see the trace output in the DebugPrint Monitor application.

A full inspection of the source code of each example will bear much fruit. In practice, one learns only by writing real code. Use these example drivers as a starting point, and add features to develop your understanding.

Device Driver Software Tools

There are two types of software available to help you with your device driver requirement as shown in Table 1.1. *Driver classes* are source code C++ class wrappers that provide much of the default functionality needed by drivers. *Generic drivers* can be tailored from user mode to talk to many straightforward types of device. OSR have a debugging aid called OSRDDK.

Chapter 4 discusses other tools that will be useful to you while developing device drivers.

Table 1.1   Device driver software tools

Company	Web site	Driver classes	Generic driver
Blue Water Systems	www.bluewatersystems.com	WinDK	WinRT
KRF Tech	www.krftech.com		WinDriver
Compuware Numega	www.vireo.com	DriverWorks	DriverAgent
Open Systems Research	www.osr.com		

In some cases, a new device driver may not be needed for new hardware. There are off-the-shelf generic products that handle simple I/O, be it memory mapped or in I/O space. Indeed, these may be a better option as NT and Windows 95 may be supported using the same interface, making your code portable. Even a simple NT device driver or Windows 95 VxD is no easy task. These generic drivers may be script-driven. Interrupt latency may be a problem if scripts have to be run when a user mode thread is scheduled. The WdmIo and PHDIo examples in the book are simple generic device drivers.

You may think that you need to write a new device driver to handle a special device attached to a parallel port. Check that you cannot achieve the desired effect using the full Win32 API interface. Perhaps issue asynchronous read or write calls and check for time-outs.

Driver Types Not Covered

This book does not cover writing VxDs for Windows 95 or Windows 98. None of the drivers in this book will run in Windows 95. Windows CE software is not described. This book does not cover printer drivers or virtual DOS drivers.

The device driver model described in this book is the basis of many types of drivers. The details of the following driver types are not covered: file system drivers, video drivers, network drivers, and SCSI drivers.

As mentioned previously, only the USB and HID system drivers are described in detail.

The drivers in this book should run in non-x86 systems. However, they have not been tested there.

A New Frame of Mind

If you have "only" written Win32 application programs, you will find that writing device drivers requires a new frame of mind. There are no windows and messages to manipulate. You do not have the protective arms of Windows to stop you from trampling over other processes and the operating system. Source level debugging is more difficult to set up. Most support libraries are not available — not even the C standard library and the C++ new operator. You even have to use makefiles to build drivers, though it is easy to control the build process from a development tool such as Visual Studio.

As a device driver is a trusted part of the operating system, you can crash the system easily. (I can assure you that you will crash the system during your driver development.) Therefore, it is your responsibility to write safe and dependable code. Comment and test your code well. Check for error return values from every kernel call you make.

Device driver problems are a constant source of difficulty for users and support staff. Please insist that your driver is fully tested before release on an unsuspecting world. Test the driver on a variety of machines.

Device Driver Environment

A device driver works in a demanding environment. More than one user application may be bombarding a driver with requests. A user program with open I/O requests may terminate suddenly. A driver may be running in a multiprocessor PC, with different pieces of the driver running on different processors. In fact, two read requests can be processed simultaneously by the same piece of driver code on two separate processors.

Low-level device drivers have to cope with hardware interrupts that may arrive at any moment. Only one part of a driver should access hardware electronics at a time. You may have to use Direct Memory Access (DMA) to transfer data from your device into memory, or vice versa.

Supporting configurable and hot-pluggable buses also adds to the burden of device driver writers. For example, a Plug and Play device might be removed by the user at any time. Also, the kernel can decide at any time that it needs to stop your device so that it can reassign all the hardware resources.

However, as mentioned previously, using the standard WDM bus and class drivers helps to reduce the amount of effort required to write drivers for certain categories of device.

Terminology and Resources

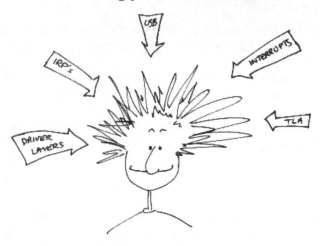

Brain Direct Memory Access

Device driver writers come face to face with a huge range of terminology. Good specification documents will help hardware and software engineers work together to achieve a common goal.

You will need to understand how your device works. While you may not need to understand the details of its electronic implementation, it certainly helps if you have a working knowledge of its technology. For example, the Universal Serial Bus places certain restrictions on maximum packet sizes. You may need to split up data transfers to meet these requirements.

The Windows Driver Model itself uses many technical terms to describe its operation. I will gradually introduce you to all the structures and concepts needed for writing device drivers.

Each technology has its own specialized terminology. For example, the USB bus refers to *interrupt transfers*. These bear no relation to hardware interrupts. However, as this is how the USB specification describes its operation, I will stick to using the correct terms. Whenever a new and important piece of terminology is introduced, it is highlighted in *italic*.

A possible source of confusion is the word "class". Windows class drivers are standard drivers that you can use to access particular types of devices, such USB, HID, or IEEE 1394. In contrast, USB devices are categorized into various device classes. There is one USB device class for printers, another for audio appliances, etc. (Thank goodness I do not use C++ classes in the source code.)

To get you started in the device driver world, Chapter 24 lists many information resources, books, newsgroups, and mailing lists that may be of help. Chapter 24 gives a summary of the PC 99 specification — the hardware and software that should be provided on new Windows computers. Finally, the Glossary explains some of the many acronyms you come across when writing drivers.

Win32 Program Interface

Before I go any further, it is worth looking at how Win32 programs call a device driver. The Win32 specification has various implications for driver writers.

Basic I/O

To Win32 programmers, a device is accessed as if it were a file, using the functions listed in Table 1.2. As well as open, read, write, and close routines, DeviceIoControl provides a driver with the option of providing any special functionality. Consult your Win32 documentation for full details of these functions.

Table 1.2   Win32 Program Interface

Win32 function	Action
CreateFile	Open device file
CloseHandle	Close device file
ReadFile ReadFileEx ReadFileScatter ReadFileVlm	Read
WriteFile WriteFileEx WriteFileGather WriteFileVlm	Write
DeviceIoControl	IOCTL
CancelIo	Cancel any overlapped operations
FlushFileBuffers	

A device driver provides one or more named devices to the Win32 programmer. If writing a device driver for a dongle that can be attached to any parallel port, the devices might be named "DongLpt1", "DongLpt2", etc. To the Win32 programmer these appear as \\.\DongLpt1, etc. (i.e., with \\.\ at the beginning). Note that when written as a C string this last device name appears as "\\\\.\\DongLpt1".

The `CreateFile` Win32 function is used to open or create a connection to a device. After the device filename, the next two parameters specify the read and write access mode and whether the file can be shared. `CloseHandle` is used to close the file handle when you have finished using the device file.

The `ReadFile` and `WriteFile` series of functions are used to issue read or write requests to the device file.

`DeviceIoControl` is used to issue special requests that can send data to a driver and read data back, all in one call. There are many predefined IOCTL codes that can be sent to standard drivers, but you can also make up your own for your driver.

When a process finishes, Win32 ensures that all handles are closed and it cancels any outstanding I/O requests.

One large set of functions deals with file systems. Similarly, serial port communication has its own set of specialised functions. Other operations, such as unloading the device driver or shutting down the system, can also result in your device driver being called.

Overlapped Asynchronous Requests

Win32 supports asynchronous overlapped I/O calls, in which a program issues a read or write request and then gets on with another task. This feature has no impact on a device driver as any user request may be processed asynchronously from the Win32 process. In Windows 98, overlapped I/O requests cannot be issued to disk file systems, but can be issued to ordinary device drivers. Chapter 14 shows how to issue Win32 overlapped I/O requests.

Environment

Any number of Win32 threads could access your device at the same time, so your driver should expect this and cope correctly, even if the action is just to allow exclusive access by one thread. The kernel I/O Manager helps considerably by providing a mechanism for processing your read and write requests one at a time.

Your driver should be prepared to run on a multiprocessor system. Many of your driver routines need to be reentrant to cope with this situation. You have to ensure that your driver can cope with being run on two different processors at the same time, usually in different parts of the driver and possibly at different interrupt levels. Techniques for achieving these goals are described in this book.

Ideally, you should provide a version of your driver for each available CPU platform. This means compiling a DEC Alpha version as well as 80x86.

The end user may not be using English. For most I/O, this is not a problem for a driver. However, if you log messages to the event log, it is nice to provide messages in a language that matches the administrator's locale. It should be easy to localize any support utilities that you provide.

Finally, your driver can determine whether Windows 2000 is running as a server or as a personal workstation. Server systems might have more memory and do more I/O.

Device Specific Restrictions

In particular cases in which you are writing both the device driver and the user mode code, you may find it useful to put more restrictions on the type of access with which your driver can cope.

For instance, if implementing a particular protocol, you might dictate that a command has to be written first using WriteFile and the results read back using ReadFile. Alternatively, you could ignore WriteFile and ReadFile completely and just use DeviceIoControl with your own IOCTL codes.

Whatever approach you use, make sure that you follow a specification. If you are interfacing your hardware to a system class driver, you will have to work to a specification laid down in the Microsoft Driver Development Kit (DDK). Otherwise, you will have to create a specification for your API that your Win32 colleagues will have to follow.

Other Win32 Access to Drivers

The file metaphor is used for most device driver interactions in Win32 programs. However, Windows calls drivers in other situations. For example, keyboard keypresses arrive in programs as Windows messages. These keypresses come from a device driver. Internally, Windows calls the system keyboard driver using the file metaphor.

Many other specialised aspects of Win32 also use device drivers, DirectInput and DirectDraw to mention two. As another example, Human Input Device (HID) user mode clients use various routines, such as HidD_SetFeature, that end up as a driver call.

All these different ways of accessing drivers from Win32 end up as calls using the file metaphor. Therefore, all WDM and NT style device drivers have the same basic structure as they process the same sorts of calls.

Conclusion

In this book, I will explain the Windows Driver Model, including how to write device drivers that work in Windows 98 and Windows 2000. I will also cover NT style drivers that work in Windows NT 3.51 and NT 4.

A driver writer has a job completely different from a standard Windows programmer. There is much terminology to learn and each type of device has its own detailed hardware and software specifications. However, in the end, a device must be made available to Win32 programs and users.

Before writing a device driver in earnest, the next couple of chapters look at the big picture: device driver design and the crucial concepts and structures needed in the driver model.

If you are itching for something to do, order your MSDN Professional subscription. Install the MSDN library, the DDKs, and the Platform SDK. If you are writing a driver for both Windows 98 and Windows 2000, either set up a dual boot machine or get another test computer to hand.

Chapter 2

The Big Picture

A device driver has many jobs to do, some mandatory and others optional. In this chapter, I will give an overview of these jobs, so you know what tasks your driver will have to implement. I will also describe the big picture to show the different types of device drivers and the environment in which drivers run. If possible, you should consider writing a driver that uses one of the Windows Driver Model system drivers, as it simplifies your work and makes it fit in.

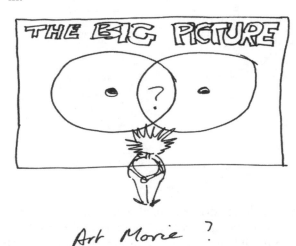

Art More ?

Choosing how to implement your driver is one of the most crucial decisions you will need to make. You need to know which jobs it must perform and which tasks it can offload to a class driver. If you are writing a driver for a Universal Serial Bus (USB) device, it soon becomes obvious that you should use the USB class driver. However, if you are going to support Power Management, it is best to decide this at the start as it may well effect the rest of your design.

Device Driver Components

Here are some of the jobs that a device driver can do:
- Initialize itself
- Create and delete devices
- Process Win32 requests to open and close a file handle
- Process Win32 Input/Output (I/O) requests
- Serialize access to hardware
- Talk to hardware
- Call other drivers
- Cancel I/O requests
- Time-out I/O requests
- Cope if a hot-pluggable device is added or removed
- Handle Power Management requests
- Report to administrators using Windows Management Instrumentation and NT events

Figure 2.1   Device driver components

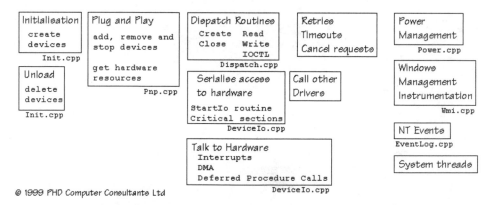

© 1999 PHD Computer Consultants Ltd

Figure 2.1 shows how I have divided up a device driver's functionality into different modules. The figure also shows the filenames that I have used for the modules in this book.

Strictly speaking, only the Initialization module is mandatory. In practice, all drivers have dispatch routines to handle user I/O requests. A WDM device driver needs a Plug and Play module, along with an installation INF file. NT style drivers will usually create their devices in their Initialization routine and delete them in an Unload routine. All other modules are optional, though in WDM drivers it is best to write minimal Power Management and Windows Management Instrumentation modules, simply to pass any requests to lower drivers.

Obviously, there will be many interactions between these different modules. Some of these interactions will be direct function calls. However, a lot of information will be passed in data structures. For example, a "device object" data structure stores information about each device.

If writing your first driver, you will no doubt be keen to know how to process reads and writes. As the figure shows, the module that handles these basic I/O requests is a depressingly small part of the whole device driver. The only consolation I can offer is this: if you base your driver on one of the examples in this book, you should be able to concentrate on your device's functionality. However, you cannot ignore what is going on in all the other modules.

Driver Entry Points and Callbacks

The kernel usually runs code in your driver by sending I/O Request Packets (IRPs). For example, a Win32 ReadFile call arrives in a device driver as a Read IRP. The size and location of the read buffer are specified as parameters within the IRP structure. The IRP structure is fundamental to device drivers. I shall be looking at IRPs more in the next chapter, and throughout the rest of the book.

A driver has one main initialization entry point — a routine that you must call DriverEntry. It has a standard function prototype. The kernel calls your DriverEntry routine when your driver is loaded, as is shown in Chapter 4.

Subsequently, the kernel may call many other routines in your driver. These routines are given the general name of *callbacks*. You tell the kernel the name of the routine and later on the kernel calls the routine back in the right circumstances. For example, if you want to handle interrupts, you must tell the kernel the name of your Interrupt Service Routine (ISR) callback. Each callback has a standard function prototype, appropriate for the circumstance in which it is called.

Table 2.1 lists all the driver entry points and callbacks. I will briefly describe these routines in this chapter and fully explain them later in the book, so do not worry about the details yet. New drivers can also provide a Common Object Model (COM) interface to the kernel, a defined series of routines that the driver implements.

Table 2.1   Standard driver entry points and callback routines

DriverEntry	Initial driver entry point. Sets up main callbacks.
I/O Request Packet (IRP) handlers	Called to process the IRPs that you wish to handle.
Unload	Unload the driver.
AddDevice	A new Plug and Play device has been added.
StartIo	A callback to handle IRPs serially.
Interrupt Service Routine (ISR)	Called to handle a hardware interrupt. Usually schedules a Deferred Procedure Call to do most interrupt servicing.

DpcForIsr	Deferred Procedure Call routine. Starts off another interrupt-driven transfer or completes an I/O request.
Critical section routine	Called to synchronize execution on one processor with no interrupts. Called by low IRQL tasks to interact with hardware
Cancel	Called to cancel an IRP
Completion	Called when a lower-level driver has completed processing an IRP. This lets the current driver do more work.
AdapterControl	Called when a DMA adapter channel is available.
ControllerControl	Called when a controller becomes free. NT and W2000 only.
Timer	A one-second timer callback.
CustomTimerDpc	For time-outs of less than one second.
CustomDpc	Usually used to handle work queues.
Reinitialize	Called if a driver takes a long time to initialize itself.
ConfigCallback	Query device hardware description callback. NT and W2000 only.
Plug and Play Notification	Called to notify you when devices have arrived, when the hardware profile changes, or when a device is being removed.
Callback	W2000 callback object handler

Dispatch Routines

A driver's DriverEntry routine must set up a series of callbacks for processing IRPs. It also sets the Unload, AddDevice, and StartIo routine callbacks, if these are needed. Table 2.2 shows the common Win32 device I/O functions and their corresponding IRPs. For example, a call to CreateFile ends up as a Create IRP sent to your driver.

Table 2.2   Dispatch routine IRPs

Win32 function	IRP
CreateFile	Create IRP
CloseHandle	Close IRP
ReadFile, etc.	Read IRP
WriteFile, etc.	Write IRP
DeviceIoControl	IOCTL IRP
	Internal IOCTL IRP

One common IRP cannot be generated from user mode code[1]. The Internal IOCTL IRP can only be generated from within the kernel. This allows drivers to expose an interface that cannot be used from Win32. These Internal IOCTLs are often made available by generic

---

1.   There are several other IRPs which are only generated by the kernel in response to kernel related events. For example, when the system decides to powers down, a Power IRP is sent all drivers.

drivers. For example, the Universal Serial Bus (USB) class drivers only accept commands in Internal IOCTLs; they do not support ordinary reads and writes.

The handlers of the Create, Close, Read, Write, IOCTL, and Internal IOCTL IRPs are commonly called *dispatch routines* because they often perform only some initial processing of the IRP, such as checking that all the parameters are valid. They then dispatch the IRP for processing elsewhere within the driver. Quite often, IRPs need to be processed serially so that the driver interacts with hardware in a safe way.

Processing in these basic routines is not quite as straightforward as you might think. Two or more IRP dispatch routines may be running "simultaneously". The problem is particularly acute in multiprocessor systems, but can easily happen when there is just one processor. For example, a dispatch routine on a single processor may block waiting for a call to a lower driver to complete. Or the dispatch routine's thread may run out of time in its execution slot. In both cases, another IRP dispatch routine may called. In due course, this second IRP will block or be completed, and work will continue on the first IRP. This is a common scenario and much of the difficult work of a driver is coping correctly with synchronization issues.

Creating Devices

How do devices come to exist in the first place? Quite simply, you have to create them, either in your DriverEntry routine or when the Plug and Play (PnP) Manager tells you to. In due course, you will delete the devices when your driver unloads or when the PnP Manager tells you that the device has been removed.

Most WDM device objects are created when the PnP Manager calls your AddDevice entry point. This routine is called when a new device has been inserted and the installation INF files indicate that your driver is the one to run. After this, a series of PnP IRPs are sent to your driver to indicate when the device should be started and to query its capabilities. Finally a *Remove Device* PnP IRP indicates that the device has been removed, so your device object must be deleted.

NT style drivers create their devices when they want to. Usually their DriverEntry routine roots around to find any hardware that can be represented as a device. For example, the system parallel port driver finds out how many parallel ports have been detected and creates an appropriate kernel device object for each one. The driver's unload routine is usually responsible for deleting any device objects.

How do user mode programs know what devices exist? You must make a symbolic link for each device object that is visible to Win32. There are two different techniques for making these symbolic links. The first is to use an explicit "hard-coded" symbolic link name. The user mode program must similarly have the device name hard-coded into its source[2]. The alternative is to use device interfaces, in which each device interface is identified by a Globally Unique Identifier (GUID). Registering your device as having a particular device interface creates a symbolic link. A user mode program can search for all devices that have a particular GUID.

---

2. In NT and W2000 you can use the QueryDosDevice Win32 function to obtain a list of all symbolic links. I am not sure how useful this list is.

Hardware Resource Assignments

Low-level drivers need to know what hardware resources have been assigned to them. The most common hardware resources are I/O ports, memory addresses, interrupts, and DMA lines. You cannot just jump straight in and access an I/O port, for example. You must be told that it is safe to use this port.

WDM drivers that handle Plug and Play (PnP) IRPs are informed of a device's resources when the *Start Device* PnP IRP is received. An NT style driver must find what resources each device needs and request use of them.

A significant number of drivers will not need any low-level hardware resources. For example, a USB client driver does not need any hardware resources. The USB bus driver does all the nitty-gritty work of talking to hardware, so only it has to know about the hardware resources that the electronics use. A USB client driver simply has to issue requests to the bus driver. The USB bus driver talks to the hardware to do your job.

Calling Other Drivers

WDM drivers spend a lot of time talking to other drivers. A Plug and Play device is in a stack of device objects. It is very common to pass IRPs to the next device down the stack.

Some types of IRP, such as Plug and Play, Power Management, and Windows Management Instrumentation IRPs, are often passed immediately to the next device. Only minimal processing is required in a driver.

In other cases, a driver's main job is achieved by calling the next device down the stack. A USB client driver often calls the USB bus drivers by passing an IRP down the stack. Indeed, a driver often creates new IRPs to do this same job. For example, it is quite common for a Read IRP handler in a USB driver to do its job by issuing many Internal IOCTL IRP requests to the USB bus drivers.

Serializing Access to Hardware

Any device that accesses hardware has to use some mechanism to ensure that different parts of the driver do not access the hardware at the same time. In a multiprocessor system, the Write IRP handler could be running at the same time on two different processors. If they both try to access hardware then very unpredictable results will occur. Similarly, if an interrupt occurs while a Write IRP handler is trying to access hardware, it is quite likely that both tasks will go seriously wrong.

There are two different mechanisms to sort out these sources of conflict. First, Critical section routines are used to ensure that code cannot be interrupted by an interrupt handler, even on another processor.

Second, you should use StartIo routines to serialize the processing of IRPs. Each device object has a built in IRP queue. A driver's dispatch routines insert the IRPs into this device queue. The kernel I/O Manager takes IRPs out of this queue one by one, and passes them to the driver's StartIo routine. The StartIo routine, therefore, processes IRPs serially, ensuring no conflict with other IRP handlers. The StartIo routine will still need to use Critical section routines to avoids conflicts with hardware interrupts.

If you hold an IRP in any sort of queue, you must be prepared to cancel it if the user thread aborts suddenly or it calls the Win32 CancelIo function. You do this by attaching a cancel callback routine to each IRP that you queue. Cancelling can be tricky, as you have to cope if the IRP has been dequeued and is being processed in your StartIo routine.

If the user mode application closes its file handle to your device with overlapped requests outstanding, you must handle the Cleanup IRP. The Cleanup IRP asks you to cancel all IRPs associated with a file handle.

Talking to Hardware

Once you have an address for an I/O Port or memory, it is straightforward to read and write hardware registers and the like. You should not hog the processor for more than 50 microseconds. Consider using system threads or system worker threads, described later, if you need prolonged access to hardware.

Handling interrupts is slightly more complicated. As mentioned earlier, you have to register your Interrupt Service Routine. This must check whether your device caused the interrupt and act on it as soon as possible.

However, this is where is gets complicated. It is not safe for an interrupt handler to call most kernel functions. If an interrupt signals that the last part of a Write request has completed, you will want to tell the I/O Manager that you have completed processing the IRP. However, interrupt handlers cannot do this job. Instead, interrupt handlers must ask for your driver's Deferred Procedure Call (DPC) routine to be run in due course. Your DPC routine can use most kernel functions, thus letting it complete IRPs, etc.

Some drivers must set up Direct Memory Access (DMA) transfers of large amounts of data from devices into memory, or vice versa. DMA is usually done using the shared system DMA controllers. However, some new devices have a built-in bus mastering capability that lets them use DMA themselves. I will not explain how to set up DMA transfers. However, Chapter 24 describes the new DMA routines for the benefit of NT 4 driver writers.

Hardware Problems

As we all know, hardware is bound to go wrong (unlike our software:-). You should be able predict the common ways in which hardware problems arise: interrupts will not arrive, buffers will overrun, printers will run out paper, and cables will be disconnected. Some of these problems are timing related. You will have to ensure, as far as possible, that your driver is available at all times to process fast I/O events.

Make sure that you check all hardware status bits (e.g., the out-of-paper indication). Further, ensure that a valid error message gets back to the user mode application.

If things go wrong, make sure you implement time-outs and retry the transfer, if appropriate. The I/O Manager provides an easy way to check time-outs with a granularity of one second. However, it is straightforward to implement a timer for smaller intervals. If a transfer still fails after a few retries, you will have to abort the IRP and signal an appropriate error.

Power Management

If a device's power consumption can be controlled, its driver should support Power Management in W98 and W2000. This applies to both WDM and NT style devices. Power Management conserves battery power in portables and reduces energy consumption and wear in desktop systems. Conversely, some people will have a sleeping or hibernating system on all the time so that it can start up again quickly.

Power Management happens on a system-wide and device-specific scale. The Power Manager can request that the whole system powers down. There are six system power states, including fully on and off, with three sleeping and one hibernating state in between. At a device level, there are four device power states, with two sleeping states in between fully on and off. A device can power itself down even if the rest of the system is running at full speed.

Drivers support Power Management by handling the Power IRP. Quite a few drivers will just pass the Power IRP down the stack of devices. However, your driver will probably be the only one that knows how to change the power usage of your device, so you will have to support the Power IRP correctly.

Windows Management Instrumentation

If possible, a driver should implement the Windows Management Instrumentation (WMI) extensions for WDM. This reports diagnostic and performance information to engineers and management. Drivers make data sets available on request and can fire events when they want. Driver methods can be invoked on demand.

Drivers support WMI by handling the System Control IRP. Again, some drivers will just pass this IRP down the device stack.

You can either use standard WMI data or event blocks or you can define your own new ones in MOF format. These must be compiled and included as a resource in your driver.

NT Event Reporting

The system event log is available in NT and Windows 2000. This is the traditional way of reporting driver problems and should still be supported, if possible. Drivers build an error log entry with an event number and possibly some strings and data. The system event log combines the event number with message strings included in a driver's resources.

System Threads

A system thread lets you do some work "in the background". A system thread could talk to very slow devices, or do some lower priority post-processing of data. Alternatively, existing system worker threads let you queue a work item for execution at lower priority.

Types of Device Driver

This section gives a picture of where device drivers fit in the general scheme of things, and the different types of driver that exist.

Windows Overview

Figure 2.2 shows a general driver schematic of Windows. User mode programs usually run in Win32 or the Virtual DOS Machine provided for DOS and Win16 applications. NT and W2000 also support other subsystems, such as OS/2 and Posix, but I think these are little used. Most user mode calls end up as requests to kernel mode.

The kernel mode executive services do any security and parameter checking that is appropriate before passing the request onto other parts of the kernel. In NT and W2000, a portion of the Win32 functionality is implemented in the kernel to achieve good performance. In particular, this Win32k.sys driver includes the Graphics Display Interface (GDI) engine. Print and display drivers live in here; these drivers are not covered by this book. Video requests are then processed by a generic video port driver. This port driver uses video miniport drivers to talk to specific display adapters.

Windows 95 and Windows 98 use Virtual Device Drivers (VxDs) to talk to hardware. These are not covered by this book. Windows NT, Windows 2000 and Windows 98 use WDM or NT style device drivers to talk to hardware. In NT and W2000, a Hardware Abstraction Layer (HAL) provides a portable interface to actual hardware.

Figure 2.2   Windows systems

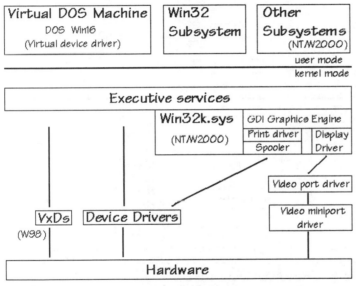

@ 1999 PHD Computer Consultants Ltd

I/O Request Processing

Figure 2.3 shows how an action by a user ends up being processed by device drivers. An application program makes a Win32 call for device I/O. This is received by the I/O System Services. The I/O Manager builds a suitable I/O Request Packet (IRP) out of the request.

In the simplest case, the I/O Manager just passes the IRP to one device driver. This interacts with hardware and, in due course, completes work on the IRP. The I/O Manager returns the data and results to Win32 and the user application.

It is very common now for an IRP to be processed by a stack of layered device drivers. Each driver breaks down the request into simpler requests. Highest level drivers, such as file system drivers, know how files are represented on disk, but not the details of how to get at the data. Intermediate level drivers process requests further (e.g., by breaking down a large request into a series of manageable chunks). Finally, lowest-level drivers actually interact with the hardware.

Wherever possible, drivers have been designed to be as generic as possible. For example, the SCSI port driver knows how to translate disk data requests into SCSI requests. However, it delegates the issuing of SCSI requests to SCSI miniport drivers that know how to talk to individual types of SCSI adapter.

One useful type of intermediate driver is a filter driver. A filter driver slips in between other driver layers to add functionality without effecting the higher or lower drivers. For example, a filter driver can be used to provide fault-tolerant disk access in which data is written to two separate physical disks to ensure that no data is lost.

Figure 2.3   Device driver calls

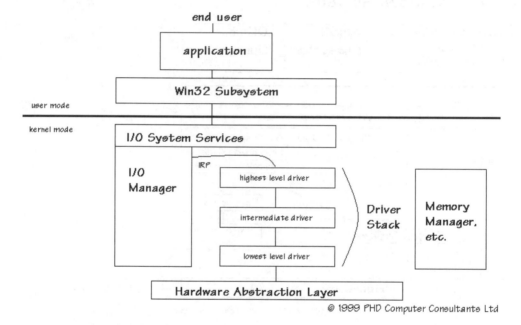

@ 1999 PHD Computer Consultants Ltd

Plug and Play Device Stacks

The Windows Driver Model (WDM) redefines driver layering to fit in with the Plug and Play system. A device stack represents the layers of drivers that process requests, as shown in Figure 2.4. A bus driver controls access to all devices on their bus. For example, you must use the USB bus drivers if you want to talk to a USB device.

The *bus* driver is responsible for enumerating its bus. This means finding all devices on the bus and detecting when devices are added or removed. The bus driver creates a Physical Device Object (PDO) to represent the device it has found. Some bus drivers simply control access to the bus; you can do what you want with the bus once you have control. In other cases, the bus driver handles all transactions on the bus for you.

A *function* driver that does know how to control a device's main features is layered above the bus driver. A function driver creates a Function Device Object (FDO) that is put in the device stack. In the USB case, the function driver must use the USB class driver to talk to its device. However, in other cases, once the bus driver has given the go ahead, the function driver can access hardware directly.

It is quite possible for more function drivers to be layered on top of the first function driver. An example of this is given shortly.

Various types of filter drivers can be slipped into the device stack. Bus driver filters are added above the bus driver for all devices on the bus. Class filter drivers are added for all function drivers of a specific class. Device filter drivers are added only for a particular device. Upper-level filter drivers go above the function driver, while lower-level drivers go below.

It is important to realize that user requests always enter at the top of a device stack. Suppose a user program has identified a function device to which it wants to talk. The I/O Manager ensures that all its requests are sent to the top of the device stack, so that any higher filter or function drivers get a chance to process the requests first.

Figure 2.4   Generic WDM driver stack

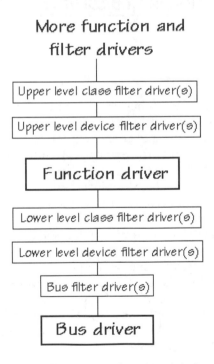

@ 1998 PHD Computer Consultants Ltd

Standard Bus and Class Drivers

Figure 2.5 shows the main class and bus drivers that are provided with Windows. These are general purpose drivers that can act as bus drivers and as function drivers. In most cases, the main system driver is accompanied by another driver that interfaces with hardware or another class driver. These helper drivers are usually called minidrivers but can be called miniclass or miniport drivers. For example, the Human Input Device (HID) class driver uses minidrivers to talk to hardware. One minidriver is provided to talk to HID devices on the USB bus. This HID minidriver translates requests from the HID class driver into requests to the USB bus driver.

I will discuss only the HID and USB class drivers in detail. Please consult the Driver Development Kit (DDK) documentation for details of the other system class and bus drivers. Any driver that uses a bus or class driver is often called a *client driver*.

Figure 2.5   Bus and class drivers

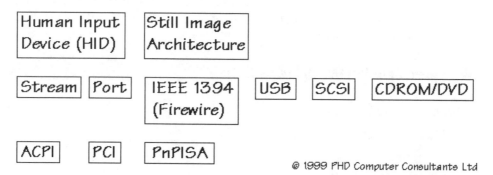

© 1999 PHD Computer Consultants Ltd

The Advanced Configuration and Power Interface (*ACPI*) bus driver interacts with the PC ACPI BIOS to enumerate the devices in the system and control their power use. The Peripheral Component Interconnect (*PCI*) bus driver enumerates and configures devices on the PCI bus. The *PnPISA* bus driver does a similar job for Industry Standard Architecture (ISA) devices that are configurable using Plug and Play.

The *Stream* class driver provides a basis for processing high bandwidth, time critical, and video and audio data. The stream class driver uses minidrivers to interface to actual hardware. Filter drivers can be written to transform data. In addition, a USB Camera helper minidriver is provided to make it easier to manage USB video and audio isochronous data streams.

The Stream class makes it easier to expose the different aspects of a piece of hardware as different subdevices. The audio *Port* class driver helps this further by using COM to identify the subdevices. Audio port and miniport are used to control the actual hardware.

The *IEEE 1394* bus driver enumerates and controls the IEEE 1394 (Firewire) high speed bus. This bus driver uses port drivers to talk to the IEEE 1394 control electronics. IEEE 1394 client drivers issue IEEE 1394 Request Blocks (IRBs) to control their device.

The Universal Serial Bus (USB) bus driver enumerates and controls the lower-speed USB bus. Host controller drivers are provided as standard to talk to the two main types of USB host controller. USB client drivers use various IOCTLs to talk to their device via the USB class driver. The most important IOCTL lets clients issue USB Request Blocks (URBs) to control their device.

The **SCSI** and **CDROM/DVD** drivers are used to access hard disks, floppies, CDs, and DVDs. As usual, a variety of port and miniport drivers are used to talk to hardware. For example, requests to a CD-ROM on the IEEE 1394 bus would be routed to the IEEE 1394 bus driver.

The *Human Input Device* (HID) class driver provides an abstract view of input devices. The actual input hardware can be connected in different ways that are shielded by HID minidrivers. Quite a few HID devices are on the USB bus, so a HID minidriver for the USB bus is provided as standard. HID clients can either be user mode applications or, if need be, kernel mode drivers. Standard parsing routines are available to make sense of HID input and output reports.

Finally, the *Still Image Architecture* (STI) is not a driver at all, but a means of obtaining scanner and still image camera using minidrivers. STI currently supports SCSI, serial, parallel, and USB devices.

Example Driver Stack

With all these different types of drivers lurking around, it is worth while having a look at one real example to see how it all fits together. You do not need to know the detailed terminology yet.

Figure 2.6 shows the hardware and software components that are involved in the handling of a HID keyboard attached to the USB bus. On power up, the PCI bus driver finds the USB controller and so loads the USB class drivers. In this case, the USB OpenHCI host controller driver is used to talk directly to the hardware.

The USB host controller electronics detects when the keyboard is plugged in. The USB root hub bus driver is informed. It tells the Plug and Play Manager that a new device has arrived. The keyboard Hardware ID is used to find the correct installation INF file. In this case, the HID class driver and its USB minidriver are loaded, along with the standard HID keyboard client driver.

The HID keyboard driver uses Read IRPs to request keyboard data. The HID class driver asks the USB minidriver to retrieve any HID input reports. The minidriver uses an Internal IOCTL to submit URBs to the USB class drivers to read data. The USB class drivers talk directly to USB host controller electronics to get any available data. The electronics finally generates the correct signals on the USB bus. These are interpreted by the USB keyboard and it returns any keypress information. The results percolate up the device stack and will eventually reach a Win32 program in message form.

Figure 2.6 also shows that a user mode HID client can access the HID keyboard using standard Win32 `ReadFile` and `WriteFile` requests.

Figure 2.6   HID USB keyboard example

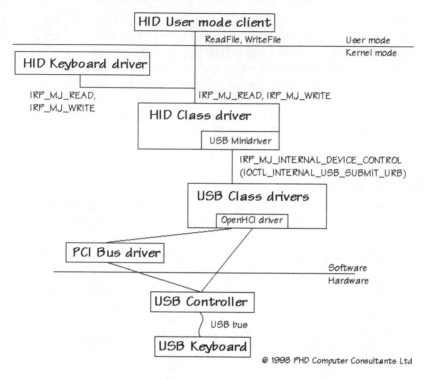

@ 1998 PHD Computer Consultants Ltd

Driver Choices

You need to evaluate all the information in this chapter so you can decide at the outset what sort of driver to write for your device. There are various factors and choices that will influence your decisions.

Off-the-Shelf Drivers

The simplest approach is to buy an off-the-shelf driver that handles your device directly. Such drivers might not be available. However, there are various commercial general purpose drivers available, as described earlier. These can be tailored with scripts or the like to interact with many simple devices. The PHDIo and WDMIo drivers described later in this book can be used to talk to devices with I/O ports and a simple interrupt handling requirement.

Use Standard Drivers

If at all possible, you should use one of the standard bus or class drivers, as these will usually have implemented a large amount of the functionality that you need. Indeed, for many types of devices, using the system driver is the only way to access it. The only drawback is that you will have to study the detailed documentation in the DDK for the class of device that you are using. However, I can assure you that it is a darn sight easier to write a USB client than having to write a huge monolithic USB driver from scratch.

Operating System

If using a standard bus or class driver, you will usually be writing a driver that supports Plug and Play. Your driver should usually, therefore, run in W98 and W2000.

Most of the standard bus and class drivers are not available in NT 4 and earlier. In fact, any driver that supports NT 4 or earlier must be an NT style driver. In some cases, it still makes sense to write an NT style driver for W98 and W2000.

Layered Device Drivers

Plug and Play drivers are always layered drivers, as they always form a layer in a device stack. However, NT style drivers can also layer themselves over other drivers. Drivers in a device stack will receive IRPs sent to any of the devices in the stack.

It is quite possible for a driver to use a device stack but not include itself in the stack. For example, drivers that use Plug and Play Notification are usually NT style drivers. Whenever a suitable device is inserted, such a driver will want to use the device. It can layer itself into the device stack, and so receive all IRPs sent to the device stack. Alternatively, it can just store a pointer to the device object at the top of the device stack. This way, it can issue requests to the device without receiving unwanted requests from above. The HID client driver described in Chapter 23 shows how to use this technique.

Monolithic Drivers

It will soon become evident that in some cases, you need to write a low-level driver to talk to your device. The question now is whether yours is a general purpose device or not. If it is has just one specialized use, then writing a monolithic driver to handle all device requests may well be the simplest solution.

However if you want your device to interface into Windows smoothly, you will usually need to write a suitable minidriver, miniclass, or miniport driver. The DDK and supporting documentation must be used to determine the exact requirements for these types of drivers. In extreme cases, you may decide to write a whole new class driver, but I would not recommend it.

Recommended and Optional Features

As mentioned earlier, your driver should support Power Management, Windows Management Instrumentation (WMI), and NT event reporting. It is best to build in Power Management support from the beginning. It is possible to bolt on the other two aspects later. However, if you wish to collect performance information for your WMI reporting, it is best to design this in from the start.

WDM Rationale

I am willing to hazard a few comments about the Windows Driver Model (WDM) design to give you my ideas of how well it achieves its goals.

It is the role of device drivers to interface the operating system to hardware devices. They must eventually support the information flows that users and their applications require. In Windows, the specification of Win32, with multiple processes and threads, has implications for device drivers. And a user requirement for fast processing of video and audio data means that device drivers now have to do much more than just receive and send data.

One Core Model

Device drivers have to process a wide variety of data. The simplest example of input data is a keyboard keypress. Keypresses do not happen very often, but they do need to get to an application with visual feedback provided quickly. The next important type of data is the blocks of data that make up file systems. Speed and data integrity are the crucial issues here. Finally, for isochronous data, being able to keep up with the data is the most important factor, even if it means dropping some samples.

The Windows Driver Model defines one basic model to handle all these types of data. In effect, the model is extended for each type of device. For example, the USB class drivers provide an abstract model for all USB devices, with a defined interface that is used by all client drivers. Having a core driver model that is common to all device types makes it easier for driver writers to move from one type of device to another. And it also means that the kernel implementation of the driver model is more likely to be solid. Having a different model for each type of device would almost certainly make life harder for us.

Is this extensible core model too complicated? If you are writing a monolithic simple low-level driver, I think that IRPs will appear to you as quite difficult beasts. Supporting Plug and Play properly is a task too onerous for simple drivers, as described in the following text.

Complexity

Any discussion of device drivers can be fairly "heavy". By their very nature, you have to interact with complicated hardware systems that may have exacting timing constraints. Any operating system interface to hardware will have a complicated job to do. Another factor is that each type of device is different, so you need to know about the details of its technology, including its hardware implementation. For example, the network device drivers field is a subject that can easily fill a whole book. All this means that any aspiring device driver writer will have a steep learning curve.

Plug and Play and Layers

When Plug and Play is first described to you, it will probably sound fairly straightforward. You are told when a device arrives. You are told when it leaves. However, some of the follow-on implications are hard work for device drivers. In a brief look, I am not convinced that any of Microsoft examples do the job fully.

The first problem is that the sudden removal of a device will interrupt a whole host of activities that your driver may be doing. However, I guess that there is no easy way round this.

A stickier problem occurs when the Plug and Play (PnP) Manager tries to juggle the system resources when a new device is inserted. The PnP Manager can ask to "stop" your device (i.e., pause it while your resource assignments are changed). Again, this may interrupt the current flow of operations. However, more importantly, you are supposed to queue up any received IRPs while your device is stopped[3]. At worst, the user is only supposed to experience a tiny delay in the functioning of a device that is paused temporarily. However, implementing an IRP queue is not a trivial exercise, especially as you have to be prepared for the worst. For example, if a device is removed while stopped, you will have to chug through all the queued IRPs completing them with an appropriate error status. (The Ks... series of Kernel Streaming IRP handling routines might provide all this functionality.)

Another potential criticism of the Plug and Play system is that too many layers of drivers are involved. Passing requests between layers can take some time. If a request has to pass through many layers of filter drivers, it could have a noticeable effect on performance.

Range of Functionality

For good reasons, device drivers cannot use many standard C or C++ functions. This makes it more complicated for driver writers as they have to learn what routines to use instead. I find the Unicode string handling routines particularly unwieldy.

I understand that Microsoft is trying remove some older areas of functionality. For example, WMI is effectively going to replace NT events. However, you have to support both in the mean time, which is more effort.

Development Environment

There is plenty of documentation available in the various DDKs. However, it is slightly confusing to have Windows 98 and Windows 2000 DDKs covering the same subject. As is usual, the documentation lags a few months behind when any new developments arise, which can be frustrating. It is worth while checking the relevant Microsoft websites regularly for appropriate articles.

The provision of lots of example source in the DDKs is great. Give us more! These show how real drivers work and show us how to use various kernel calls and techniques.

The driver build environment is still essentially command line. Although it is easy enough to provide a Visual Studio wrapper for this build environment, a fully integrated system would be appreciated. The provision of a nice debugger that can be run on the same computer as the driver you are testing would be a boon.

Developer Support

You will have to make do with the DDK documentation and any colleagues that you can find to help out. First, scour the DDKs, the DDK source, documents in the DDK source, the Microsoft website, and the Platform SDK. Your next port of call might well be the newsgroups and mailing lists. Finally, you can pay Microsoft for help.

Conclusion

In this chapter, I have tried to give the device driver big picture, showing what components your driver will need and how it fits into the general scheme of things. There are different types of device driver and different ways of implementing the same driver functionality. Some driver components are not compulsory, but it is certainly recommended that you support Plug and Play, Power Management, and Windows Management Instrumentation, if possible.

---

3. You are also supposed to queue IRPs that arrive while your device is sleeping. The simple solution to this problem is to wake up your device when an IRP arrives.

A crucial part of the Windows Driver Model is the provision of several bus and class drivers. These make it considerably easier to write many types of driver.

In the next chapter, I will look at the crucial background concepts that are needed before device driver development in earnest.

Chapter 3

Device Driver Design

Introduction

This chapter introduces the basic concepts and structures that you will need to write drivers. Some ideas are sufficiently important and deserve to be mentioned now. All the other techniques, routines, and objects are covered in the following chapters.

Windows uses a layered design for passing requests from the end user to a driver stack and eventually to the hardware. As a trusted part of the operating system, a driver must be written well to fit into the kernel. It should use the kernel's abstract model of the processor and use memory carefully. Wherever possible, a driver should use the recommended techniques, routines, and structures.

Driver Design Guide

A good device driver needs to be well written. Remember that it is a trusted part of the operating system. The PC will not get the protection that a normal Win32 application has. If anything goes wrong, it is likely that the system will crash, or worse — destroy valuable data before crashing.

For important projects, my ultimate design philosophy is to design code three times. At the outset, it is definitely worth while knowing where I am going and what features I am going to implement. I then prefer an incremental approach, in which each individual section of code is designed as I go along. However, as with many programs, it often only when I have finished a first-cut prototype of the whole driver that I really understand what is going on.

Doing some real hardware interactions might invalidate some of my initial design decisions. Similarly, I might decide to expose an IOCTL interface instead of reads and writes. In the end, it might be best to scrap some sections of code and start again[1].

Documentation

It is particularly important that you document your driver well and provide ample comments in your code. Device drivers are complicated enough at the best of times. Having to look at someone else's code without any explanation of what is going on can be difficult.

Your hardware and firmware engineering colleagues should have provided you with a suitable interface specification for the device you have to control. You will bring these requirements together with the specification of any bus or class drivers you intend to use. At the end of the day you will have specified your driver's *lower edge* — the hardware and software interactions that you use to get your job done.

Similarly, you should define your *upper edge*. This is the interface that your driver presents to user mode applications or higher-level drivers. You need to define what devices you make available and which Win32 functions can be used to access the devices. Specify the parameters in detail and the contents of all buffers. Give your Win32 colleagues a suitable header file.

Good Design

Whatever your approach, a good design needs to be configurable, portable, preemptive, interruptible, and multiprocessor-safe.

A driver should be as configurable as possible — at least as configurable as the hardware allows. A device on a configurable bus ought to have a Plug and Play driver. Use the registry and the control panel to control how a driver runs.

Making a driver portable between platforms is usually fairly straightforward. Drivers should be easily recompiled to support non-x86 processors. WDM drivers need to run on W98 and W2000. Make sure that you only use routines that are available on both platforms. Write the driver in C or C++ and avoid using assembly language. Use the appropriate kernel routines to access hardware. Beware of data types that have different sizes on different processors. Other processors may have different virtual memory page sizes.

When your driver is running at a low interrupt level (see the following text), the operating system may perform a context switch at any time. Remember that higher priority interrupts can occur at any time, so your driver must bear this in mind. Even your ISR can be interrupted.

In an NT/W2000 multiprocessor system, your driver could be called simultaneously on different processors, so the code needs to be reentrant. As NT/W2000 supports symmetric multiprocessing, an interrupt may be handled on any processor. Even on a single processor system, your code needs to be reentrant. Your code could block while processing one IRP, or a context switch could occur. This could allow another IRP to start processing before the first IRP has completed.

---

1. I used this approach when writing this book. My initial contents were completely different from the end result, and I rewrote these initial chapters after the rest of the book was complete.

Various techniques and routines are used to handle interrupts and multiprocessor systems. For example, critical sections ensure that a device operation is not interrupted. Spin lock tokens can guard access to work queues. Various dispatcher objects (events, Mutexes, semaphores, timers, and threads) can be used to signal conditions and synchronize with the other parts of your driver.

The kernel also provides mechanisms to arbitrate access to shared hardware devices such as the system DMA controllers. In NT/W2000, if your own hardware has a controller that is shared between devices, use the Controller object to arbitrate access, rather than rolling your own.

Look to see if the kernel already provides you with a feature (e.g., string manipulation facilities and management of a pool of memory blocks). These facilities are discussed as appropriate throughout the book.

A good driver checks each and every return value from any routines that it uses (e.g., in the kernel or other drivers). If an error occurs, make sure you rollback your transaction correctly before returning the error.

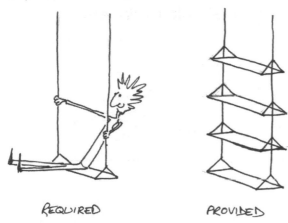

REQUIRED   PROVIDED

The layered device driver design

By now you will, of course, be writing Y2K-compliant code. However, this problem may not occur all that often in drivers. If appropriate, your driver should include accessibility options for people who have difficulty performing certain tasks. Bear in mind that many of your users may not use English as their first language, so internationalize as necessary. For example, provide different language event messages. Note that the strings that drivers must use are in Unicode.

Finally, it is best if you follow the driver routine naming conventions described later. Document and test your work as you go along. At the end, go through all the comments and make sure they are up to date.

Kernel Calls

Device drivers cannot access any standard C, C++, or Win32 functions. You cannot use the C++ new and delete operators. Instead, you must make use of a large number of kernel routines. The main categories of kernel routines are shown in Table 3.1.

The book describes each kernel routine when it is first used. However, I decided not to provide a full list of each function's parameters because it would take up too much room. You may find it useful to have the Driver Development Kit (DDK) documentation to hand. However, I hope that you can follow the thread of the argument without having to sit beside your computer.

If you are using one of the proprietary toolkits listed in Chapter 1, these will provide a set of classes that make drivers easier to write. This includes routines that make it easier to call the kernel. For example, drivers must use Unicode strings when talking to the rest of the kernel. The kernel routines that manipulate Unicode strings are fairly laborious to use. Most proprietary toolkits provide a class wrapper for Unicode strings that makes them easier to use.

There are various Win32 function calls that are used only in connection with device drivers. In particular, the SetupDi... routines are used to find devices that implement a specified device interface.

Table 3.1   Kernel Mode Routines available to drivers

Ex...	Executive Support
Hal...	Hardware Abstraction Layer (NT/W2000 only)
Io...	I/O Manager (including Plug and Play routines)
Ke...	Kernel
Ks..	Kernel stream IRP management routines
Mm...	Memory Manager
Ob...	Object Manager
Po...	Power Management
Ps...	Process Structure
Rtl...	Runtime Library
Se...	Security Reference Monitor
Zw...	Other Routines

Table 3.2 shows routines that are used only by specific types of driver. Most of these are provided for miniclass drivers, miniport drivers, and minidrivers. Others, such as the Hid... and Usb... functions are used by client drivers.

In addition to those routines listed, audio miniport drivers define several standard interfaces. The IEEE 1394 bus driver makes many routines available, with no common initial name.

Table 3.2   Bus driver and class specific routines

BatteryClass...	Battery class routines for miniclass drivers
Hid...	Human Input Device routines
Pc...	Port class driver routines

ScsiPort...	SCSI port driver routines for miniport drivers
StreamClass...	Stream class driver functions for stream minidrivers
TapeClass...	SCSI Tape class routines for miniclass drivers
Usb...	Universal Serial Bus Driver Interface routines for USB client drivers

Kernel Macros

If you look in the DDK header files, you will find that a few of the kernel functions are implemented as macros. The definition of one or two of these macros is quite poor. For example, RemoveHeadList is defined as follows:

```
#define RemoveHeadList(ListHead)
    (ListHead)->Flink;
    {RemoveEntryList((ListHead)->Flink)}
```

If you call RemoveHeadList in the following way, the wrong code is compiled.

```
if( SomethingInList)
    Entry = RemoveHeadList(list);
```

The only way to make this safe is to use braces.

```
if( SomethingInList)
{
    Entry = RemoveHeadList(list);
}
```

Therefore, to be on the safe side, it is best to use braces in all if, for, and while statements, etc.

Kernel Objects

The DDK documentation makes much use of the word *object* when describing kernel structures. This does not mean object in the C++ sense of the word. However, it means the same in principle. In the purest definition, a kernel object should only be accessed using kernel function calls. In practice, each kernel structure usually has many fields that can be accessed directly. However, it is definitely true that some fields should be considered private and not touched by your driver. This book describes which fields can be used safely in each kernel object.

Driver Routine Names

There is a useful convention for naming driver routines, similar to the kernel API naming scheme. Each routine name should have a small prefix based on the driver's name (e.g. Par for a Parallel port driver). This prefix should be followed by verbs or nouns as necessary. This naming scheme makes it easier to identify a driver when looking at a debugger trace. A driver initially has just one exposed routine that must be called DriverEntry so it can found.

Many driver routines have standard names that most people use. For example, the Create IRP handler in the Wdm1 driver is called `Wdm1Create`; the Read IRP is handled in `Wdm1Read`, etc.

Processor Model

Windows 98, NT, and Windows 2000 all primarily run in x86 systems. However, NT and W2000 can also run on Alpha processors. To handle any differences, WDM uses a model of a general processor and provides an abstract view of the resources available on the supported processors.

Always use kernel routines to access hardware. For some types of drivers, you do not need to talk to hardware directly. Instead, make the appropriate calls to the relevant bus driver to get it to perform the low-level I/O for you.

Processor Modes

The operating system requires that the processor support only two modes: a user mode for one or more applications to run in and a kernel mode for the bulk of the operating system. User mode programs are protected so that they cannot easily damage each other or the kernel. Conversely, the kernel can do anything it wants to and can access all memory.

Interrupts and exceptions are ways of stopping the processor from doing something and asking it to do something else. A hardware interrupt is an input signal to the processor from external hardware. Software interrupts are instructions that interrupt the processor to switch it to kernel mode. Exceptions are interrupts generated by the processor, usually when something goes wrong.

User programs make kernel function calls using software interrupts. In fact, a Win32 call initially goes to a user mode Win32 subsystem DLL. If appropriate, this then calls the main operating system in kernel mode.

The kernel uses a regular timer hardware interrupt to switch between the available user program threads of execution. This makes it appear that the threads are running simultaneously. On multiprocessor systems, the threads really can be running simultaneously. The kernel can also have its own threads running, which are switched in the same way as user threads.

The result is that drivers can be called in several ways. A user application can issue a Win32 device I/O request. Eventually, a driver is called to process this call. The driver might then start its hardware. If its hardware generates an interrupt, then the driver is called to process it (i.e., stop the interrupt), flag that it has happened, and, if necessary, start another operation. In both these cases, the driver does not necessarily operate in the context of the user application that called it.

Drivers can also be called in the context of a kernel thread. For example, the Plug and Play calls to a driver are usually called in the context of a system thread. As will be shown, this gives the driver more scope to do things. Finally drivers can set up their own kernel mode threads that can run as long as needed, getting their own share of the processor time.

Interrupt Levels

Hardware or software interrupts stop the processor from doing one task and force it to run some interrupt handling code. A processor prioritizes interrupts so that lower priority interrupts can themselves be interrupted by higher priority interrupts. This makes sure that very important tasks are not interrupted by jobs that can be done at a later stage.

Table 3.3   Abstract Processor Interrupt Levels

	IRQL	Description
No interrupts	PASSIVE_LEVEL	Normal thread execution
Software interrupts	APC_LEVEL	Asynchronous Procedure Call execution
	DISPATCH_LEVEL	Thread scheduling Deferred Procedure Call execution
Hardware interrupts	DIRQLs	Device Interrupt Request Level handler execution
	PROFILE_LEVEL	Profiling timer
	CLOCK2_LEVEL	Clock
	SYNCH_LEVEL	Synchronization level
	IPI_LEVEL	Interprocessor interrupt level
	POWER_LEVEL	Power failure level

Table 3.3 shows the interrupt levels that Windows uses to provide an abstract view of the actual processor interrupt levels. The lowest priority interrupt levels are at the top. Hardware-generated interrupts always take priority over software interrupts.

Drivers only use three of these interrupt levels. Somewhat confusingly, driver dispatch routines are called at PASSIVE_LEVEL[2] (i.e., with no interrupts). Many driver callbacks run at DISPATCH_LEVEL. Finally, driver hardware interrupt service routines operate at Device Interrupt Request Level (DIRQL).

The DIRQLs level in fact represents the many hardware interrupt levels that are available on the specific processor. A disk hardware interrupt level may well have higher priority than a parallel port hardware interrupt, so the disk may interrupt a parallel port interrupt. Usually a driver has just one DIRQL, though there is no reason why it cannot support more, if that is how its hardware works. The relevant driver interrupt level is referred to as its DIRQL.

The interrupt level at which a driver is operating is very important, as it determines the types of operation that can be undertaken. For example, driver hardware interrupt service routines cannot access memory that might be paged out to a swap file.

A driver has to be aware that it could be interrupted by a higher priority task at any stage. This means that its own interrupt service routine could run in the middle of its own dispatch routines. For this reason, if a normal driver routine is about to interact with its own hardware, it uses a kernel function to raise its interrupt level temporarily to its DIRQL. This stops its own interrupt handler from running at the same time. In addition, in a multiprocessor system, another driver routine might be attempting the same task on another processor. The

---

2.   If a file system driver is fetching a page back into memory then the IRQL may be APC_LEVEL.

KeSynchronizeExecution routine is used to run such critical sections in a multiprocessor-safe way.

Runtime Priorities

The Interrupt Level should not be confused with the scheduling priority. All threads normally run at the lowest interrupt level, PASSIVE_LEVEL. The scheduler uses the priority values to determine which thread to run next. Any other interrupt takes priority over a thread.

Drivers need to be aware of scheduling priorities. When it completes an IRP, it can give the calling thread a temporary priority boost so that, for example, programs that interact with the mouse can continue their run more quickly. This technique generally improves the perceived responsiveness of the system.

Deferred Procedure Calls

Any routine servicing a hardware interrupt stops normal program execution. For this reason, it is best to make an interrupt service routine as quickly as possible.

Any nonessential interrupt processing should be deferred until later. Windows lets driver writers use Deferred Procedure Call (DPC) routines, which are called at DISPATCH_LEVEL. The interrupt routine requests that the driver's DPC post-interrupt routine (DpcForIsr) is called when things have calmed down. A typical job in this routine is to indicate that the current I/O request is complete; the relevant kernel call can only be carried out at DISPATCH_LEVEL.

When writing your driver, be careful to comment the interrupt level at which each routine can be called. Similarly, make only kernel calls that are appropriate for the current interrupt level.

Using Memory

A device driver has to be very careful when allocating or accessing memory. However, it is not as gruesome to access memory as some other types of driver. For example, you do not need to worry about the murky x86 world of segments and selectors, as the kernel handles all this stuff.

Pool Memory

Windows implements *virtual* memory, where the system pretends that it has more memory than it really has. This allows more applications (and the kernel) to keep running than would otherwise be the case.

Virtual memory is implemented by breaking each application's potential address space into fixed size chunks called *pages*. (x86 processors have a 4KB-page size, while Alpha processors use 8KB.) A page can either be *resident* in physical memory, or not present and so swapped to hard disk.

Drivers can allocate memory that can be paged out, called *paged* memory. Alternatively, it can allocate memory that is permanently resident, called *nonpaged* memory. If you try to access paged memory at DISPATCH_LEVEL or above, you will cause a page fault and the kernel will crash. If you access nonresident paged memory at PASSIVE_LEVEL, the kernel will block your thread until the memory manager loads the page back into memory.

Please do not make extravagant use of nonpaged memory. However, you will find that most of the memory your driver uses will be in nonpaged memory. You must use nonpaged memory if it is going to be accessed at DISPATCH_LEVEL or above. Your driver StartIo and ISRs, for example, can access only nonpaged memory.

Table 3.4 shows how to allocate both paged and nonpaged memory using the kernel ExAllocatePool function. The table uses the DDK convention of listing IN or OUT before each parameter. IN parameters contain information that is passed to the function, and vice versa. Instances of IN and OUT in DDK header files are removed using macros.

Specify the ExAllocatePool PoolType parameter as PagedPool if you want to allocate paged memory or NonPagedPool if you want to allocate nonpaged memory. The other Pool-Type values are rarely used. Do not forget to check whether a NULL error value was returned by ExAllocatePool. The ExAllocatePool return type is PVOID, so you will usually need to cast the return value to the correct type.

When you are finished using the memory, you must release it using ExFreePool, passing the pointer you obtained from ExAllocatePool. Use ExFreePool for all types of memory. If you forget to free memory, it will be lost forever, as the kernel does not pick up the pieces after your driver has unloaded.

The ExAllocatePoolWithTag function associates a four-letter tag with the allocated memory. This makes it easier to analyze memory allocation in a debugger or in a crash dump.

Table 3.4   ExAllocatePool **function**

PVOID ExAllocatePool	(IRQL<=DISPATCH_LEVEL) If at DISPATCH_LEVEL, use one of the NonPagedXxx values
Parameter	**Description**
IN POOL_TYPE PoolType	PagedPool PagePoolCacheAligned NonPagedPool NonPagedPoolMustSucceed NonPagedPoolCacheAligned NonPagedPoolCacheAlignedMustS
IN ULONG NumberOfBytes	Number of bytes to allocate
Returns	Pointer to allocated memory or NULL

Lookaside Lists

If your driver keeps on allocating and deallocating small amounts of pool memory then it will be inefficient and the available heap will fragment. The kernel helps in this case by providing *lookaside lists*[3] for fixed-size chunks of memory. A lookaside list still gets memory from the pool. However, when you free a chunk of memory, it is not necessarily returned to the pool. Instead, some chunks are kept in the lookaside list, ready to satisfy the next allocation request. The number of chunks kept is determined by the kernel memory manager.

---

3. NT 3.51 drivers should use zone buffers, which are now obsolete.

Lookaside lists can contain either paged or nonpaged memory. Use `ExInitializeNPaged-LookasideList` to initialize a lookaside list for nonpaged memory and `ExDeleteNPaged-LookasideList` to delete the lookaside list. When you want to allocate a chunk of memory, call `ExAllocateFromNPagedLookasideList`. To free a chunk, call `ExFreeToNPaged-LookasideList`. A similar set of functions is used for paged memory. Consult the DDK for full details of these functions.

Other Memory Considerations

The kernel stack is nonpaged memory for local variables[4]. There is not room for huge data structures on the kernel stack because it is only 8KB–12KB long.

Drivers need to be reentrant, so they can be called simultaneously on different processors. The use of global variables is, therefore, strongly discouraged. However, you might read in some registry settings into globals, as they are effectively constant for all the code. Local static variables should also not normally be used for the same reason.

Finally, you can reduce your driver's memory usage in other ways. Once the driver initialization routines have completed, they will not be needed again, so they can be discarded. Similarly, some routines may be put into a pageable code segment. However, routines running at `DISPATCH_LEVEL` or above need to be in nonpageable nondiscardable memory.

Accessing User Application Memory

There are two main techniques for accessing user data buffers. If you use *Buffered I/O*, the I/O Manager lets you use a nonpaged buffer visible in system memory for I/O operations. The I/O Manager copies write data into this buffer before your driver is run, and copies back read data into user space when the request has completed.

The alternative technique, *Direct I/O*, is preferable, as it involves less copying of data. However, it is slightly harder to use and is usually only used by DMA drivers that transfer large amounts of data. The I/O Manager passes a Memory Descriptor List (MDL) that describes the user space buffer. While a driver can make the user buffer visible in the system address space, the MDL is usually passed to the DMA handling kernel routines.

DMA

Direct Memory Access (DMA) hardware controllers perform I/O data transfers direct from a device to main memory (or vice versa) without going through the processor. The first implication is that DMA memory cannot be paged. Secondly, the DMA controller has to be programmed with a physical memory address, not a processor virtual address. The kernel provides routines to help with both these tasks.

IRP Processing

I/O Request Packets (IRPs) are central to the operation of drivers. It is helpful to look at them briefly at this point, but they are described in full later. In particular, this section looks at how

---

4. Actually, the kernel stack may be paged if a driver issues a user-mode wait (e.g. in `KeWaitForSing-leObject`).

IRP parameters are stored and how the *IRP stack* lets an IRP be processed by a stack of drivers.

An IRP is a kernel "object", a predefined data structure with a set of I/O Manager routines that operate on it. The I/O Manager receives an I/O request and then allocates and initializes an IRP before passing it to the highest driver in the appropriate driver stack.

An IRP has a fixed header part and a variable number of IRP stack location blocks, as shown in Figure 3.1. Each I/O request has a major function code (such as IRP_MJ_CREATE corresponding to a file open) and possibly a minor function code. For example, the IRP_MJ_PNP Plug and Play IRP has several minor functions (e.g., IRP_MN_START_DEVICE). Table 3.5 lists the common IRP major function codes.

Table 3.5   Common IRP Major Function Codes

IRP_MJ_CREATE	Create or open device file
IRP_MJ_CLOSE	Close handle
IRP_MJ_READ	Read
IRP_MJ_WRITE	Write
IRP_MJ_CLEANUP	Cancel any pending IRPs on a file handle
IRP_MJ_DEVICE_CONTROL	Device I/O control
IRP_MJ_INTERNAL_DEVICE_CONTROL This is also called IRP_MJ_SCSI	Device I/O control from a higher driver
IRP_MJ_SYSTEM_CONTROL	Windows Management Instrumentation
IRP_MJ_POWER	Power Management request
IRP_MJ_PNP	Plug and Play message
IRP_MJ_SHUTDOWN	Shutdown notification

The fixed part of an IRP (the IRP structure itself) contains fixed attributes of the IRP. Each stack location — an IO_STACK_LOCATION structure — in fact contains most of the pertinent IRP parameters.

More than one IRP stack location is used when an IRP might be processed by more than one driver. Each driver gets its IRP parameters from the **current IRP stack location**. If you pass an IRP down the stack of drivers for the current device, you must set up the next stack location with the correct parameters. The parameters that you pass down may be different from the ones you are working with.

Figure 3.1   IRP Overview

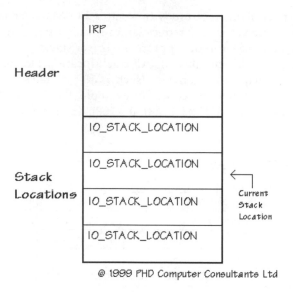

@ 1999 PHD Computer Consultants Ltd

IRP Parameters

When a write I/O request is made into an IRP, the I/O Manager fills in the main IRP header and builds the first IRP stack location. For a write, the IRP header contains the user buffer information. If you use Buffered I/O, the IRP *AssociatedIrp.SystemBuffer* field contains a pointer to the nonpaged copy of the user's buffer. For Direct I/O, the IRP *MdlAddress* field has a pointer to the user buffer MDL.

The IRP stack location has the main write request parameters. The stack *MajorFunction* field has the IRP_MJ_WRITE major function code that indicates that a write has been requested. The *Parameters.Write.Length* field has the byte transfer count and *Parameters.Write.ByteOffset* has the file pointer. The stack also has other important fields that are described later.

As I said earlier, you have to set up the next stack location if you call another driver. This means that you can theoretically change the IRP major function code to something else. While this might work in some cases, it is generally not a good idea, as the parameters in the fixed part of the IRP might not be correct.

However, it might be appropriate to change the number of bytes to transfer and the file pointer. You might do this if you know that the lower driver can only handle short transfers. To handle a large transfer, you might, therefore, send the IRP down several times. Each transfer request you send down will be within the capabilities of the lower driver.

Processing IRPs in a Driver Stack

In practice, the I/O stack locations are not usually used to alter these fundamental IRP parameters. Instead, the IRP stack is normally used to let an IRP be processed by all the drivers in a device stack.

Figure 3.2 shows how an IRP might be processed by four drivers in the device stack. The first IRP arrives at the highest driver, Driver 1. This uses the function `IoGetCurrentIrpStack-Location` to obtain a pointer to the current stack location. The figure shows that this returns the topmost IRP stack location.

Driver 1 decides that it needs to pass the IRP down the stack for processing. The IRP might be a Power Management IRP that the lowest bus driver needs to see. Driver 1 might not do anything with this IRP, but it still needs to pass the IRP down the stack.

Driver 1 therefore sets up the stack location for the next driver. In many cases, it simply copies the current stack location to the next using the `IoCopyCurrentIrpStackLocation-ToNext` or `IoSkipCurrentIrpStackLocation` functions. If you need to alter the next stack location, then use `IoGetNextIrpStackLocation` to get a pointer to it.

Driver 1 then calls the next driver down the stack using the `IoCallDriver` function. The I/O Manager now changes the "current IRP stack location" pointer so that Driver 2 sees the second IRP stack location down (the one that Driver 1 set up for it). This process continues until the lowest driver, Driver 4, receives the IRP.

Driver 4 now processes the IRP. When it has finished with the IRP, Driver 4 calls `IoCompl-eteRequest` to indicate that it has finished processing the IRP. The IRP travels back up the device stack until it eventually pops out the top and is returned to the user.

Each of the drivers in the stack is given an opportunity to work with the IRP again as it travels up the stack, if they wish to do so. To do this, a driver must attach a *completion routine* to an IRP using the `IoSetCompletionRoutine` function. The completion routine information is stored in each driver's IRP stack location. This lets each of the drivers numbered 1 to 3 work on the IRP as it travels back up the stack in the order: Driver 3, Driver 2, and then Driver 1. A driver does not need to attach a completion routine. In this case, the I/O Manager does not call the driver as the IRP passes up the stack.

Figure 3.2   How all drivers in a stack process an IRP

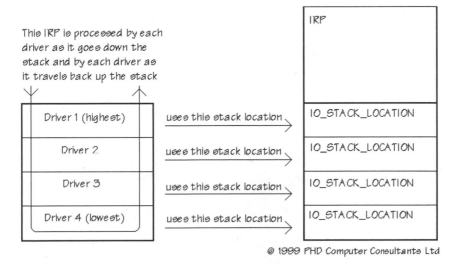

@ 1999 PHD Computer Consultants Ltd

A driver does not have to pass an IRP down the stack. If it detects an error in a parameter, or is able to process the IRP itself, then it should do its job and complete the IRP with IoCompleteRequest.

When a driver processes an IRP as it travels back up the device stack, it does not necessarily let the IRP progress further up the stack straightaway. If Driver 2 for example, splits Write IRPs into small chunks that Driver 3 can digest, then it may send the IRP back down the stack again, with changed parameters.

A further possibility is that a driver can build a new IRP and send it down the stack. Driver 2 might be processing a read request. To satisfy this request, it has to send an IOCTL request to the lower drivers. When the IOCTL request returns, Driver 2 checks that it worked satisfactorily and carries on processing its Read IRP.

To reiterate, an IRP includes a stack of I/O request operations. A driver only looks at the current IRP stack location and does not have to worry if there are higher-level drivers above.

Conclusion

Windows provides a well-defined hierarchy and structure for drivers. Go with the grain and learn how to fit in nicely with the rest of the operating system. Do not forget to cope with multiprocessor machines.

Windows provides a generic model of different processors that may be running. Drivers have routines that run at PASSIVE_LEVEL, DISPATCH_LEVEL, and DIRQL interrupt levels. Use system memory carefully and access user memory using the relevant kernel routines.

The kernel calls a driver in many ways. Most driver processing is in response to I/O Request Packets (IRPs). Conversely, a driver can make use of a whole host of kernel routines. Many system drivers have specific interfaces defined to handle particular I/O control codes. Drivers should use these interfaces wherever possible.

Enough prevarication, let's start on the first real device driver. The following chapters describe the Wdm1 virtual device driver in detail, how it is implemented and used, and how to test and debug drivers.

Chapter 4

WDM Driver Environment

This chapter writes a simple WDM device driver, called Wdm1. It shows how Wdm1 is built and installed in Windows 98 and Windows 2000. The basics of the driver are explained, but a full explanation of the guts has to wait until the following chapters.

The Wdm1 driver is for a virtual WDM device that does not correspond to any real hardware. For now, the Plug and Play and Power Management support is minimal. The Wdm1 driver implements a shared memory buffer for all Wdm1 devices.

First, I describe how to set up your computer for WDM driver development by installing the various development kits and by setting up Visual Studio. I describe the other useful tools that you will need.

The book software provides a Visual Studio workspace called WDM Book that you can use to compile the drivers and any associated user mode Win32 applications. Be careful if editing within this workspace as the different projects in this workspace often have files of the same name.

System Set Up

The section details what you will have to do to set up your development computer or computers for WDM driver development. This task is laborious, especially as you have to do it at least twice, once for Windows 2000 and once for Windows 98.

You will have to be an Administrative user to install the W2000 Driver Development Kit (DDK) and drivers in Windows 2000.

These instructions assume that you are using Visual C++ for development. While other compilers can almost certainly do the job, most Windows driver writers will be firmly in the

Visual Studio camp. Visual C++ also has various useful tools that you will need such as *rebase* and *guidgen*. I used VC++ version 5.

You will need a Microsoft Developer Network (MSDN) Professional (or Universal) subscription to get the necessary development kits. While some of these kits are available free online, it is best to get an MSDN subscription to ensure that you receive the most recent releases and beta versions. The Installable File System (IFS) DDK costs extra and is only available currently in the USA and Canada.

DDKs

Install the W98 DDK in Windows 98. Install the W2000 DDK in Windows 2000. Install the Platform Software Development Kit (SDK) and MSDN Library in both versions of Windows. The Platform SDK is not vital, but it has quite a few useful tools.

The DDK tools and documentation are listed in the Start+Programs+Development kits menu. The Platform SDK tools are listed in the Start+Programs+Platform SDK menu.

The SETENV.BAT file in the DDK bin directory is used to set the environment variables for the build process[1]. Shortcuts to this batch file are set up in the Start menu. It calls other batch files in the bin directory. The current batch files install DosKey; I decided to comment these lines out as I already have a similar command processor installed.

The Platform SDK may have more up to date tools than are supplied with VC++ (e.g., the *rebase* utility). Put the MSSDK bin directory before the Visual Studio directories on your path.

You may need to add the DDK bin directory to the directories list in VC++.

Currently, the documentation is released in HTML Help format. Each help file is viewed in its own incarnation of the HTML Help viewer, and is not merged into the Visual Studio 97 documentation. The help files are probably merged together in Visual Studio 6.

Each function found in the W2000 DDK states which header files you can use. If wdm.h is listed, the routine can be used in a WDM driver for Windows 98 or Windows 2000. There are some routines that compile correctly in Windows 98 but have no effect, such as the event log writing routines. If the W2000 DDK lists only ntddk.h for a routine, it can be used only in Windows 2000 drivers, not Windows 98. The W98 DDK lists only routines that run in Windows 98.

The W98 DDK and the W2000 DDK are largely the same for WDM drivers. At the time of writing, the W98 DDK does not include the Windows Management Instrumentation libraries. Some header files are also different (e.g., SETUPAPI.H). For some reason, the checked build produced by W2000 seems quite a bit smaller than that produced by the W98 DDK.

Book Software Installation

The source code for the Wdm1 driver is supplied on the book's accompanying disk. Copy the source code directory tree to a fresh directory on your hard disk. These instructions assume that you have installed the software to C:\WDMBook. However, you can change the directory if you wish. If you use a different drive, you will need to alter the settings of every VC++ project, so it will be much easier for you if you can install the software on drive C:.

---

1. You will probably need to ensure that the Windows 98 environment space size is at least 2048 for device driver development.

The `WDM Book` Visual Studio workspace in the book software base directory contains projects for each of the drivers in the book.

Three steps are needed before the book software drivers can be compiled.

- Ensure that the `DDKROOT` environment variable is set to the W2000 DDK or W98 DDK base directory.

- Set up an environment variable called `WDMBOOK` to point to the book software base directory (e.g., `C:\WDMBook`). In Windows 98, set environment variables by adding them to `AUTOEXEC.BAT`. You will probably need to reboot for the new definitions to come into effect. In Windows 2000, set environment variables in the Control Panel System properties applet "Advanced tab" Environment variables editor. You will need to restart Visual Studio and any DOS boxes for these changes to take effect.

- Ensure that the `MakeDrvr.bat` batch file in the book software base directory is available to be run by Visual Studio, as it is used to invoke the build make process. Either add `MakeDrvr.bat` to a directory that is on your path, or add the book software base directory to the list of "Executable files" directories in the Visual Studio Tools+Options Directories tab.

Recompiling the Book Software

You can check that you have installed the development kits and book software correctly by compiling the book software drivers[2]. From the `Start+Programs+Dvelopment Kits+Windows XX DDK` menu, select either the free or checked build environment. Then move to the book software base directory and enter `build -nmake /a`. If you want to recompile the Win32 applications, you will have to do this in the Visual Studio workspace.

Check that the drivers are actually updated. In the worst case, it will appear as if the files have compiled but nothing will in fact have been done. If the example drivers do not compile, then there are two usual causes. The first possible problem is that the environment variables have not been set up correctly.

The second problem is that some of the source code explicitly includes header files from the Windows 98 DDK, which I assume has been installed in the `C:\98DDK` directory. Some of the linker settings have been also been set up to refer to this same directory. If you have used a different directory, or have only installed the Windows 2000 DDK, you will need to change each instance where `C:\98DDK` is mentioned, either in the source code or in the `SOURCES` file.

The `Wdm3` example specifically refers to a library in the W2000 DDK in its `SOURCES` files. I assume that this DDK is installed in `C:\NTDDK`, so change this if need be.

The *Wdm2Power* Win32 application may need some special project settings before it will compile. See Chapter 10 for details.

Shortcuts

I find it useful to have shortcuts to common tools available, more quickly than through the Start menu.

One possible place for these shortcuts is in the Visual Studio Tools menu. Another option is to place the shortcuts in the *Quick Launch* folder on the Taskbar.

---

2.   The `PHDIo` and `Wdm3` drivers are not included in the compilation list as they do not compile in Windows 98.

The option I prefer is to use the free *QuickStart* utility, available from my company, PHD Computer Consultants Ltd., website at www.phdcc.com. This displays an icon in the Taskbar tray. Clicking this icon shows a menu of various useful options, including submenus for each Desktop folder that you have. Make sure that you put QuickStart in the Startup folder so that it runs from start up.

I therefore suggest making a *Tools* Desktop folder with shortcuts copied from the Start menu. Table 4.1 shows the most useful tools that I put in this folder.

Table 4.1   Useful Tools

Command line prompt	`cmd` or `command` (or equivalent, e.g. 4NT)	
Free build environment	DOS box set up for driver free builds	
Checked build environment	DOS box set up for driver checked builds	
Registry Editor	`%windir%\regedit.exe`	
W2000 Registry Editor	`%windir%\System32\regedt32.exe`	NT and W2000
WinObj	`C:\MSSDK\bin\winnt\Winobj.Exe`	NT and W2000
Computer Management	`%SystemRoot%\system32\compmgmt.msc /s` (from Start+Programs+Administrative tools)	W2000 only
Servicer	`%WDMBOOK%\Servicer\Release\Servicer.exe`	NT and W2000
WBEM Object Browser	`<WBEM>\Applications\browser.htm`	
DebugPrint Monitor	`%WDMBOOK%\DebugPrint\exe\Release\DebugPrintMonitor.exe`	
Wdm2Power	`%WDMBOOK%\Wdm2\Power\Release\Wdm2Power.exe`	
Driver Verifier	`Start+Programs+Development Kits+Windows 2000 DDK+Driver Verifier`	W2000 only

You might find a *Documentation* Desktop folder useful for all the DDK and SDK help shortcuts.

Utilities

This section gives a brief overview of the most useful tools for driver development and how they might best be used. Their detailed use will be covered later.

DOS Boxes

Although you can do most of your development work in Visual Studio, you will occasionally need to work at a command line prompt.

The supplied DDK *build* tool runs at a command prompt. The environment variables have to be set up appropriately as free or *checked* driver development. The DDKs provide two shortcuts to get these command prompt boxes.

Computer Management Console

The Windows 2000 Computer Management Console (from `Start+Programs+Administrative tools`) has a useful System Tools section with Event Viewer and Device Manager tools.

The *Event Viewer* displays the Windows 2000 events. Drivers should write to the System Log to inform administrators of informational, warning, or error events.

The *Device Manager* shows all your Windows 98 or Windows 2000 devices and lets you change their properties (e.g., change port settings and reallocate resources). You can change (i.e., update) the driver for a device, and uninstall a device. The Device Manager is also available in the Control Panel System applet.

NT Devices Applet

In NT 4 and NT 3.51, the Control Panel *Devices* applet lets you see what drivers are running. You can stop or start drivers and change their start up characteristics. In W2000, the Device Manager does not initially list the non-WDM drivers. Selecting the "Show hidden devices" checkbox brings up a list of these drivers. You can start or stop them and change their startup characteristics. The book *Servicer* program also lets you do some of these functions.

Hardware Wizard

Some drivers are loaded automatically when Windows detects a new device in the system. For others, you need to use the Control Panel Add New Hardware wizard to prompt the loading of the driver.

Registry Editors

You will probably need to inspect the registry at some point during driver development. The basic Registry Editor *regedit* is useful for most jobs, particularly as it can search the registry. However, for some jobs in NT and Windows 2000, you will need to use **RedEdt32**, as it can handle all the registry types, such as `REG_MULTI_SZ` strings.

This book assumes that you have a reasonable working knowledge of the registry. `HKLM` is used as abbreviation for `HKEY_LOCAL_MACHINE`. The DDKs sometimes refer to `HKLM` as `Registry\MACHINE`.

INF Editor

The *InfEdit* tool can be used (in Windows 98 only) to create and edit INF files. However, it seems to clobber some INF file information and does not display Windows 2000 specific sections correctly.

WBEM

The WBEM SDK includes the **WBEM Object Browser**. Use this to inspect the Windows Management Instrumentation data that drivers produce.

Debuggers

There are various tools available to help you test and debug drivers. Unfortunately, the Visual Studio debugger cannot be used on kernel mode code. The Microsoft **WinDbg** kernel mode debugger must be used between two NT or W2000 computers.

NuMega Compuware sells the SoftICE debugger, which can run on the same PC as the driver under test. Use the NuMega utility **nmsym** utility to build the necessary symbols for SoftICE.

Finally, some tools let you include trace statements in your driver code that can be viewed in a user mode application on the same PC. The Compuware Numega DriverWorks software includes its **Driver Monitor** tool for this job. The **DebugPrint** software from PHD Computer Consultants Ltd. provides a similar facility. DebugPrint is used extensively in this book and its innards are described in detail as an example of a fully working driver. The **OSR DDK** software from Open Systems Resources logs all your DDK function calls so that they can be viewed by their OSRTracer tool.

NT and Windows 2000 Utilities

WinObj is an NT and W2000 utility that displays various Windows objects, including device names and symbolic links.

The \device branch displays the device names. These device names are not directly visible to user applications. Non-WDM kernel mode device drivers must provide a symbolic link between a Win32 visible name and the underlying device name. The \?? branch displays these symbolic links. Double-clicking on \??\COM1 shows that it is a symbolic link to the underlying device \Device\Serial1. (The \?? branch used to be called \DosDevices, so this name appears in some of the DDK documentation).

Double-clicking most other **WinObj** entries yields no useful information.

ObjDir is a command line version of **WinObj**.

Drivers is a command line utility that lists all drivers and their memory usage.

Book Software Tools

DebugPrint

The **DebugPrint** software is used by test drivers to produce formatted print statements. The trace messages are viewed in the **DebugPrint Monitor** Win32 application. This tool is explained in full in Chapter 6, because the book software makes extensive use of this debugging technique. Chapter 14 describes the source for the DebugPrint driver.

Wdm2Power

The **Wdm2Power** tool firstly lets you inspect the AC and battery status of your system. It also displays any system power events. Finally, it lets you suspend or hibernate your computer.

MakeDrvr

This batch file is used from Visual Studio to initiate each build, as described in the following text.

Servicer

The *Servicer* utility lets you inspect the status of any running Windows 2000 service, including drivers. You can see if a driver is running or not, and attempt to start or stop it.

Driver Targets

Drivers can be built in free or checked versions, and in NT and Windows 2000 for different processors.

The *free* target is the final release retail version, optimized as necessary with all debug symbols removed.

The *checked* target is an unoptimised debug version that includes symbols to make debugging easier.

NT and Windows 2000 come in free and checked versions. If you use the Microsoft *WinDbg* debugger then you need two computers running NT or W2000. The development PC should be the faster computer running the free version of Windows. The driver should be running under test on the other target PC that is running the checked build. The fact that there are fewer resources available on the target system is good as it makes it easier to check that your driver will work in stressful situations.

Windows 2000 also runs on the Dec Alpha platform, so you can also build for the Alpha platform free and checked targets. This book only discusses the x86 platform.

In this book, the emphasis is on writing drivers that work in both Windows 98 and Windows 2000. However, a few features are present in only one operating system. The following preprocessor directives can be used to determine whether you are using the W2000 DDK or the W98 DDK. If you have separate versions of your driver, the installation files will have to be slightly different. As Chapter 11 shows, a single installation INF file can include separate instructions for W2000 and W98.

```
#if _WIN32_WINNT>=0x0500
    // W2000+ code
#else
    // W98 code
#endif
```

Driver Language and Libraries

A driver is a Dynamic Link Library (DLL) with the file extension .sys. It is usually written in C or C++ and can include resources, such as a version block, event messages, and Windows Management Instrumentation (WMI) class definitions. In Windows 98, a driver executable must have an 8.3 filename.

Although drivers were traditionally written just in C, it is quite straightforward to use C++. The main requirement is to use the extern "C" directive in a couple of important places. However, do not use the new keyword in C++. The new keyword may be implemented using malloc, which is not available to kernel mode drivers.

In fact, most standard libraries and classes are not available to driver writers in either language, because they make inappropriate use of memory. If you are using one of the proprietary driver development kits, these provide various useful classes, including safe memory allocators.

Instead, you can use any of the routines provided by the operating system to kernel mode devices, as described in Chapter 3. While these are useful, it takes a while to get used to the different set of routines that are available.

Assembly code can be used if absolutely necessary. Obviously, this makes for more work if you port the driver to the Windows 2000 Alpha platform.

Resources

A resource .rc file should include a standard version block. Increment the version numbers as new builds are released. Make sure that you keep a full source backup of each version you release. Many version control packages can help you manage this task.

If you generate NT or Windows 2000 events, you should write these in an .mc file that is compiled using the *mc* utility and included in a driver's resource file. More details of this process are in Chapter 13.

Similarly, if you generate custom WMI classes, you need to write a .mof file that is compiled using the *mofcomp* tool and included in the driver's resource file. See Chapter 12 for more information.

A sophisticated driver might need to download microcode to its device and so would include the microcode as a binary resource in its executable.

Good Code

A driver is an integral part of the operating system, so it can easily crash Windows if it goes wrong. You do not have the protection of a Win32 address space to stop you from overwriting memory that does not belong to you.

Please be especially careful when you write your driver. Keep it as simple as possible and document it well.

Treat all compiler warnings as errors that need to be fixed. For example, whether an integer is signed or not can make all the difference.

Make sure that you check the return values of all kernel functions that you call, and act on them accordingly. For example, if you get an error after you create a device, make sure that the clean-up code deletes the device.

Make sure that you use the kernel resources carefully, particularly memory. Some resources are scarce and overuse may degrade system functioning.

build Utility

The DDK *build* command line utility is the primary tool for building drivers. It invokes the *nmake* make utility to build your driver using the correct compiler and linker settings. If necessary, *build* can be used to build standard user mode Win32 executables, etc.

The next section describes how to invoke *build* from within Visual Studio. However, you must still set up *build* so that it can be run at the command line. As well as your source code, you must specify a SOURCES file, a standard makefile, the directory structure, and optionally a makefile.inc file and a dirs file. All these steps are described in the following text.

build displays progress details and error results to its standard output. In addition, it lists the errors in a file called build.err, the warnings in build.wrn, and a log in build.log. In

W2000 there are free and checked build versions of each of these files, i.e., `buildfre.log`, `buildchk.log`, etc.

makefiles

Younger readers may not have come across `makefiles`. In the days before Integrated Development Environments (IDEs) such as Visual Studio, you had to use `makefiles` to determine which files in a project needed recompiling. If you changed only one module in a project consisting of eight modules, you want to recompile only that one module and then link the whole lot together.

The *nmake* utility uses instructions in a file called `makefile` to determine what commands to run to update a project. The following `makefile` shows that if `haggis.cpp` has been updated, it is compiled into `haggis.obj` using the *cl* compiler. `haggis.obj` is linked to make `haggis.exe` using the link tool.

```
haggis.exe:  haggis.obj
             link -o haggis.exe haggis.obj

haggis.obj:  haggis.cpp
             cl haggis
```

Most `makefiles` are a good deal more complicated than this, which is why they were happily forgotten when IDEs came along. However, setting up the compiler and linker settings for drivers is quite a complicated task. Therefore, Microsoft has stuck with `makefiles`. See the Visual Studio *nmake* documentation for more details of makefiles.

SOURCES

build looks for an *nmake* macro file called `SOURCES` in the current directory for details of what to build.

Listing 4.1 shows the `SOURCES` file for the Wdm1 project. It specifies that the driver target name is `Wdm1.sys`, that it is a WDM driver, and that it should be built in the `OBJ` subdirectory. Source browser information should be generated. The DDK `inc` directory is added to the search list for header files. The `SOURCES` macro specifies a list of files to compile. `NTTARGET-FILES` specifies some post build steps, as described in the following text.

Other less common `SOURCES` macro definitions can be found in the DDK documentation. Note that there must be no spaces between the `SOURCES` macro and its equal sign.

Listing 4.1 Wdm1 **project** SOURCES **file**

```
TARGETNAME=Wdm1
TARGETTYPE=DRIVER
DRIVERTYPE=WDM
TARGETPATH=OBJ
BROWSER_INFO=1


INCLUDES=$(BASEDIR)\inc;
```

Listing 4.1   Wdm1 **project** SOURCES **file (continued)**

```
SOURCES=init.cpp \
        dispatch.cpp \
        pnp.cpp \
        DebugPrint.c \
        version.rc


NTTARGETFILES=PostBuildSteps
```

makefile **File**

You must provide a standard file called makefile as shown in Listing 4.2. This invokes the standard make file makefile.def in the DDK inc directory.

As it says, do not edit this file at all. If you want to add to the list of files to compile, add them to the SOURCES macro in the SOURCES file.

Listing 4.2   Wdm1 **project** makefile

```
#
# DO NOT EDIT THIS FILE!!!  Edit .\sources. if you want to add a new source
# file to this component.  This file merely indirects to the real make file
# that is shared by all the driver components of the Windows NT DDK
#

!INCLUDE $(NTMAKEENV)\makefile.def
```

build Directories

Windows 98 and NT

build always puts the compiled object files in the OBJ\i386 subdirectory (for x86 targets). The SOURCES TARGETPATH macro specifies where the final executables go. If you specify OBJ for this, then the driver executables go in the OBJ\i386\free and OBJ\i386\checked subdirectories. There is a long-standing bug in *build* that means that you have to create these last two directories by hand for *build*. If you were making Wdm1 from scratch, then you would have to make subdirectories OBJ\i386\free and OBJ\i386\checked.

    Both the free and checked builds put their object files in the same directory. If you switch between these build types, make sure that you rebuild all the files in the project (by putting "-nmake /a" on the *build* command line).

    The checked build output file OBJ\i386\checked\Wdm1.sys contains the debug symbols. The build process also generates a file called Wdm1.dbg with the debug symbols in the OBJ\i386\free directory.

Windows 2000

In W2000, *build* keeps the free and checked build object files separate. If the TARGETPATH is OBJ, the free build x86 object files and the final driver go in the OBJFRE\i386 directory. The checked build object files and driver go in the OBJCHK\i386 directory.

Wdm1 Directories

The end result is the following series of subdirectories for Wdm1.

Directory	Contents
OBJ	Has *build* list of files to build in _objects.mac
OBJ\i386	W98 Compiled object files
OBJ\i386\free	W98 Free build Wdm1.sys
OBJ\i386\checked	W98 Checked build Wdm1.sys
OBJFRE\i386	W2000 Free build objects and Wdm1.sys
OBJCHK\i386	W2000 Checked build objects and Wdm1.sys

Other *build* Steps

Another makefile called makefile.inc is invoked if you use certain optional macros in the SOURCES file. Table 4.2 shows the macro name and when the make target is invoked.

Table 4.2   SOURCES **optional macros**

SOURCES macro name	When invoked
NTTARGETFILE0	after dependency scan
NTTARGETFILE1	before linking
NTTARGETFILES	during link

Listing 4.3 shows the standard makefile.inc that is used in all the book software projects.

Listing 4.3   **Book software projects** makefile.inc

```
PostBuildSteps: $(TARGET)
!if "$(DDKBUILDENV)"=="free"
    rebase -B 0x10000 -X . $(TARGET)
!endif
    copy $(TARGET) $(WINDIR)\system32\drivers
```

The line in the Wdm1 SOURCES file that says NTTARGETFILES=PostBuildSteps ensures that the target PostBuildSteps is built during the link process. As the PostBuildSteps target depends on $(TARGET) — the driver executable — the build commands for PostBuildSteps are carried out after the driver is built. The PostBuildSteps output is displayed in the build.log file. Note that problems in PostBuildSteps may not evident unless you inspect this file.

The actual post-build steps in makefile.inc do two jobs. First, the *rebase* utility is run on the driver executable for free builds, and the driver is copied to the Windows system32\drivers directory. Copying the driver does not mean that it is installed.

rebase strips any remaining debug symbols that are left, even in the free driver executable. The base load address is kept at 0x10000 and the symbols are put in the .dbg file in the current directory.

DIRS **File**

The final main feature of *build* is that it can recursively build files in other directories. If a DIRS file is present, build looks at the DIRS directive in there for a list of directories to build. These directories may themselves contain further DIRS files.

The book software base directory has a DIRS file with the following line:

```
DIRS=Wdm1 Wdm2 WdmIo UsbKbd HidKbd DebugPrint PassThru
```

Running *build* in the book software base directory will compile the drivers in all these directories. The companion Win32 user mode applications are not built by this process.

The directories listed in the DIRS directive must be only one level below the current directory. Therefore, the Wdm1 directory has a DIRS file that instructs *build* to go to the SYS subdirectory.

VC++ **Projects**

You can set up VC++ to build drivers from within Visual Studio. It is possible to configure a project's settings so that Visual Studio can compile your driver directly. However, it is laborious changing all the settings and is error prone. The final problem is that Microsoft might change driver compile or link requirements in the DDK standard makefile. Such changes would not be reflected automatically in your settings.

The best way therefore is to use a *Makefile* project. This invokes a command line utility to build the driver. The downside of this approach is that some common tasks, such as adding a file to the compile list, have to be done in a different way, as described in the following text.

The book software projects are set up already to use the *Makefile* technique. This eventually invokes the *build* command as described previously. All the necessary build files must be set up correctly: SOURCES, makefile, the target directories, and possibly the makefile.inc and DIRS files.

You might find it useful to select the Visual Studio Tools+Options menu Editor tab "Automatic reload of externally modified files" checkbox so that changes to the build log files are loaded with no fuss.

Makefile **Build Environment**

When you make a new *Makefile* project, Visual Studio gives you two *build* configurations by default, "Win32 Debug" and "Win32 Release". I prefer to use the Build+Configurations menu to remove these and have configurations named "Win32 Checked" and "Win32 Free", instead.

For the free configuration, set the project settings as shown in Table 4.3. For the checked build, change "free" to "checked" in the build command line and the browse info filename.

If you installed the book software to a driver other than C: you will need to change the drive letter in the build command line.

The build command line runs the MakeDrvr.bat batch file, using the DDKROOT and WDMBOOK environment variables. The options -nmake /a are added to this command line if you request a complete rebuild in Visual Studio. The output filename is set so that the correct name is displayed in the *build* menu.

Table 4.3   Win32 Free configuration settings

Build command line	MakeDrvr %DDKROOT% c: %WDMBook%\wdm1\sys free
Rebuild all options	-nmake /a
Output file name	Wdm1.sys
Browse info file name	obj\i386\free\Wdm1.bsc (W98/NT) objfre\i386\Wdm1.bsc (W2000)

MakeDrvr

When you ask Visual Studio to build your driver, the batch file MakeDrvr.bat, listed in Listing 4.4, is run. This is always passed at least four parameters: the DDK base directory, the source drive, the source directory, and the build type ("free" or "checked"). Any further arguments are passed straight to build.

MakeDrvr first does some basic checks on the parameters that are passed. It then calls the DDK setenv command to set up the environment variables correctly for the build target, changes directory to the source drive and directory, and finally calls build. The -b build option ensures that the full error text is displayed. The -w option ensures that the warnings appear on the screen output, so that Visual Studio can find them in the *build* Output window.

The screen output of the MakeDrvr command file appears in the Visual Studio Output window. You can then use F4 as usual to go the next error or warning.

Listing 4.4   MakeDrvr.bat

```
@echo off
if "%1"=="" goto usage
if "%3"=="" goto usage
if not exist %1\bin\setenv.bat goto usage
call %1\bin\setenv %1 %4
%2
cd %3
build -b -w %5 %6 %7 %8 %9
goto exit

:usage
echo usage    MakeDrvr DDK_dir Driver_Drive Driver_Dir
              free/checked [build_options]
echo eg       MakeDrvr %%DDKROOT%% C: %%WDMBOOK%% free -cef
:exit
```

Directories

The *build* output goes into the OBJ, OBJFRE, or OBJCHK subdirectories.

In Windows 98, the free driver is in OBJ\i386\free\Wdm1.sys, with debug symbols in OBJ\i386\free\Wdm1.dbg. The checked build products are in OBJ\i386\checked. The intermediate object files for both builds are in the OBJ\i386 directory. If you change from the free to checked targets, make sure that you do a "Rebuild all" to recompile all the source files with the correct debug preprocessor defines.

In Windows 2000, the free build object files and driver are in OBJFRE\i386 and the checked build products are in OBJCHK\i386. A complete rebuild is not required if you switch between free and checked builds.

Common Tasks

As you are using a Visual Studio *Makefile* project, these common tasks must be done is a different way.

Add File to Project

If you add a file to your project, you must add it to the SOURCES file for it to be built.

Make Browse Information

Source browser information is generated alongside the target executable if the SOURCES file contains this line.

```
BROWSER_INFO=1
```

Build Steps

Various additional build steps can be defined in a makefile.inc file if the SOURCES file contains one or more of the NTTARGETFILE0, NTTARGETFILE1, or NTTARGETFILES macros.

The results of any additional build steps are not shown in the Visual Studio output window, but only in the build.log file. The build does not necessarily stop if one of these steps fail.

You could also alter MakeDrvr.bat to do tasks that are common to all your driver projects.

Compiling a Single File

You cannot compile a single file in *Makefile* projects.

The Wdm1 Driver Code

Table 4.4 lists all the source files used by the first driver called Wdm1. Table 4.5 lists all the build files that have already been described. In W2000 the *build* output files have slightly different names. These files are on the book CD-ROM.

This chapter looks at only some of the source files. As far as possible, the minimum possible functionality has been implemented in Wdm1. For example, a stub function has been written to handle Win32 create file requests. These stub functions usually make each request succeed.

If you were to put this Wdm1 driver to the test, it would not work in some circumstances. The succeeding chapters explain how the driver works and how its functionality has been enhanced to make it work better, as well as showing how to call the driver from Win32 code.

Table 4.4   Wdm1 source files

Wdm1.h	Driver header
Init.cpp	Entry and unload code
Pnp.cpp	Plug and Play and Power handling code
Dispatch.cpp	Main IRP dispatch routines
DebugPrint.c	DebugPrint code
DebugPrint.h	DebugPrint header
Wdm1.rc	Version resource
GUIDs.h	GUID definition
Ioctl.h	IOCTL definition
resource.h	Visual Studio resource editor header
Wdm1free.inf	Free build installation instructions
Wdm1checked.inf	Checked build installation instructions

Table 4.5   Wdm1 build files

SOURCES	*build* instructions
makefile.inc	Post build steps for makefile
makefile	Standard makefile
MakeDrvr.bat	*Makefile* project batch file
build.log	*build* results log output
build.err	*build* errors output
build.wrn	*build* warnings output

Compiler Options

The code includes some directives to the compiler that need some explaining.

The extern "C" directive is used to ensure that the compiler uses the correct linkage to reference kernel routines. The driver entry point, DriverEntry, must have extern "C" to ensure it is found.

The #pragma code_seg preprocessor directive is used to force routines into certain code segments. The INIT code segment is discarded after the driver has initialized itself, and the PAGE segment contains code that can be paged out of kernel memory. Using segments helps to lower a driver's memory usage.

Only routines that run at PASSIVE_LEVEL IRQL can be paged from memory. All the dispatch routines in the Wdm1 driver operate at PASSIVE_LEVEL so they can be put in the PAGE segment. Routines that are not given a segment are never paged from memory.

If writing C code, you can use the alloc_text pragma instead to set the code segment of named routines.

Header Files

The Wdm1.h header file is included in all the source files. Wdm1.h first includes the main DDK header file for WDM projects, wdm.h. If you were writing NT style drivers, you would use the similar ntddk.h. You may find that you need to include some other DDK header files for particular types of drivers.

Next, a device extension structure is defined. This structure is where a driver can hold any information it needs about a device (more on this later).

The GUIDs.H header defines a Globally Unique Identifier (GUID) for the Wdm1 device interface. The next chapter shows how this GUID is used by Win32 user mode applications to find Wdm1 devices. I used the *guidgen* utility to generate this GUID in the DEFINE_GUID format. I defined several consecutive GUIDs at once for all the examples in this book.

Finally, the IOCTL.H header defines the IOCTL codes that Wdm1 supports. These are explained in Chapter 7.

Driver Entry Module

Init.cpp contains the driver entry point. Listing 4.5 shows this routine, which must be called DriverEntry and use C linkage.

The Plug and Play Manager locates the correct driver and calls DriverEntry to initialize the driver (at PASSIVE_IRQL). In DriverEntry, the main job is to store a series of call back routine pointers in the passed DriverObject. This DRIVER_OBJECT structure is used by the operating system to store any information relevant to the driver. A separate structure is used later to store information about each device.

In Wdm1, DriverEntry sets a whole series of callback routines. These routines are called by the kernel when a device is added and when IRPs need to be sent to the driver. The WdmUnload routine, later in Init.cpp, does nothing at this stage.

Finally, DriverEntry returns an NTSTATUS value of STATUS_SUCCESS. Almost all driver routines have to return a NTSTATUS value, from the list in the NTSTATUS.H DDK header file. Note that these error codes do not correspond to Win32 error codes. The kernel does the necessary mapping between the two types of error code.

Listing 4.5   DriverEntry **routine**

```
extern "C"
NTSTATUS DriverEntry( IN PDRIVER_OBJECT DriverObject,
                      IN PUNICODE_STRING RegistryPath)
{
    NTSTATUS status = STATUS_SUCCESS;

    // ...
```

Listing 4.5   DriverEntry routine (continued)

```
    // Export other driver entry points...
    DriverObject->DriverExtension->AddDevice = Wdm1AddDevice;
    DriverObject->DriverUnload = Wdm1Unload;

    DriverObject->MajorFunction[IRP_MJ_CREATE] = Wdm1Create;
    DriverObject->MajorFunction[IRP_MJ_CLOSE] = Wdm1Close;

    DriverObject->MajorFunction[IRP_MJ_PNP] = Wdm1Pnp;
    DriverObject->MajorFunction[IRP_MJ_POWER] = Wdm1Power;

    DriverObject->MajorFunction[IRP_MJ_READ] = Wdm1Read;
    DriverObject->MajorFunction[IRP_MJ_WRITE] = Wdm1Write;
    DriverObject->MajorFunction[IRP_MJ_DEVICE_CONTROL] = Wdm1DeviceControl;

    DriverObject->MajorFunction[IRP_MJ_SYSTEM_CONTROL] = Wdm1SystemControl;

    // ...

    return status;
}
```

Version Resource

Wdm1.rc simply defines a version resource block with version and copyright information.

Accessing the Registry

The RegistryPath parameter to DriverEntry contains the registry key of the driver. The Wdm1 code does not currently use its RegistryPath parameter. However, it is common for drivers to use its registry key to store parameters for the whole driver. Therefore, I present the ReadReg routine shown in Listing 4.6. This reads two values from the driver's registry path Parameters subkey. The first value is a ULONG obtained from the value named UlongValue. The second value is a string obtained from the default value for the Parameters key.

Eventually ReadReg needs to call the RtlQueryRegistryValues kernel routine to read both the registry values in one fell swoop. One of the parameters is the absolute registry path, as a NULL-terminated wide string. The driver registry path is supplied in a UNICODE_STRING structure. Although this contains a wide string buffer, it may not necessarily be NULL-terminated. Unfortunately, this means that we have to laboriously make a copy of the string, simply to add on that dratted NULL-terminating character.

The first section of ReadReg does this job. It works out the length of buffer required and uses ExAllocatePool to allocate the memory from the paged pool. The RtlCopyMemory function is used to copy the bulk of the string over and RtlZeroMemory zeroes that all-important last character. You can do the copy and zero by hand if you want, though it should be more efficient to call the kernel functions.

Listing 4.6   ReadReg

```
void ReadReg( IN PUNICODE_STRING DriverRegistryPath)
{
    // Make zero terminated copy of driver registry path
    USHORT FromLen = DriverRegistryPath->Length;
    PUCHAR wstrDriverRegistryPath =
        (PUCHAR)ExAllocatePool( PagedPool, FromLen+sizeof(WCHAR));
    if( wstrDriverRegistryPath==NULL) return;
    RtlCopyMemory( wstrDriverRegistryPath, DriverRegistryPath->Buffer,
        FromLen);
    RtlZeroMemory( wstrDriverRegistryPath+FromLen, sizeof(WCHAR));

    // Initialise our ULONG and UNICODE_STRING values
    ULONG UlongValue = -1;
    UNICODE_STRING UnicodeString;
    UnicodeString.Buffer = NULL;
    UnicodeString.MaximumLength = 0;
    UnicodeString.Length = 0;

    // Build up our registry query table
    RTL_QUERY_REGISTRY_TABLE QueryTable[4];
    RtlZeroMemory( QueryTable, sizeof(QueryTable));

    QueryTable[0].Name  = L"Parameters";
    QueryTable[0].Flags = RTL_QUERY_REGISTRY_SUBKEY;
    QueryTable[0].EntryContext = NULL;
    QueryTable[1].Name  = L"UlongValue";
    QueryTable[1].Flags = RTL_QUERY_REGISTRY_DIRECT;
    QueryTable[1].EntryContext = &UlongValue;
    QueryTable[2].Name  = L""; // Default value
    QueryTable[2].Flags = RTL_QUERY_REGISTRY_DIRECT;
    QueryTable[2].EntryContext = &UnicodeString;

    // Issue query
    NTSTATUS status =
        RtlQueryRegistryValues(
            RTL_REGISTRY_ABSOLUTE, (PWSTR)wstrDriverRegistryPath,
            QueryTable, NULL, NULL);
```

Listing 4.6   ReadReg **(continued)**

```
    // Print results
    DebugPrint( "ReadReg %x: UlongValue %x UnicodeString %T",
            status, UlongValue, &UnicodeString);


    // Do not forget to free buffers
    if( UnicodeString.Buffer!=NULL)
        ExFreePool(UnicodeString.Buffer);
    ExFreePool(wstrDriverRegistryPath);
}
```

The UNICODE_STRING **Structure**

This is how the UNICODE_STRING type is defined.

```
typedef struct _UNICODE_STRING {
    USHORT Length;
    USHORT MaximumLength;
    PWSTR  Buffer;
} UNICODE_STRING, *PUNICODE_STRING;
```

The *Buffer* field points to a wide 16-bit character buffer. The character string is not usually NULL-terminated. Instead, the *Length* field gives the current size of the string in bytes. The *MaximumLength* field gives the maximum size of the string that can fit in the buffer in bytes. This design is used to avoid reallocating string buffers too often. However, it does make manipulating Unicode strings a bit awkward. Table 4.6 shows all the kernel routines that you can use with Unicode strings.

Just in case you were interested, you do not have to use these kernel routines to access Unicode strings. You can fiddle with a string structure however you like, as long as it is in a valid format when passed to the kernel.

Table 4.6   UNICODE_STRING **functions**

RtlAnsiStringToUnicodeString	Converts an ANSI string to a Unicode string, optionally allocating a buffer.
RtlAppendUnicodeStringToString	Append one Unicode string to another, up to the length of the destination buffer.
RtlAppendUnicodeToString	Append a wide string to a Unicode string, up to the length of the destination buffer.
RtlCompareUnicodeString	Compares two Unicode strings, optionally case-insensitive.
RtlCopyUnicodeString	Copies one Unicode string to another, up to the length of the destination buffer.
RtlEqualUnicodeString	Returns TRUE if the two Unicode strings are equal, optionally case-insensitive.
RtlFreeUnicodeString	Frees the Unicode string buffer memory from the pool.

RtlInitUnicodeString	Sets the Unicode string buffer to point to the given wide string, and sets the length fields to match.
RtlIntegerToUnicodeString	Converts a ULONG value to a Unicode string in the specified base. The string buffer must have been initialized beforehand.
RtlPrefixUnicodeString[†]	Sees if one Unicode string is a prefix of another, optionally case-insensitive.
RtlUnicodeStringToAnsiString	Converts a Unicode string into ANSI. If you ask for the destination ANSI buffer to be allocated, free it eventually with RtlFreeAnsiString.
RtlUnicodeStringToInteger	Converts a Unicode string to an integer.
RtlUpcaseUnicodeString[†]	Converts a Unicode string into uppercase, optionally allocating a buffer.
[†]NT/W2000 only	

ReadReg declares the variables that will receive the values from the registry. For a UNICODE_STRING, its *Buffer*, *Length*, and *MaximumLength* fields have to be initialized. In this case, these fields are initialized to NULL and zero. The call to RtlQueryRegistryValues will allocate a *Buffer* and set the length fields.

In most cases, a UNICODE_STRING's buffer needs to be set up correctly. If you want to store an unchanging wide string value in a Unicode string, use the RtlInitUnicodeString function. The Unicode string *Buffer* is set to point to the passed string and the *Length* and *Maximum-Length* strings are set to the length of the string[3].

```
RtlInitUnicodeString( &UnicodeString, L"\\Device\\Wdm1");
```

If you wish to work with the contents of a Unicode string, you need to provide the wide string buffer. Use code like this.

```
const int MAX_CHARS = 30;
UNICODE_STRING UnicodeString;
WCHAR UnicodeStringBuffer[MAX_CHARS];
UnicodeString.Buffer = UnicodeStringBuffer;
UnicodeString.MaximumLength = MAX_CHARS*2;
UnicodeString.Length = 0;
```

In this case, the buffer is on the stack. If you want to use the Unicode string after this routine has completed, you must allocate the buffer from pool memory. Do not forget to free this memory once you have finished using the string.

Calling RtlQueryRegistryValues

You must send a query table array to RtlQueryRegistryValues. This array details the actions that you want to do. The last entry in the query table must be zeroed to indicate the end of

---

3. Be careful if initializing strings in paged code segments or code segments that are discarded after initialization. The *Buffer* data has the attributes of the underlying code segment and so may not be available when you want to access it. Chapter 14 gives an example in which I initially got this wrong.

the list. This is achieved when the entire query table is zeroed using RtlZeroMemory in ReadReg.

There is a wide range of options available when querying the registry. The *Flags* field in each query table element indicates the action you want to do. The first query listed previously uses RTL_QUERY_REGISTRY_SUBKEY, which means that the *Name* field contains the subkey for subsequent queries.

The following queries both set the *Flags* field to RTL_QUERY_REGISTRY_DIRECT. In this case, the *Name* field contains the registry value name, and the *EntryContext* field contains a pointer to the variable to receive the value. You have to trust that no one has fiddled with the value types so that a string is returned when you were expecting a ULONG.

A safer approach (not used here) is to pass the name of a callback routine in the *QueryRoutine* field and set the *Flags* field to zero. Your routine is called for each value found, and indicates the type of the found value.

The remaining parameters to RtlQueryRegistryValues give even more flexibility. RTL_REGISTRY_ABSOLUTE indicates that the Path parameter is an absolute registry path. Various other useful options can be given (e.g., if the Path is relative to the HKLM\System\CurrentControlSet\Services key).

RtlQueryRegistryValues only returns STATUS_SUCCESS if all the queries were processed correctly and all the registry values were found. The code in ReadReg simply displays the return status and the values retrieved. If using this routine for real, you will probably want to store the values in global variables.

Operating system version

You can use a registry setting to determine at run time whether you are running in W98 or not. In NT and W2000 the following registry value is available, HKLM\System\CurrentControlSet\Control\ProductOptions\ProductType. The value is "WinNT" for the Workstation/Professional Windows version and either "LanmanNT" or "ServerNT" for the Server/Enterprise version. In W98 this registry value should not be available.

Installing Wdm1

That's enough on the Wdm1 code so far. The compiled driver, Wdm1.sys, is provided on the CD in free and checked build versions. However, you can recompile it if you wish.

Normally, Windows detects when a device is installed and prompts for the necessary drivers if they cannot be found already in the system. Full details of the driver selection process are given later.

For the virtual Wdm1 device, use the Control Panel "Add New Hardware" applet wizard. The process is basically the same for Windows 98 and Windows 2000.

Click Next two times, select "No, I want to select hardware from a list" and click Next. Select "Other devices" and click Next. Click "Have Disk..." and browse to the path of the Wdm1 driver (e.g., C:\WDMBook\Wdm1\Sys) and click "OK".

Two models are listed from the found installation INF files, one for the Wdm1 checked build and one for the free build. Select the model you want to install and select Next. Select "Finish" to complete installation.

The installation process does whatever the INF file specifies. The Wdm1 INF files copy the relevant driver to the Windows system32\drivers directory, adds registry settings, etc.

Windows should now have created a Wdm1 device for you. Check that it appears in the Device Manager "Other devices" category (e.g., named "WDM Book: WDM1 Example, free build"). The next chapter describes a Win32 user mode program that you can use to test that the driver is working.

In the Wdm1 code, the DriverEntry routine has been called along with various other Plug and Play callback routines, as described later.

Installation Details

You may be interested to know precisely what happens as a result of installing the Wdm1 driver and one Wdm1 device.

INF Files

Windows 98

Windows 98 copies the INF file to the Windows INF\OTHER directory. The INF file is renamed, after the manufacturer name, to "WDM BookWDM1.INF".

Windows 98 remembers that the INF file has been installed in the registry. The HKLM\ Software\Microsoft\Windows\CurrentVersion\Setup\SetupX\INF\OEM Name key has a value C:\W98\INF\OTHER\WDM BookWDM1.INF set to "WDM BookWDM1.INF". This entry is not deleted if you remove the Wdm1 device.

Windows 98 also keeps a note of the most recently used install locations in the registry.

Windows 2000

Windows 2000 copies the INF file and renames it to the next available OEM filename in the Windows INF directory. For example, WDM1.INF might be copied and renamed as C:\WINNT\ INF\OEM1.INF.

Windows 2000 also remembers install locations in the registry.

Registry

The device interface for Wdm1 devices has a registry key at HKLM\System\CurrentControlSet\ Control\DeviceClasses\{C0CF0640-5F6E-11d2-B677-00C0DFE4C1F3} along with a series of subkeys for each device and their associated symbolic link(s). This key includes the WDM1_GUID {C0CF0640...}.

Windows 98

The HKLM\Enum\Root\Unknown\0000 key is the entry for the Wdm1 device. The digits 0000 will increment for each Unknown device. The HKLM\System\CurrentControlSet\Services\ Class\Unknown\0000 key is the entry for the Wdm1 driver. This key is passed to DriverEntry in the RegistryPath string.

These entries are removed if you remove the Wdm1 device.

The *ClassGUID* value in the Wdm1 device key is {4D36E97E...} for "Unknown" type devices. This has a registry entry at HKLM\System\CurrentControlSet\Services\Class\ {4D36E97E...}.

Windows 2000

The `HKLM\System\CurrentControlSet\Control\Class\{4D36E97E...}\0000` key is the entry for the Wdm1 device. {4D36E97E...} is the GUID for "Unknown" type devices.

The driver service entry key is `HKLM\System\CurrentControlSet\Services\Wdm1`. This key is passed to `DriverEntry` in the `RegistryPath` string. The `Enum` subkey has values named 0 onwards such as `Root\UNKNOWN\0000` for each device instance.

Windows 2000 Objects

The *winobj* program lets you see what Windows 2000 objects are present for the driver and the device.

The driver has an entry `\driver\wdm1`.

There is a symbolic link object `\??\Root#UNKNOWN#0000#{C0CF0640...}`. If you double-click this, you will see that this links to one of the listed devices (e.g., to `\device\004059`). Symbolic links are covered in the next chapter.

Managing Devices and Drivers

Having installed Wdm1, there are various device and driver management jobs that you can perform.

Add Another Device

You can open another device with the same driver using the Add New Hardware wizard again. Windows made a copy of the INF file so you can select the device from the list without having to specify your driver location. Additional entries are made in the registry for the second Wdm1 device and the device interface to it.

Removing a Device

In the Device Manager, select the device you want to remove. Click Uninstall or Remove.

Most of the device and driver registry entries are removed if the Wdm1 device is removed. However, the registry entry for the device interface and the Windows 2000 `Services` entry persist after a device is removed.

When you remove a device, the INF file is not removed from the Windows INF directory structure. Similarly, the Wdm1.sys driver file remains in place in the `System32\Drivers` directory. This makes it easy to reinstall a Wdm1 device.

If all the driver's devices have been removed, the driver is unloaded from memory. At this stage, if you update the driver in the Windows `System32\drivers` directory and reinstall the device, the new driver is used.

Updating the Driver

The simplest way to update a driver is to use the Update/Change driver option in the Device Manager properties for a device. The book software projects always make a copy of the latest build of a driver in the Windows `System32\Drivers` directory so you can reinstall a new driver without having to select the Have disk option. W2000 uses the files in this directory while W98 requires you to specify the location of the new files.

Alternatively, remove all devices and invoke the Add New Hardware wizard again.

NT Style Drivers

NT style (non-Plug and Play) drivers must use a special installation process, as described in Chapter 11.

Updating an NT style driver happens in a different way, as well. First, you must get your driver into the Windows `System32\drivers` directory.

In NT 3.51 and NT 4, run the Control Panel Devices applet. Find your driver, stop it, and start it again. In Windows 2000 you must opt to show hidden devices in the Device Manager before you can start or stop NT style drivers. Alternatively, you can run the book software *Servicer* program; type in the name of your driver; press Stop and then Start. If you run an NT style driver in W98, you must reboot the computer to use an updated driver.

Conclusion

This chapter has shown how to set up a development computer for device driver development. A very basic WDM driver has been written and installed in Windows 98 and Windows 2000.

The following chapters explain how to access this driver from a user program and enhance this driver to implement the correct Plug and Play and Power Management handling.

Listing 4.7   Wdm1.h

```
/////////////////////////////////////////////////////////////////////////////
//Copyright © 1998 Chris Cant, PHD Computer Consultants Ltd
//WDM Book for R&D Books, Miller Freeman Inc
//
//Wdm1 example
/////////////////////////////////////////////////////////////////////////////
//wdm1.hCommon header
/////////////////////////////////////////////////////////////////////////////
//Version history
//27-Apr-991.0.0CCcreation
/////////////////////////////////////////////////////////////////////////////


/////////////////////////////////////////////////////////////////////////////
//Include WDM standard header with C linkage

#ifdef __cplusplus
extern "C"
{
#endif
```

```c
#include "wdm.h"
#ifdef __cplusplus
}
#endif

//////////////////////////////////////////////////////////////////////
//DebugPrint and Guid headers

#include "DebugPrint.h"

#include "GUIDs.h"

//////////////////////////////////////////////////////////////////////
//Spin lock to protect access to shared memory buffer

extern KSPIN_LOCK BufferLock;
extern PUCHAR Buffer;

//////////////////////////////////////////////////////////////////////
//Our device extension

typedef struct _WDM1_DEVICE_EXTENSION
{
PDEVICE_OBJECTfdo;
PDEVICE_OBJECTNextStackDevice;
UNICODE_STRINGifSymLinkName;

} WDM1_DEVICE_EXTENSION, *PWDM1_DEVICE_EXTENSION;

//////////////////////////////////////////////////////////////////////
// Forward declarations of global functions

VOID Wdm1Unload(IN PDRIVER_OBJECT DriverObject);

NTSTATUS Wdm1Power(IN PDEVICE_OBJECT fdo,
IN PIRP Irp);

NTSTATUS Wdm1Pnp(IN PDEVICE_OBJECT fdo,
IN PIRP Irp);

NTSTATUS Wdm1AddDevice(IN PDRIVER_OBJECT DriverObject,
IN PDEVICE_OBJECT pdo);
```

```
NTSTATUS Wdm1Create(IN PDEVICE_OBJECT fdo,
IN PIRP Irp);

NTSTATUS Wdm1Close(IN PDEVICE_OBJECT fdo,
IN PIRP Irp);

NTSTATUS Wdm1Write(IN PDEVICE_OBJECT fdo,
IN PIRP Irp);

NTSTATUS Wdm1Read(IN PDEVICE_OBJECT fdo,
IN PIRP Irp);

NTSTATUS Wdm1DeviceControl(IN PDEVICE_OBJECT fdo,
IN PIRP Irp);

NTSTATUS Wdm1SystemControl(IN PDEVICE_OBJECT fdo,
IN PIRP Irp);


/////////////////////////////////////////////////////////////////////////////

NTSTATUS CompleteIrp( PIRP Irp, NTSTATUS status, ULONG info);


/////////////////////////////////////////////////////////////////////////////
```

Listing 4.8   Init.cpp

```
/////////////////////////////////////////////////////////////////////////////
//Copyright © 1998 Chris Cant, PHD Computer Consultants Ltd
//WDM Book for R&D Books, Miller Freeman Inc
//
//Wdm1 example
/////////////////////////////////////////////////////////////////////////////
//init.cpp:Driver initialization code
/////////////////////////////////////////////////////////////////////////////
//DriverEntryInitialisation entry point
//Wdm1UnloadUnload driver routine
/////////////////////////////////////////////////////////////////////////////
//Version history
//27-Apr-991.0.0CCcreation
/////////////////////////////////////////////////////////////////////////////
```

```
#include "wdm1.h"

#pragma code_seg("INIT") // start INIT section


/////////////////////////////////////////////////////////////////////////////
//DriverEntry:
//
//Description:
//This function initializes the driver, and creates
//any objects needed to process I/O requests.
//
//Arguments:
//Pointer to the Driver object
//Registry path string for driver service key
//
//Return Value:
//This function returns STATUS_XXX

extern "C"
NTSTATUS DriverEntry(IN PDRIVER_OBJECT DriverObject,
IN PUNICODE_STRING RegistryPath)
{
NTSTATUS status = STATUS_SUCCESS;

#if DBG
DebugPrintInit("Wdm1 checked");
#else
DebugPrintInit("Wdm1 free");
#endif

DebugPrint("RegistryPath is %T",RegistryPath);

// Export other driver entry points...
DriverObject->DriverExtension->AddDevice = Wdm1AddDevice;
DriverObject->DriverUnload = Wdm1Unload;

DriverObject->MajorFunction[IRP_MJ_CREATE] = Wdm1Create;
DriverObject->MajorFunction[IRP_MJ_CLOSE] = Wdm1Close;

DriverObject->MajorFunction[IRP_MJ_PNP] = Wdm1Pnp;
DriverObject->MajorFunction[IRP_MJ_POWER] = Wdm1Power;
```

```
DriverObject->MajorFunction[IRP_MJ_READ] = Wdm1Read;
DriverObject->MajorFunction[IRP_MJ_WRITE] = Wdm1Write;
DriverObject->MajorFunction[IRP_MJ_DEVICE_CONTROL] = Wdm1DeviceControl;

DriverObject->MajorFunction[IRP_MJ_SYSTEM_CONTROL] = Wdm1SystemControl;

//Initialise spin lock which protects access to shared memory buffer
KeInitializeSpinLock(&BufferLock);

DebugPrintMsg("DriverEntry completed");

return status;
}
#pragma code_seg() // end INIT section


/////////////////////////////////////////////////////////////////////////
//Wdm1Unload
//
//Description:
//Unload the driver by removing any remaining objects, etc.
//
//Arguments:
//Pointer to the Driver object
//
//Return Value:
//None

#pragma code_seg("PAGE") // start PAGE section

VOID Wdm1Unload(IN PDRIVER_OBJECT DriverObject)
{
// Free buffer (do not need to acquire spin lock)
if( Buffer!=NULL)
ExFreePool(Buffer);

DebugPrintMsg("Wdm1Unload");
DebugPrintClose();
}          .


/////////////////////////////////////////////////////////////////////////
#pragma code_seg() // end PAGE section
```

Listing 4.9   Pnp.cpp

```
/////////////////////////////////////////////////////////////////////////
//Copyright © 1998 Chris Cant, PHD Computer Consultants Ltd
//WDM Book for R&D Books, Miller Freeman Inc
//
//Wdm1 example
/////////////////////////////////////////////////////////////////////////
//pnp.cpp:Plug and Play and Power IRP handlers
/////////////////////////////////////////////////////////////////////////
//Wdm1AddDeviceAdd device routine
//Wdm1PnpPNP IRP dispatcher
//Wdm1PowerPOWER IRP dispatcher
/////////////////////////////////////////////////////////////////////////
//Version history
//27-Apr-991.0.0CCcreation
/////////////////////////////////////////////////////////////////////////

#define INITGUID// initialize WDM1_GUID in this module

#include "wdm1.h"

#pragma code_seg("PAGE")// start PAGE section

/////////////////////////////////////////////////////////////////////////
//Wdm1AddDevice:
//
//Description:
//Cope with a new Pnp device being added here.
//Usually just attach to the top of the driver stack.
//Do not talk to device here!
//
//Arguments:
//Pointer to the Driver object
//Pointer to Physical Device Object
//
//Return Value:
//This function returns STATUS_XXX

NTSTATUS Wdm1AddDevice(IN PDRIVER_OBJECT DriverObject,
IN PDEVICE_OBJECT pdo)
{
DebugPrint("AddDevice");
```

```
NTSTATUS status;
PDEVICE_OBJECT fdo;

// Create our Functional Device Object in fdo
status = IoCreateDevice (DriverObject,
sizeof(WDM1_DEVICE_EXTENSION),
NULL,// No Name
FILE_DEVICE_UNKNOWN,
0,
FALSE,// Not exclusive
&fdo);
if( !NT_SUCCESS(status))
return status;

// Remember fdo in our device extension
PWDM1_DEVICE_EXTENSION dx = (PWDM1_DEVICE_EXTENSION)fdo->DeviceExtension;
dx->fdo = fdo;
DebugPrint("FDO is %x",fdo);

// Register and enable our device interface
status = IoRegisterDeviceInterface(pdo, &WDM1_GUID, NULL, &dx->ifSymLinkName);
if( !NT_SUCCESS(status))
{
IoDeleteDevice(fdo);
return status;
}
IoSetDeviceInterfaceState(&dx->ifSymLinkName, TRUE);
DebugPrint("Symbolic Link Name is %T",&dx->ifSymLinkName);

// Attach to the driver stack below us
dx->NextStackDevice = IoAttachDeviceToDeviceStack(fdo,pdo);

// Set fdo flags appropriately
fdo->Flags &= ~DO_DEVICE_INITIALIZING;
fdo->Flags |= DO_BUFFERED_IO;

return STATUS_SUCCESS;
}

////////////////////////////////////////////////////////////////////////
//Wdm1Pnp:
//
//Description:
```

```
//Handle IRP_MJ_PNP requests
//
//Arguments:
//Pointer to our FDO
//Pointer to the IRP
//Various minor codes
//IrpStack->Parameters.QueryDeviceRelations
//IrpStack->Parameters.QueryInterface
//IrpStack->Parameters.DeviceCapabilities
//IrpStack->Parameters.FilterResourceRequirements
//IrpStack->Parameters.ReadWriteConfig
//IrpStack->Parameters.SetLock
//IrpStack->Parameters.QueryId
//IrpStack->Parameters.QueryDeviceText
//IrpStack->Parameters.UsageNotification
//
//Return Value:
//This function returns STATUS_XXX

NTSTATUS Wdm1Pnp(IN PDEVICE_OBJECT fdo,
IN PIRP Irp)
{
DebugPrint("PnP %I",Irp);
PWDM1_DEVICE_EXTENSION dx=(PWDM1_DEVICE_EXTENSION)fdo->DeviceExtension;

// Remember minor function
PIO_STACK_LOCATION IrpStack = IoGetCurrentIrpStackLocation(Irp);
ULONG MinorFunction = IrpStack->MinorFunction;

// Just pass to lower driver
IoSkipCurrentIrpStackLocation(Irp);
NTSTATUS status = IoCallDriver( dx->NextStackDevice, Irp);

// Device removed
if( MinorFunction==IRP_MN_REMOVE_DEVICE)
{
DebugPrint("PnP RemoveDevice");
// disable device interface
IoSetDeviceInterfaceState(&dx->ifSymLinkName, FALSE);
RtlFreeUnicodeString(&dx->ifSymLinkName);

// unattach from stack
```

```
if (dx->NextStackDevice)
IoDetachDevice(dx->NextStackDevice);

// delete our fdo
IoDeleteDevice(fdo);
}


return status;
}


///////////////////////////////////////////////////////////////////////////
//Wdm1Power:
//
//Description:
//Handle IRP_MJ_POWER requests
//
//Arguments:
//Pointer to the FDO
//Pointer to the IRP
//IRP_MN_WAIT_WAKE:IrpStack->Parameters.WaitWake.Xxx
//IRP_MN_POWER_SEQUENCE:IrpStack->Parameters.PowerSequence.Xxx
//IRP_MN_SET_POWER:
//IRP_MN_QUERY_POWER:IrpStack->Parameters.Power.Xxx
//
//Return Value:
//This function returns STATUS_XXX

NTSTATUS Wdm1Power(IN PDEVICE_OBJECT fdo,
IN PIRP Irp)
{
DebugPrint("Power %I",Irp);
PWDM1_DEVICE_EXTENSION dx = (PWDM1_DEVICE_EXTENSION)fdo->DeviceExtension;

// Just pass to lower driver
PoStartNextPowerIrp( Irp);
IoSkipCurrentIrpStackLocation(Irp);
return PoCallDriver( dx->NextStackDevice, Irp);
}

#pragma code_seg()// end PAGE section
```

Listing 4.10 Dispatch.cpp

```cpp
//////////////////////////////////////////////////////////////////////////
//Copyright © 1998 Chris Cant, PHD Computer Consultants Ltd
//WDM Book for R&D Books, Miller Freeman Inc
//
//Wdm1 example
//////////////////////////////////////////////////////////////////////////
//dispatch.cpp:Other IRP handlers
//////////////////////////////////////////////////////////////////////////
//Wdm1CreateHandle Create/Open file IRP
//Wdm1CloseHandle Close file IRPs
//Wdm1ReadHandle Read IRPs
//Wdm1WriteHandle Write IRPs
//Wdm1DeviceControlHandle DeviceIoControl IRPs
//Wdm1SystemControlHandle WMI IRPs
//////////////////////////////////////////////////////////////////////////
//Version history
//27-Apr-991.0.0CCcreation
//////////////////////////////////////////////////////////////////////////


#include "wdm1.h"
#include "Ioctl.h"


//////////////////////////////////////////////////////////////////////////
//////////////////////////////////////////////////////////////////////////
//Buffer and BufferSize and guarding spin lock globals (in unpaged memory)

KSPIN_LOCK BufferLock;
PUCHARBuffer = NULL;
ULONGBufferSize = 0;


//////////////////////////////////////////////////////////////////////////
//Wdm1Create:
//
//Description:
//Handle IRP_MJ_CREATE requests
//
//Arguments:
//Pointer to our FDO
//Pointer to the IRP
//IrpStack->Parameters.Create.xxx has create parameters
//IrpStack->FileObject->FileName has file name of device
```

```
//
//Return Value:
//This function returns STATUS_XXX

NTSTATUS Wdm1Create(IN PDEVICE_OBJECT fdo,
IN PIRP Irp)
{
PIO_STACK_LOCATION IrpStack = IoGetCurrentIrpStackLocation(Irp);
DebugPrint( "Create File is %T", &(IrpStack->FileObject->FileName));

// Complete successfully
return CompleteIrp(Irp,STATUS_SUCCESS,0);
}


/////////////////////////////////////////////////////////////////////////////
//Wdm1Close:
//
//Description:
//Handle IRP_MJ_CLOSE requests
//
//Arguments:
//Pointer to our FDO
//Pointer to the IRP
//
//Return Value:
//This function returns STATUS_XXX

NTSTATUS Wdm1Close(IN PDEVICE_OBJECT fdo,
IN PIRP Irp)
{
DebugPrintMsg("Close");

// Complete successfully
return CompleteIrp(Irp,STATUS_SUCCESS,0);
}


/////////////////////////////////////////////////////////////////////////////
//Wdm1Read:
//
//Description:
//Handle IRP_MJ_READ requests
//
```

```
//Arguments:
//Pointer to our FDO
//Pointer to the IRP
//IrpStack->Parameters.Read.xxx has read parameters
//User buffer at:AssociatedIrp.SystemBuffer(buffered I/O)
//MdlAddress(direct I/O)
//
//Return Value:
//This function returns STATUS_XXX

NTSTATUS Wdm1Read(IN PDEVICE_OBJECT fdo,
  IN PIRP Irp)
{
PIO_STACK_LOCATION IrpStack = IoGetCurrentIrpStackLocation(Irp);
NTSTATUS status = STATUS_SUCCESS;
LONG BytesTxd = 0;

// Get call parameters
LONGLONG FilePointer = IrpStack->Parameters.Read.ByteOffset.QuadPart;
ULONG ReadLen = IrpStack->Parameters.Read.Length;
DebugPrint("Read %d bytes from file pointer %d",(int)ReadLen,(int)FilePointer);

// Get access to the shared buffer
KIRQL irql;
KeAcquireSpinLock(&BufferLock,&irql);

// Check file pointer
if( FilePointer<0)
status = STATUS_INVALID_PARAMETER;
if( FilePointer>=(LONGLONG)BufferSize)
status = STATUS_END_OF_FILE;

if( status==STATUS_SUCCESS)
{
// Get transfer count
if( ((ULONG)FilePointer)+ReadLen>BufferSize)
{
BytesTxd = BufferSize - (ULONG)FilePointer;
if( BytesTxd<0) BytesTxd = 0;
}
else
BytesTxd = ReadLen;
```

```
// Read from shared buffer
if( BytesTxd>0 && Buffer!=NULL)
RtlCopyMemory( Irp->AssociatedIrp.SystemBuffer, Buffer+FilePointer, BytesTxd);
}

// Release shared buffer
KeReleaseSpinLock(&BufferLock,irql);

DebugPrint("Read: %d bytes returned",(int)BytesTxd);

// Complete IRP
return CompleteIrp(Irp,status,BytesTxd);
}

////////////////////////////////////////////////////////////////////////////////
//Wdm1Write:
//
//Description:
//Handle IRP_MJ_WRITE requests
//
//Arguments:
//Pointer to our FDO
//Pointer to the IRP
//IrpStack->Parameters.Write.xxx has write parameters
//User buffer at:AssociatedIrp.SystemBuffer(buffered I/O)
//MdlAddress(direct I/O)
//
//Return Value:
//This function returns STATUS_XXX

NTSTATUS Wdm1Write(IN PDEVICE_OBJECT fdo,
IN PIRP Irp)
{
PIO_STACK_LOCATION IrpStack = IoGetCurrentIrpStackLocation(Irp);
NTSTATUS status = STATUS_SUCCESS;
LONG BytesTxd = 0;

// Get call parameters
LONGLONG FilePointer = IrpStack->Parameters.Write.ByteOffset.QuadPart;
ULONG WriteLen = IrpStack->Parameters.Write.Length;
DebugPrint("Write %d bytes from file pointer %d",(int)WriteLen,(int)FilePointer);
```

```
if( FilePointer<0)
status = STATUS_INVALID_PARAMETER;
else
{
// Get access to the shared buffer
KIRQL irql;
KeAcquireSpinLock(&BufferLock,&irql);

BytesTxd = WriteLen;

// (Re)allocate buffer if necessary
if( ((ULONG)FilePointer)+WriteLen>BufferSize)
{
ULONG NewBufferSize = ((ULONG)FilePointer)+WriteLen;
PVOID NewBuffer = ExAllocatePool(NonPagedPool,NewBufferSize);
if( NewBuffer==NULL)
{
BytesTxd = BufferSize - (ULONG)FilePointer;
if( BytesTxd<0) BytesTxd = 0;
}
else
{
RtlZeroMemory(NewBuffer,NewBufferSize);
if( Buffer!=NULL)
{
RtlCopyMemory(NewBuffer,Buffer,BufferSize);
ExFreePool(Buffer);
}
Buffer = (PUCHAR)NewBuffer;
BufferSize = NewBufferSize;
}
}

// Write to shared memory
if( BytesTxd>0 && Buffer!=NULL)
RtlCopyMemory( Buffer+FilePointer, Irp->AssociatedIrp.SystemBuffer, BytesTxd);

// Release shared buffer
KeReleaseSpinLock(&BufferLock,irql);
}
```

```
DebugPrint("Write: %d bytes written",(int)BytesTxd);

// Complete IRP
return CompleteIrp(Irp,status,BytesTxd);
}


///////////////////////////////////////////////////////////////////////////
//Wdm1DeviceControl:
//
//Description:
//Handle IRP_MJ_DEVICE_CONTROL requests
//
//Arguments:
//Pointer to our FDO
//Pointer to the IRP
//Buffered:AssociatedIrp.SystemBuffer (and IrpStack->Parameters.DeviceIoCon-
trol.Type3InputBuffer)
//Direct:MdlAddress
//
//IrpStack->Parameters.DeviceIoControl.InputBufferLength
//IrpStack->Parameters.DeviceIoControl.OutputBufferLength
//
//Return Value:
//This function returns STATUS_XXX

NTSTATUS Wdm1DeviceControl(IN PDEVICE_OBJECT fdo,
IN PIRP Irp)
{
PIO_STACK_LOCATION IrpStack = IoGetCurrentIrpStackLocation(Irp);
NTSTATUS status = STATUS_SUCCESS;
ULONG BytesTxd = 0;

ULONG ControlCode = IrpStack->Parameters.DeviceIoControl.IoControlCode;
ULONG InputLength = IrpStack->Parameters.DeviceIoControl.InputBufferLength;
ULONG OutputLength = IrpStack->Parameters.DeviceIoControl.OutputBufferLength;

DebugPrint("DeviceIoControl: Control code %x InputLength %d OutputLength %d",
ControlCode, InputLength, OutputLength);

// Get access to the shared buffer
KIRQL irql;
KeAcquireSpinLock(&BufferLock,&irql);
switch( ControlCode)
```

```
{
///////Zero Buffer
case IOCTL_WDM1_ZERO_BUFFER:
// Zero the buffer
if( Buffer!=NULL && BufferSize>0)
RtlZeroMemory(Buffer,BufferSize);
break;

///////Remove Buffer
case IOCTL_WDM1_REMOVE_BUFFER:
if( Buffer!=NULL)
{
ExFreePool(Buffer);
Buffer = NULL;
BufferSize = 0;
}
break;

///////Get Buffer Size as ULONG
case IOCTL_WDM1_GET_BUFFER_SIZE:
if( OutputLength<sizeof(ULONG))
status = STATUS_INVALID_PARAMETER;
else
{
BytesTxd = sizeof(ULONG);
RtlCopyMemory(Irp->AssociatedIrp.SystemBuffer,&BufferSize,sizeof(ULONG));
}
break;

///////Get Buffer
case IOCTL_WDM1_GET_BUFFER:
if( OutputLength>BufferSize)
status = STATUS_INVALID_PARAMETER;
else
{
BytesTxd = OutputLength;
RtlCopyMemory(Irp->AssociatedIrp.SystemBuffer,Buffer,BytesTxd);
}
break;

///////Invalid request
default:
```

```
        status = STATUS_INVALID_DEVICE_REQUEST;
    }
    // Release shared buffer
    KeReleaseSpinLock(&BufferLock,irql);

    DebugPrint("DeviceIoControl: %d bytes written",(int)BytesTxd);

    // Complete IRP
    return CompleteIrp(Irp,status,BytesTxd);
}


/////////////////////////////////////////////////////////////////////////////
//Wdm1SystemControl:
//
//Description:
//Handle IRP_MJ_SYSTEM_CONTROL requests
//
//Arguments:
//Pointer to our FDO
//Pointer to the IRP
//Various minor parameters
//IrpStack->Parameters.WMI.xxx has WMI parameters
//
//Return Value:
//This function returns STATUS_XXX

NTSTATUS Wdm1SystemControl(IN PDEVICE_OBJECT fdo,
IN PIRP Irp)
{
DebugPrintMsg("SystemControl");

// Just pass to lower driver
IoSkipCurrentIrpStackLocation(Irp);
PWDM1_DEVICE_EXTENSION dx = (PWDM1_DEVICE_EXTENSION)fdo->DeviceExtension;
return IoCallDriver( dx->NextStackDevice, Irp);
}


/////////////////////////////////////////////////////////////////////////////
//Wdm1Cleanup:
//
//Description:
//Handle IRP_MJ_CLEANUP requests
```

```
//Cancel queued IRPs which match given FileObject
//
//Arguments:
//Pointer to our FDO
//Pointer to the IRP
//IrpStack->FileObject has handle to file
//
//Return Value:
//This function returns STATUS_XXX


//Not needed for Wdm1


////////////////////////////////////////////////////////////////////////
//CompleteIrp:Sets IoStatus and completes the IRP

NTSTATUS CompleteIrp( PIRP Irp, NTSTATUS status, ULONG info)
{
Irp->IoStatus.Status = status;
Irp->IoStatus.Information = info;
IoCompleteRequest(Irp,IO_NO_INCREMENT);
return status;
}


////////////////////////////////////////////////////////////////////////
```

Listing 4.11 Wdm1.rc

```
//Microsoft Developer Studio generated resource script.
//
#include "resource.h"

#define APSTUDIO_READONLY_SYMBOLS
////////////////////////////////////////////////////////////////////////
//
// Generated from the TEXTINCLUDE 2 resource.
//
#include "afxres.h"


////////////////////////////////////////////////////////////////////////
#undef APSTUDIO_READONLY_SYMBOLS


////////////////////////////////////////////////////////////////////////
// English (U.S.) resources
```

```
#if !defined(AFX_RESOURCE_DLL) || defined(AFX_TARG_ENU)
#ifdef _WIN32
LANGUAGE LANG_ENGLISH, SUBLANG_ENGLISH_US
#pragma code_page(1252)
#endif //_WIN32

#ifndef _MAC
/////////////////////////////////////////////////////////////////////////////
//
// Version
//

VS_VERSION_INFO VERSIONINFO
 FILEVERSION 1,0,5,0
 PRODUCTVERSION 1,0,0,0
 FILEFLAGSMASK 0x3fL
#ifdef _DEBUG
 FILEFLAGS 0x1L
#else
 FILEFLAGS 0x0L
#endif
 FILEOS 0x4L
 FILETYPE 0x2L
 FILESUBTYPE 0x0L
BEGIN
    BLOCK "StringFileInfo"
    BEGIN
        BLOCK "080904b0"
        BEGIN
            VALUE "Comments", "Chris Cant\0"
            VALUE "CompanyName", "PHD Computer Consultants Ltd\0"
            VALUE "FileDescription", "Wdm1\0"
            VALUE "FileVersion", "1, 0, 5, 0\0"
            VALUE "InternalName", "Wdm1 driver\0"
           VALUE "LegalCopyright", "Copyright © 1998,1999 PHD Computer Consultants Ltd\0"
            VALUE "OriginalFilename", "Wdm1.sys\0"
            VALUE "ProductName", "WDM Book\0"
            VALUE "ProductVersion", "1, 0, 0, 0\0"
        END
    END
    BLOCK "VarFileInfo"
```

```
    BEGIN
        VALUE "Translation", 0x809, 1200
    END
END


#endif    // !_MAC

#endif    // English (U.S.) resources
/////////////////////////////////////////////////////////////////////////////



/////////////////////////////////////////////////////////////////////////////
// English (U.K.) resources

#if !defined(AFX_RESOURCE_DLL) || defined(AFX_TARG_ENG)
#ifdef _WIN32
LANGUAGE LANG_ENGLISH, SUBLANG_ENGLISH_UK
#pragma code_page(1252)
#endif //_WIN32

#ifdef APSTUDIO_INVOKED
/////////////////////////////////////////////////////////////////////////////
//
// TEXTINCLUDE
//

1 TEXTINCLUDE DISCARDABLE
BEGIN
    "resource.h\0"
END

2 TEXTINCLUDE DISCARDABLE
BEGIN
    "#include ""afxres.h""\r\n"
    "\0"
END

3 TEXTINCLUDE DISCARDABLE
BEGIN
    "\r\n"
    "\0"
END
```

```
#endif    // APSTUDIO_INVOKED

#endif    // English (U.K.) resources
//////////////////////////////////////////////////////////////////////////


#ifndef APSTUDIO_INVOKED
//////////////////////////////////////////////////////////////////////////
//
// Generated from the TEXTINCLUDE 3 resource.
//


//////////////////////////////////////////////////////////////////////////
#endif    // not APSTUDIO_INVOKED
```

Listing 4.12   Ioctl.h

```
//DeviceIoControl IOCTL codes supported by Wdm1

#define IOCTL_WDM1_ZERO_BUFFER CTL_CODE(\
FILE_DEVICE_UNKNOWN,\
0x801,\
METHOD_BUFFERED,\
FILE_ANY_ACCESS)

#define IOCTL_WDM1_REMOVE_BUFFER CTL_CODE(\
FILE_DEVICE_UNKNOWN,\
0x802,\
METHOD_BUFFERED,\
FILE_ANY_ACCESS)

#define IOCTL_WDM1_GET_BUFFER_SIZE CTL_CODE(\
FILE_DEVICE_UNKNOWN,\
0x803,\
METHOD_BUFFERED,\
FILE_ANY_ACCESS)
```

```
#define IOCTL_WDM1_GET_BUFFER CTL_CODE(\
FILE_DEVICE_UNKNOWN,\
0x804,\
METHOD_BUFFERED,\
FILE_ANY_ACCESS)

#define IOCTL_WDM1_UNRECOGNISED CTL_CODE(\
FILE_DEVICE_UNKNOWN,\
0x805,\
METHOD_BUFFERED,\
FILE_ANY_ACCESS)
```

Listing 4.13   GUIDs.h

```
////////////////////////////////////////////////////////////////////
//Wdm1 device interface GUID

// {C0CF0640-5F6E-11d2-B677-00C0DFE4C1F3}
DEFINE_GUID(WDM1_GUID, 0xc0cf0640, 0x5f6e, 0x11d2, 0xb6, 0x77, 0x0, 0xc0, 0xdf, 0xe4,
0xc1, 0xf3);

////////////////////////////////////////////////////////////////////
//Wdm2 device interface GUID

// {C0CF0641-5F6E-11d2-B677-00C0DFE4C1F3}
DEFINE_GUID(WDM2_GUID, 0xc0cf0641, 0x5f6e, 0x11d2, 0xb6, 0x77, 0x0, 0xc0, 0xdf, 0xe4,
0xc1, 0xf3);

////////////////////////////////////////////////////////////////////
//Wdm3 device interface GUID

// {C0CF0642-5F6E-11d2-B677-00C0DFE4C1F3}
DEFINE_GUID(WDM3_GUID, 0xc0cf0642, 0x5f6e, 0x11d2, 0xb6, 0x77, 0x0, 0xc0, 0xdf, 0xe4,
0xc1, 0xf3);

////////////////////////////////////////////////////////////////////
//Wdm3 WMI data block GUID

// {C0CF0643-5F6E-11d2-B677-00C0DFE4C1F3}
DEFINE_GUID(WDM3_WMI_GUID, 0xc0cf0643, 0x5f6e, 0x11d2, 0xb6, 0x77, 0x0, 0xc0, 0xdf,
0xe4, 0xc1, 0xf3);
```

```
//////////////////////////////////////////////////////////////////////////
//Wdm3 WMI event block GUID


// {C0CF0644-5F6E-11d2-B677-00C0DFE4C1F3}
DEFINE_GUID(WDM3_WMI_EVENT_GUID, 0xc0cf0644, 0x5f6e, 0x11d2, 0xb6, 0x77, 0x0, 0xc0,
0xdf, 0xe4, 0xc1, 0xf3);


/*
// {C0CF0645-5F6E-11d2-B677-00C0DFE4C1F3}
DEFINE_GUID(<<name>>,
0xc0cf0645, 0x5f6e, 0x11d2, 0xb6, 0x77, 0x0, 0xc0, 0xdf, 0xe4, 0xc1, 0xf3);

// {C0CF0646-5F6E-11d2-B677-00C0DFE4C1F3}
DEFINE_GUID(<<name>>,
0xc0cf0646, 0x5f6e, 0x11d2, 0xb6, 0x77, 0x0, 0xc0, 0xdf, 0xe4, 0xc1, 0xf3);

// {C0CF0647-5F6E-11d2-B677-00C0DFE4C1F3}
DEFINE_GUID(<<name>>,
0xc0cf0647, 0x5f6e, 0x11d2, 0xb6, 0x77, 0x0, 0xc0, 0xdf, 0xe4, 0xc1, 0xf3);

// {C0CF0648-5F6E-11d2-B677-00C0DFE4C1F3}
DEFINE_GUID(<<name>>,
0xc0cf0648, 0x5f6e, 0x11d2, 0xb6, 0x77, 0x0, 0xc0, 0xdf, 0xe4, 0xc1, 0xf3);
*/


//////////////////////////////////////////////////////////////////////////
```

Listing 4.14 resource.h

```
//{{NO_DEPENDENCIES}}
// Microsoft Developer Studio generated include file.
// Used by Wdm1.rc
//

// Next default values for new objects
//
#ifdef APSTUDIO_INVOKED
#ifndef APSTUDIO_READONLY_SYMBOLS
#define _APS_NO_MFC                    1
#define _APS_NEXT_RESOURCE_VALUE       101
#define _APS_NEXT_COMMAND_VALUE        40001
```

```
#define _APS_NEXT_CONTROL_VALUE          1000
#define _APS_NEXT_SYMED_VALUE            101
#endif
#endif
```

Listing 4.15 Wdm1free.inf

```
; Wdm1free.Inf - install information file
; Copyright © 1998,1999 Chris Cant, PHD Computer Consultants Ltd

[Version]
Signature="$Chicago$"
Class=Unknown
Provider=%WDMBook%
DriverVer=04/26/1999,1.0.5.0

[Manufacturer]
%WDMBook% = WDM.Book

[WDM.Book]
%Wdm1%=Wdm1.Install, *wdmBook\Wdm1

[DestinationDirs]
Wdm1.Files.Driver=10,System32\Drivers
Wdm1.Files.Driver.NTx86=10,System32\Drivers

[SourceDisksNames]
1="Wdm1 build directory"...

[SourceDisksFiles]
Wdm1.sys=1,obj\i386\free

[SourceDisksFiles.x86]
Wdm1.sys=1,objfre\i386


;;;;;;;;;;;;;;;;;;;;;;;;;;;;;;;;;;;;;;
; Windows 98

[Wdm1.Install]
CopyFiles=Wdm1.Files.Driver
AddReg=Wdm1.AddReg
```

```
[Wdm1.AddReg]
HKR,,DevLoader,,*ntkern
HKR,,NTMPDriver,,Wdm1.sys


[Wdm1.Files.Driver]
Wdm1.sys


;;;;;;;;;;;;;;;;;;;;;;;;;;;;;;;;;;;;;;;;
; Windows 2000


[Wdm1.Install.NTx86]
CopyFiles=Wdm1.Files.Driver.NTx86


[Wdm1.Files.Driver.NTx86]
Wdm1.sys,,,%COPYFLG_NOSKIP%


[Wdm1.Install.NTx86.Services]
AddService = Wdm1, %SPSVCINST_ASSOCSERVICE%, Wdm1.Service


[Wdm1.Service]
DisplayName    = %Wdm1.ServiceName%
ServiceType    = %SERVICE_KERNEL_DRIVER%
StartType      = %SERVICE_DEMAND_START%
ErrorControl   = %SERVICE_ERROR_NORMAL%
ServiceBinary  = %10%\System32\Drivers\Wdm1.sys


;;;;;;;;;;;;;;;;;;;;;;;;;;;;;;;;;;;;;;;;
; Strings


[Strings]
WDMBook="WDM Book"
Wdm1="WDM Book: Wdm1 Example, free build"
Wdm1.ServiceName="WDM Book Wdm1 Driver"


SPSVCINST_ASSOCSERVICE=0x00000002; Driver service is associated with device being
installed
COPYFLG_NOSKIP=2; Do not allow user to skip file
SERVICE_KERNEL_DRIVER=1
SERVICE_AUTO_START=2
SERVICE_DEMAND_START=3
SERVICE_ERROR_NORMAL=1
```

Listing 4.16 SOURCES

```
TARGETNAME=Wdm1
TARGETTYPE=DRIVER
DRIVERTYPE=WDM
TARGETPATH=OBJ

INCLUDES=$(BASEDIR)\inc;

SOURCES=init.cpp \
dispatch.cpp \
pnp.cpp \
DebugPrint.c \
Wdm1.rc

NTTARGETFILES=PostBuildSteps
```

Listing 4.17 makefile.inc

```
PostBuildSteps: $(TARGET)
!if "$(DDKBUILDENV)"=="free"
rebase -B 0x10000 -X . $(TARGET)
!endif
copy $(TARGET) $(WINDIR)\system32\drivers
```

Listing 4.18 makefile

```
#
# DO NOT EDIT THIS FILE!!!  Edit .\sources. if you want to add a new source
# file to this component.  This file merely indirects to the real make file
# that is shared by all the driver components of the Windows NT DDK
#

!INCLUDE $(NTMAKEENV)\makefile.def
```

Listing 4.19 MakeDrvr.bat

```
@echo off
if "%1"=="" goto usage
if "%3"=="" goto usage
if not exist %1\bin\setenv.bat goto usage
call %1\bin\setenv %1 %4
```

```
%2
cd %3
build -b -w %5 %6 %7 %8 %9
goto exit

:usage
echo usage    MakeDrvr DDK_dir Driver_Drive Driver_Dir free/checked [build_options]
echo eg       MakeDrvr %%DDKROOT%% C: %%WDMBOOK%% free -cef
:exit
```

5

Chapter 5

Device Interfaces

This chapter explains how to use device interfaces to make a driver available to the kernel and Win32 user-mode applications. The device interface code in the Wdm1 driver is described.

The example *Wdm1Test* user-mode program shows how to use the Win32 functions to find device interfaces. *Wdm1Test* then sends some basic I/O requests to the Wdm1 driver.

Devices

A driver implements one or more *devices*. Each device represents an instance of a piece of real or virtual hardware. User-mode programs and the kernel access a driver's device through a device object.

Once a WDM driver's DriverEntry routine has completed, the driver has usually not created any devices. Instead, it has provided the kernel with various callback routine pointers. The Plug and Play (PnP) Manager calls the driver AddDevice callback routine each time the driver should create a device. The PnP Manager makes further calls to configure the device using the PnP IRP, as described in Chapter 8.

This chapter considers only two Plug and Play messages: when a device is added using AddDevice, and when it is removed. The *Remove Device* message is a PnP IRP with a minor function code of IRP_MN_REMOVE_DEVICE.

Device Access

User mode programs access drivers using the Win32 CreateFile function, passing a string in the lpFileName parameter. This routine is normally used to access files on disk. However, it can also be used to access a variety of other devices, such as pipes, mailslots, communication resources, or the console. At the simplest level, specifying COM1 opens the first serial port.

To access other devices, use \\.\ at the start of the lpFileName string. The following characters specify the device to open. For example, COM1 can also be opened as \\.\COM1. This form allows devices beyond COM9 to be opened. For example, opening COM10 will not work but \\.\COM10 will work (assuming such a device exists).

To access a device created by a driver, you must know the *symbolic link name* of the device. A symbolic link name is simply the name of the device exposed by the driver. Creating these names is described later.

Suppose a driver called *Dongle* has created a device with a symbolic link of Dongle1. A Win32 program could open it using \\.\Dongle1 as the filename parameter passed to Create-File.

There are two aspects to opening files that are important to understand. The first is that a device can be opened more than once, either in the same process or by different processes or threads. A driver can state that a Win32 thread gets exclusive access to the device until the handle is closed. Or the Win32 program can request exclusive access by setting the Create-File dwShareMode parameter to zero. Alternative settings of dwShareMode permit other open requests if they specify read and/or write access. Even if a Win32 thread has exclusive access to a device, it can use overlapped I/O to issue more than one request to a driver at the same time.

Your device should cope with access through multiple handles, if it permits nonexclusive access. For example, if your device supports the notion of a file pointer, it has to survive "simultaneous" read or write requests at the same file pointer location. The usual technique is to serialize all requests so that they are processed in full, one by one, as described in Chapter 16.

The second important aspect of opening a device is that Win32 programs can use file or directory names after the device name. A driver must decide whether it supports the concept of "files" when a process opens a device handle. For example, \\.\Dongle1\dir1\file1 could be used in the call to CreateFile. It is up to you whether your driver interprets such information.

Figure 5.1 illustrates a possible device access situation. The Dongle driver has created two devices, called "Dongle1" and "Dongle2". Process A has opened handles to each of the Dongle devices. Process B has opened a handle to "file" on the "Dongle2" device.

Figure 5.1   Device access

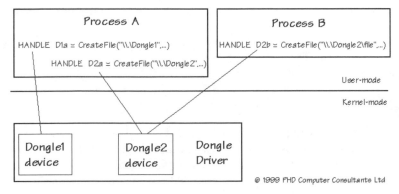

© 1999 PHD Computer Consultants Ltd

Subsequent I/O

Once a user mode program has opened a handle to your device, it can use various Win32 routines to access it, such as ReadFile, WriteFile, DeviceIoControl, and CloseHandle. Each of these calls results in a request being passed to your driver in the form of an I/O Request Packet (IRP). If a Win32 program terminates abruptly and finishes without calling CloseHandle, Windows ensures that this call is made (after cleaning up any pending I/O requests).

The Wdm1 handlers for these IRPs are described later, along with an introduction to the AddDevice and IRP_MJ_PNP handlers.

Device Objects and Device Extensions

The kernel stores information about devices in *device objects*, DEVICE_OBJECT structures. The relevant DEVICE_OBJECT is passed to your driver callback routines for each interaction with a device.

For example, a Win32 call to ReadFile results in a Read IRP (i.e., with major function code IRP_MJ_READ) being sent to your device. The DriverEntry routine must set up a handler for the relevant callback. In the Wdm1 driver, this routine, Wdm1Read, is passed a pointer to the relevant device object and a pointer to the IRP.

The device object is "owned" by the kernel, but has a few fields that should be used by a device driver, as described later. However, a driver can define a block of memory called a *device extension*, which it can use for whatever it wants. The device extension structure for each Wdm1 device is defined in Wdm1.h as shown in Listing 5.1. Your device extension structures will probably start with the same fields as Wdm1.

Listing 5.1   Wdm1 **Device Extension definition**

```
typedef struct _WDM1_DEVICE_EXTENSION
{
    PDEVICE_OBJECT  fdo;
    PDEVICE_OBJECT  NextStackDevice;
    UNICODE_STRING  ifSymLinkName;
} WDM1_DEVICE_EXTENSION, *PWDM1_DEVICE_EXTENSION;
```

A device is created using the IoCreateDevice kernel call. The DeviceExtensionSize parameter gives the size of the device extension required. IoCreateDevice allocates the memory for you from the nonpaged pool. Access the device extension through the device object *DeviceExtension* field. For example, Wdm1 accesses the its device extension using the following code.

```
PWDM1_DEVICE_EXTENSION dx = (PWDM1_DEVICE_EXTENSION)fdo->DeviceExtension;
```

A device object is deleted using IoDeleteDevice. This routine deallocates the device extension memory. Make sure that you clean up all your device-related objects before calling IoDeleteDevice.

Creating and Deleting Device Objects

Wdm1 creates and deletes devices in the Pnp.cpp module as shown in Listing 5.2. The chapter on Plug and Play explains when the Wdm1AddDevice and Wdm1Pnp routines are called.

The NT_SUCCESS macro is used to test whether the IoCreateDevice kernel call has completed. Do not forget to check that all calls to the kernel succeed. The NT_ERROR macro is not equivalent to !NT_SUCCESS. It is better to use !NT_SUCCESS, as it catches warning status values as well as errors.

Listing 5.2   Wdm1 Pnp.cpp **Create and Delete device code**

```
NTSTATUS WdmlAddDevice( IN PDRIVER_OBJECT DriverObject, IN PDEVICE_OBJECT pdo)
{
    ...
    NTSTATUS status;
    PDEVICE_OBJECT fdo;

    // Create our Functional Device Object in fdo
    status = IoCreateDevice( DriverObject,
        sizeof(WDM1_DEVICE_EXTENSION),
        NULL,     // No Name
        FILE_DEVICE_UNKNOWN,
        0,
        FALSE,     // Not exclusive
        &fdo);
    if( !NT_SUCCESS(status))
        return status;

    // Remember fdo in our device extension
    PWDM1_DEVICE_EXTENSION dx = (PWDM1_DEVICE_EXTENSION)fdo->DeviceExtension;
    dx->fdo = fdo;

    ...
    // Attach to the driver stack below us
    dx->NextStackDevice = IoAttachDeviceToDeviceStack(fdo,pdo);

    // Set fdo flags appropriately
    fdo->Flags &= ~DO_DEVICE_INITIALIZING;
    fdo->Flags |= DO_BUFFERED_IO;
    ...
}

NTSTATUS WdmlPnp( IN PDEVICE_OBJECT fdo, IN PIRP Irp)
{
    ...
    // Device removed
    if( MinorFunction==IRP_MN_REMOVE_DEVICE)
    {
```

Listing 5.2   Wdm1 Pnp.cpp **Create and Delete device code (continued)**

```
            // unattach from stack
            if (dx->NextStackDevice)
                IoDetachDevice(dx->NextStackDevice);


            // delete our fdo
            IoDeleteDevice(fdo);
        }
    return status;
}
```

Creating Devices

Wdm1AddDevice is called with two parameters, a pointer to the driver object and a pointer to a device object called the Physical Device Object (PDO).

Wdm1AddDevice calls IoCreateDevice to create the Wdm1 device using the parameters described in Table 5.2. The DeviceName parameter can be used to give a name to the device. However, as Wdm1 uses device interfaces, no name needs to be given, so NULL is passed here. Wdm1 devices do not correspond to any of the hardware device types, so FILE_DEVICE_UNKNOWN is given for DeviceType.

Table 5.1   IoCreateDevice function

NTSTATUS IoCreateDevice	(IRQL==PASSIVE_LEVEL)
Parameter	Description
IN PDRIVER_OBJECT DriverObject	Driver object
IN ULONG DeviceExtensionSize	Size of device extension required
IN PUNICODE_STRING DeviceName	Name of device, or NULL
IN DEVICE_TYPE DeviceType	The type of the device: one of the FILE_DEVICE_xxx values listed in the standard header WDM.H or NTDDK.H
IN ULONG DeviceCharacteristics	Various constants ORed together to indicate removable media, read only, etc.
IN BOOLEAN Exclusive	TRUE if only a single thread can access the device at a time
OUT PDEVICE_OBJECT* DeviceObject	The returned device object.

Table 5.2   IoRegisterDeviceInterface function

NTSTATUS IoRegisterDeviceInterface	(IRQL==PASSIVE_LEVEL)
Parameter	**Description**
IN PDEVICE_OBJECT PhysicalDeviceObject	The device PDO
IN CONST GUID *InterfaceClassGuid	The GUID being registered
IN PUNICODE_STRING ReferenceString	Usually NULL. A reference string becomes part of the interface name and so can be used to distinguish between different interfaces to the same device.
OUT PUNICODE_STRING SymbolicLinkName	The output interface symbolic link name. Do not forget to free the unicode string buffer using RtlFree-UnicodeString when finished with it.

If IoCreateDevice returns successfully, the DeviceObject parameter is set to point to the new device object. Wdm1 calls this its Functional Device Object (FDO). The Plug and Play chapter explains FDOs and PDOs in detail. Wdm1AddDevice returns an error if IoCreateDevice fails.

The Wdm1 device extension is now available through the FDO *DeviceExtension* field. Wdm1AddDevice stores a pointer to the FDO in its device extension.

Wdm1AddDevice now attaches the Wdm1 device to the device stack using IoAttachDeviceToDeviceStack, which must be called at PASSIVE_LEVEL[1]. The FDO and the given PDO are passed to this function. It returns a pointer to yet another device object that is also stored in the Wdm1 device extension. All this shenanigans are explained in the Plug and Play chapters.

Finally, two bits in the FDO *Flags* field must be changed. AddDevice routines must clear the DO_DEVICE_INITIALIZING flag. You do not need to clear this flag if you call IoCreateDevice in your DriverEntry routine. The DO_BUFFERED_IO flag is set in the *Flags* field. Again, this is covered later.

Deleting Devices

The Wdm1 driver must delete its FDO if it receives a *Remove Device* request. This is an IRP_MJ_PNP IRP with an IRP_MN_REMOVE_DEVICE minor version code. The code in Wdm1Pnp first calls IoDetachDevice to detach the FDO from the device stack. Finally, IoDeleteDevice is called to delete the FDO and its device extension.

IoDetachDevice and IoDeleteDevice must be called at IRQL PASSIVE_LEVEL.

Device Names

The Wdm1 AddDevice code calls IoCreateDevice to create a Wdm1 device object. There are two ways to provide a name that is available to Win32 programs. The old way is to provide an explicit symbolic link name. The newfangled approach is to use a device interface to identify the devices that support a defined API.

---

1. If you call IoAttachDeviceToDeviceStack at a higher IRQL, your driver may crash.

Symbolic Links

The IoCreateDevice call has a DeviceName parameter that you can use to give your device a name. This name identifies the device to the kernel, not to Win32.

You must create a symbolic link to make the kernel device name available to Win32. To do this the old way, call IoCreateSymbolicLink at IRQL PASSIVE_LEVEL, passing the desired symbolic link name and the kernel device name as parameters.

Explicit device names created using this technique usually have a device number at the end. By convention, kernel device name numbers increment from zero, while symbolic link names increment from one.

The best way to illustrate device naming is using the *WinObj* tool, which only runs in NT and Windows 2000. *WinObj* is supplied in the Platform SDK, not the W2000 DDK. The standard serial ports have kernel device names \device\Serial0, \device\Serial1, etc. In *WinObj*, these names appear in the \device folder.

Windows provides symbolic links "COM1", "COM2", etc., to make these serial ports available to Win32. Symbolic links appear in the \?? folder in *WinObj*. You can double-click on each symbolic link to find out to which kernel device it refers. The screenshot in Figure 5.2 shows that COM3, \??\com3, is a symbolic link to kernel device Serial0.

The \?? folder used to be called \DosDevices, which you might come across in some old documentation.

Figure 5.2   *WinObj* tool screenshot

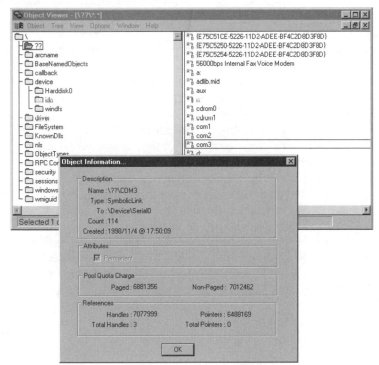

@ 1999 PHD Computer Consultants Ltd

The Wdm1 driver could have given its first device object a kernel name of \device\ Wdm1device0 and called IoCreateSymbolicLink to create a symbolic link with the name \??\ Wdm1dev1. A Win32 program would then have been able to open \\.\Wdm1dev1 using Create-File.

Note that C strings represent each \ character as "\\", so you would have to pass "\\\\.\ \Wdm1dev1" to CreateFile.

If this technique is used, you must keep track of how many devices are in use, so you can generate the correct names. If you use a global variable to keep the device count, you must use it carefully to ensure that simultaneous accesses do not corrupt its value. Achieve this using the InterlockedIncrement function that atomically increments a LONG value. This can be run at any IRQL and on paged memory. There are companion decrement, exchange, and compare functions.

Explicit kernel and symbolic link names are used most often in NT style drivers, which create devices in their DriverEntry routine. In this case, counting devices is easy, as you can be sure that no other part of your code will be trying to increment the device count.

You can use the QueryDosDevice Win32 function to get a list of all the symbolic link names currently available. For each symbolic link, you can call QueryDosDevice again to find the kernel name. QueryDosDevice is supposed to run in Windows 98, but I did not succeed in getting it to work. In NT and W2000, you can use the DefineDosDevice Win32 function to change symbolic links. You can experiment with both these commands using the *dosdev* utility in the W2000 DDK.

Device Interfaces

The Wdm1 driver uses a *device interface* to make its devices visible to Win32 programs. The idea is that each Wdm1 device makes a defined Application Programmer Interface (API) available. A Globally Unique Identifier (GUID) is used to identify this interface.

The device interface that each Wdm1 device exposes is the set of commands that a Win32 programmer can use to access the device. A Wdm1 device responds to Win32 CreateFile, ReadFile, WriteFile, DeviceIoControl, and CloseHandle calls in a defined way. For example, a Wdm1 device is nonexclusive, so more than one thread can open it at the same time.

The Wdm1 driver's job is to implement a memory buffer that is shared by all the Wdm1 devices. Any writes extend the buffer, if necessary. Reads do not extend the buffer. Four IOCTLs are supported: Zero buffer, remove buffer, get buffer size, and get buffer.

One benefit of device interfaces is that another driver could be written that also implements the Wdm1 API. As long as it does this faithfully, there is no reason why a Win32 program should notice the difference between the new driver and Wdm1.

While it seems unlikely that the Wdm1 device interface will be copied by others, this technique is very useful for other types of drivers. For example, audio miniport drivers should implement the IMiniport COM interface. This is identified using the IID_IMiniport GUID and means that the driver implements two functions, GetDescription and DataRangeIntersection.

Wdm1 **Device Interface**

Implementing a device interface for Wdm1 devices is straightforward. First, a new 16-byte GUID must be generated using the *guidgen* tool. Then, include code to register the Wdm1 device interface for each Wdm1 FDO device object.

Generating GUIDs

To get a suitable new GUID, run the *guidgen* tool in the Platform SDK or the VC++ executables directory. Figure 5.3 shows the window that appears. *guidgen* can generate GUIDs in formats suitable for four different uses. For driver header file definitions, choose the DEFINE_GUID(..) option. Pressing the *Copy* button copies the necessary code to the clipboard. Paste it into your header and amend the text <<name>> to define your own identifier.

Microsoft says that all GUIDs that *guidgen* generates are unique.

If you need to allocate a series of identifiers, press the *New GUID* button as often as necessary and copy each one to the relevant header file. Each new GUID increments the last digit of the first eight characters of the GUID. For example, the next GUID after the one shown in the screenshot starts {F4886541...}. In this book, I often use the form {F4886540...} as an abbreviation for the full GUID {F4886540-749E-11d2-B677-00C0DFE4C1F3}.

Figure 5.3   *guidgen* tool

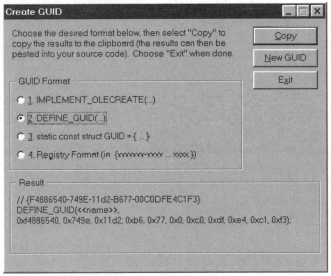

@ 1999 PHD Computer Consultants Ltd

The definition of WDM1_GUID in GUIDs.h is shown in Listing 5.3. There is a final twist in the tail for GUID definitions. One of the driver modules, Pnp.cpp in Wdm1, must have the following preprocessor directive before including GUIDs.h to formally declare WDM1_GUID.[2]

```
#define INITGUID
```

---

2.   If you are only compiling in W2000, you can include initguid.h instead.

Listing 5.3   `WDM1_GUID` **definition in** `GUIDs.h`

```
// Wdm1 device interface GUID


// {C0CF0640-5F6E-11d2-B677-00C0DFE4C1F3}
DEFINE_GUID(WDM1_GUID, 0xc0cf0640, 0x5f6e, 0x11d2, 0xb6, 0x77, 0x0, 0xc0, 0xdf,
    0xe4, 0xc1, 0xf3);
```

Device Interface Registration

Wdm1 does not give a kernel device name to its device object and does not use IoCreateSymbolicLink to create a device name accessible to Win32 programs. Instead, the Wdm1 AddDevice routine calls IoRegisterDeviceInterface to register its interface as shown in Listing 5.4. It does this after creating its device but before attaching it to the device stack. IoRegisterDeviceInterface creates a Win32 symbolic link name connection to the Wdm1 device object. This is stored in the device extension *ifSymLinkName* field.

The Wdm1AddDevice code checks the return value from IoRegisterDeviceInterface. If this call fails, it tidies up properly by deleting the FDO and returning the status error code.

The final step is to enable the device interface by calling IoSetDeviceInterfaceState, at IRQL PASSIVE_LEVEL. In subsequent drivers, the device interface is only enabled when the PnP *Start Device* message has been received.

Listing 5.4   **Wdm1** `Pnp.cpp` **Device Interface code**

```
// Register and enable our device interface
status = IoRegisterDeviceInterface(pdo, &WDM1_GUID, NULL, &dx->ifSymLinkName);
if( !NT_SUCCESS(status))
{
    IoDeleteDevice(fdo);
    return status;
}
IoSetDeviceInterfaceState(&dx->ifSymLinkName, TRUE);
```

Table 5.3   `IoRegisterDeviceInterface` **function**

NTSTATUS IoRegisterDeviceInterface	
Parameter	**Description**
IN PDEVICE_OBJECT PhysicalDeviceObject	The device PDO
IN CONST GUID *InterfaceClassGuid	The GUID being registered
IN PUNICODE_STRING ReferenceString	Usually NULL. A reference string becomes part of the interface name and so can be used to distinguish between different interfaces to the same device.
OUT PUNICODE_STRING SymbolicLinkName	The output interface symbolic link name. Do not forget to free the unicode string buffer using RtlFreeUnicodeString when finished with it.

The code in Wdm1Pnp to handle device removal has to be altered, as well. First, the device interface has to be disabled. Then, it must free the memory buffer allocated for the interface symbolic link name. It does not have to unregister the device interface.

```
IoSetDeviceInterfaceState(&dx->ifSymLinkName, FALSE);
RtlFreeUnicodeString(&dx->ifSymLinkName);
```

Device Interface Notes

As mentioned in the last chapter, registering a device interface creates a registry entry in the HKLM\System\CurrentControlSet\Control\DeviceClasses key. A subkey named after the GUID and further subkeys for each device implements that GUID. Drivers can use IoOpenDeviceInterfaceRegistryKey to open a handle to this registry key. For example, you might want to add a FriendlyName value to this key using an INF file AddInterface section, as described in Chapter 11. Win32 applications can use SetupDiOpenDeviceInterfaceRegKey to open this same key and retrieve the friendly name.

Device interface registrations can be removed from user mode, if necessary.

A driver can register that a device supports more than one device interface if desired.

IoGetDeviceInterfaces can be used by a driver to find a list of symbolic links to devices that support a specific device interface.

Behind the scenes, Windows generates a kernel device name for your device. This might be something like \device\004059. The symbolic link name generated by IoRegisterDeviceInterface looks like \DosDevices\000000000000001c#{c0cf0640...}. The *WinObj* entry for the symbolic link looks like \??\Root#UNKNOWN#0000#{C0CF0640...}.

Win32 Device Interface Access

We are ready — at last — to access a Wdm1 device. The *Wdm1Test* program opens a connection to the Wdm1 driver and puts it through its paces. Remember that all the calls in this section are to Win32 routines, not to the kernel.

The source code for the *Wdm1Test* program is in the Wdm1\exe directory. The book software has a Visual Studio WDM Book workspace, which includes a project for the *Wdm1Test* program. The *Wdm1Test* code in Wdm1Test.cpp is listed at the end of the chapter.

Wdm1Test is a standard Win32 console application. When run, it appears in a DOS box. As it is a standard Win32 program, you can debug it in Visual Studio as normal (e.g., step through the code).

There are two points to note about the *Wdm1Test* project. The first is that it includes the c:\98ddk\inc\win98\setupapi.h header file. The VC++ 5 version of this file is seriously out of date, so the code specifically includes the Windows 98 DDK version. The second special setting required is to ensure that c:\98ddk\lib\i386\free\setupapi.lib is listed in the Link property page Output/library modules section of the Project settings.

Getting a Device's Interface Name

The GetDeviceViaInterface routine in Wdm1Test.cpp opens a handle to a device, given the device interface's GUID, as shown in Listing 5.5. GetDeviceViaInterface has a second parameter, instance, which is the zero-based index into the count of available devices. The total number of devices cannot be determined in advance; simply call GetDeviceViaInterface until NULL is returned.

The *Wdm1Test* main function calls GetDeviceViaInterface with WDM1_GUID and 0 as parameters to open the first available Wdm1 device. Special steps must be taken again to ensure that WDM1_GUID is declared properly. In one module #include "initguid.h" before including the GUID header file.

Listing 5.5   GetDeviceViaInterface

```
HANDLE GetDeviceViaInterface( GUID* pGuid, DWORD instance)
{
    // Get handle to relevant device information set
    HDEVINFO info = SetupDiGetClassDevs( pGuid, NULL, NULL, DIGCF_PRESENT |
        DIGCF_INTERFACEDEVICE);
    if( info==INVALID_HANDLE_VALUE)
    {
        printf("No HDEVINFO available for this GUID\n");
        return NULL;
    }


    // Get interface data for the requested instance
    SP_INTERFACE_DEVICE_DATA ifdata;
    ifdata.cbSize = sizeof(ifdata);
    if( !SetupDiEnumDeviceInterfaces( info, NULL, pGuid, instance, &ifdata))
    {
        printf("No SP_INTERFACE_DEVICE_DATA available for this GUID instance
            \n");
        SetupDiDestroyDeviceInfoList(info);
        return NULL;
    }


    // Get size of symbolic link name
    DWORD ReqLen;
    SetupDiGetDeviceInterfaceDetail( info, &ifdata, NULL, 0, &ReqLen, NULL);
    PSP_INTERFACE_DEVICE_DETAIL_DATA ifDetail =
        (PSP_INTERFACE_DEVICE_DETAIL_DATA)(new char[ReqLen]);
    if( ifDetail==NULL)
    {
        SetupDiDestroyDeviceInfoList(info);
        return NULL;
    }

    // Get symbolic link name
    ifDetail->cbSize = sizeof(SP_INTERFACE_DEVICE_DETAIL_DATA);
```

Listing 5.5   `GetDeviceViaInterface` **(continued)**

```
    if( !SetupDiGetDeviceInterfaceDetail( info, &ifdata, ifDetail, ReqLen, NULL,
        NULL))
    {
        SetupDiDestroyDeviceInfoList(info);
        delete ifDetail;
        return NULL;
    }

    printf("Symbolic link is %s\n",ifDetail->DevicePath);
    // Open file
    HANDLE rv = CreateFile( ifDetail->DevicePath,
        GENERIC_READ | GENERIC_WRITE,
        FILE_SHARE_READ | FILE_SHARE_WRITE,
        NULL, OPEN_EXISTING, FILE_ATTRIBUTE_NORMAL, NULL);

    delete ifDetail;
    SetupDiDestroyDeviceInfoList(info);
    return rv;
}
```

The Win32 call to `SetupDiGetClassDevs` opens a "device information set" about devices with the specified GUID. The `DIGCF_PRESENT` and `DIGCF_INTERFACEDEVICE` flags ensure that only devices that are present are found.

`SetupDiEnumDeviceInterfaces` is then called to retrieve a `SP_INTERFACE_DEVICE_DATA` context structure of information about the device instance in which you are interested. `GetDeviceViaInterface` tries to retrieve information about only one instance, but you might want to find all instances by incrementing the `MemberIndex` parameter from 0 until `SetupDiEnumDeviceInterfaces` fails and `GetLastError` returns `ERROR_NO_MORE_ITEMS`.

The next task is to obtain the symbolic link name for the instance that has been found. This string is in the `PSP_INTERFACE_DEVICE_DETAIL_DATA` `ifDetail` structure returned by a call to `SetupDiGetDeviceInterfaceDetail`. `SetupDiGetDeviceInterfaceDetail` must be called twice. The first call retrieves the required size for `ifDetail`, while the second actually gets the structure.

Eventually `ifDetail->DevicePath` contains the filename that is used to open a handle to the relevant Wdm1 device. This filename is passed to `CreateFile` and then `ifDetail` is deleted. Do not forget to call `SetupDiDestroyDeviceInfoList` on all paths to close the device information set.

The call to `CreateFile` in `GetDeviceViaInterface` uses a pretty standard set of parameters. As noted before, you can change the share and access modes, if necessary. Further, the device can be opened for overlapped nonblocking access. Overlapped access works for devices in Windows 98, NT, and Windows 2000 and lets an application issue an I/O request and get on with other work while the request is being processed. The detailed explanation of the DebugPrint Monitor program in Chapter 14 gives an example of this technique.

Running *Wdm1Test*

Simply run Wdm1Test.exe in the Wdm1\exe\Release directory. Make sure that the Wdm1 driver is installed. *Wdm1Test* is a console application running in a DOS box. Press the Enter key to exit the program.

If you run *Wdm1Test* in W2000, the output should look like that given in Listing 5.6. The first time round, Test 2 is designed to fail; subsequent tests should succeed. In Windows 98, the output is different, as it does not seem to support the SetFilePointer function on devices, such as Wdm1.

The rest of the *Wdm1Test* main function performs each of these tests in turn.

1. Open the first Wdm1 device.
2. Read the first DWORD stored in the shared memory buffer.

   This will fail first time round, as there will be nothing in the buffer.
3. Write 0x12345678 to the start of the buffer.

   0x78 is written at file pointer zero, 0x56 at one, 0x34 at two, and 0x12 at three.
4. Set the file pointer to position 3.
5. Read one byte from the file. This should be 0x12.
6. Write 0x12345678 to the buffer starting at file position 3.
7. Use an IOCTL to get the buffer size, which should be 7 (i.e., 3+4).
8. Use an IOCTL to get the entire buffer. The first word is printed and should be 0x78345678.
9. Check that issuing an IOCTL with an invalid parameter fails. The IOCTL asks for the entire buffer with a request size that is too big.
10. Use an IOCTL to zero all the bytes in the buffer. Get the entire buffer again. The first word is printed which should be 0x00000000.
11. Use an IOCTL to remove the buffer. Check that the buffer size is now zero.
12. Try to issue an invalid IOCTL code. Confirm that this fails.
13. Write 0xabcdef01to the start of the buffer. When *Wdm1Test* is run next, Test 2 should find this value.

Listing 5.6   *Wdm1Test* output on W2000

```
Test 1
Symbolic link is \\?\root#unknown#0003#{c0cf0640-5f6e-11d2-b677-00c0dfe4c1f3}
     Opened OK


Test 2
     Read successfully read stored value of 0xABCDEF01


Test 3
     Write 0x12345678 succeeded


Test 4
     SetFilePointer worked
```

Listing 5.6   *Wdm1Test* output on W2000 (continued)

```
Test 5
      Read successfully read stored value of 0x12


Test 6
      Write at new file pointer succeeded


Test 7
      Buffer size is 7 (4 bytes returned)


Test 8
      First DWORD of buffer is 78345678 (7 bytes returned)


Test 9
      Too big get buffer failed correctly 87


Test 10
      Zero buffer succeeded
      First DWORD of buffer is 00000000 (7 bytes returned)


Test 11
      Remove buffer succeeded
      Buffer size is 0 (4 bytes returned)


Test 12
      Unrecognised IOCTL correctly failed 1


Test 13
      SetFilePointer worked
      Write 0xabcdef01 succeeded


Test 14
      CloseHandle worked


Press enter please
```

Wdm1Test exercises all the functions of the Wdm1 driver. It checks that things that should work do work, and things that should fail do fail.

Writing this test program showed that the Wdm1 device does behave differently in Windows 98 and Windows 2000. However, the problem does not lie in the driver itself, but is due to Windows 98 not supporting SetFilePointer for device files.

Conclusion

This chapter shows how to open a connection to a Wdm1 device using device interfaces. The *Wdm1Test* user mode program puts the Wdm1 driver through its paces.

Chapter 7 shows how Wdm1 implements the read, write, and IOCTL calls. However, before I continue looking at the driver, the next chapter describes how to test and debug drivers. It explains how this book uses the *DebugPrint* software to see what is going on in a driver.

Listing 5.7   Wdm1Test.cpp

```
/////////////////////////////////////////////////////////////////////////
//Copyright © 1998 Chris Cant, PHD Computer Consultants Ltd
//WDM Book for R&D Books, Miller Freeman Inc
//
//Wdm1Test example
/////////////////////////////////////////////////////////////////////////
//Wdm1Test.cpp:Win32 console application to exercise Wdm1 devices
/////////////////////////////////////////////////////////////////////////
//mainProgram main line
//GetDeviceViaInterfaceOpen a handle via a device interface
/////////////////////////////////////////////////////////////////////////
//Version history
//27-Apr-991.0.OCCcreation
/////////////////////////////////////////////////////////////////////////

#include "windows.h"
#include "c:\98ddk\inc\win98\setupapi.h"// VC++ 5 one is out of date
#include "stdio.h"
#include "initguid.h"
#include "..\sys\GUIDs.h"
#include "winioctl.h"
#include "..\sys\Ioctl.h"

HANDLE GetDeviceViaInterface( GUID* pGuid, DWORD instance);

int main(int argc, char* argv[])
{
int TestNo = 1;

/////////////////////////////////////////////////////////////////////
// Open device
printf("\nTest %d\n",TestNo++);
HANDLE hWdm1 = GetDeviceViaInterface((LPGUID)&WDM1_GUID,0);
```

```
if( hWdm1==NULL)
{
printf("XXX  Could not find open Wdm1 device\n");
return 1;
}
printf("    Opened OK\n");


//////////////////////////////////////////////////////////////////////////
// Read first ULONG that's left in buffer
printf("\nTest %d\n",TestNo++);
DWORD TxdBytes;
ULONG Rvalue = 0;
if( !ReadFile( hWdm1, &Rvalue, 4, &TxdBytes, NULL))
printf("XXX  Could not read value %d\n". GetLastError());
else if( TxdBytes==4)
printf("    Read successfully read stored value of 0x%X\n",Rvalue);
else
printf("XXX  Wrong number of bytes read: %d\n",TxdBytes);


//////////////////////////////////////////////////////////////////////////
// Write 0x12345678
printf("\nTest %d\n",TestNo++);
ULONG Wvalue = 0x12345678;
if( !WriteFile( hWdm1, &Wvalue, 4, &TxdBytes, NULL))
printf("XXX  Could not write %X\n",Wvalue);
else if( TxdBytes==4)
printf("    Write 0x12345678 succeeded\n");
else
printf("XXX  Wrong number of bytes written: %d\n",TxdBytes);


//////////////////////////////////////////////////////////////////////////
// Set file pointer
printf("\nTest %d\n",TestNo++);
DWORD dwNewPtr = SetFilePointer( hWdm1, 3, NULL, FILE_BEGIN);
if( dwNewPtr==0xFFFFFFFF)
printf("XXX  SetFilePointer failed %d\n",GetLastError());
else
printf("    SetFilePointer worked\n");


//////////////////////////////////////////////////////////////////////////
// Read
printf("\nTest %d\n",TestNo++);
```

```
Rvalue = 0;
if( !ReadFile( hWdm1, &Rvalue, 1, &TxdBytes, NULL))
printf("XXX  Could not read value\n");
else if( TxdBytes==1)
printf("     Read successfully read stored value of 0x%X\n",Rvalue);
else
printf("XXX  Wrong number of bytes read: %d\n",TxdBytes);


/////////////////////////////////////////////////////////////////////
// Write
printf("\nTest %d\n",TestNo++);
if( !WriteFile( hWdm1, &Wvalue, 4, &TxdBytes, NULL))
printf("XXX  Could not write %X\n",Wvalue);
else if( TxdBytes==4)
printf("     Write at new file pointer succeeded\n");
else
printf("XXX  Wrong number of bytes written: %d\n",TxdBytes);


/////////////////////////////////////////////////////////////////////
// Get buffer size
printf("\nTest %d\n",TestNo++);
ULONG BufferSize;
DWORD BytesReturned;
if( !DeviceIoControl( hWdm1, IOCTL_WDM1_GET_BUFFER_SIZE,
NULL, 0,// Input
&BufferSize, sizeof(ULONG),// Output
&BytesReturned, NULL))
printf("XXX  Could not get buffer size\n");
else
printf("     Buffer size is %i (%d bytes returned)\n",BufferSize,BytesReturned);


/////////////////////////////////////////////////////////////////////
// Get buffer size
printf("\nTest %d\n",TestNo++);
char* Buffer = new char[BufferSize+1];
if( !DeviceIoControl( hWdm1, IOCTL_WDM1_GET_BUFFER,
NULL, 0,// Input
Buffer, BufferSize,// Output
&BytesReturned, NULL))
printf("XXX  Could not get buffer\n");
else
printf("     First DWORD of buffer is %08X (%d bytes returned)\
n",*((DWORD*)Buffer),BytesReturned);
```

```
///////////////////////////////////////////////////////////////////
// Get too big a buffer size
printf("\nTest %d\n",TestNo++);
if( !DeviceIoControl( hWdm1, IOCTL_WDM1_GET_BUFFER,
NULL, 0,// Input
Buffer, BufferSize+1,// Output
&BytesReturned, NULL))
printf("    Too big get buffer failed correctly %d\n",GetLastError());
else
printf("XXX  Too big get buffer unexpectedly succeeded\n");


///////////////////////////////////////////////////////////////////
// Zero all buffer bytes
printf("\nTest %d\n",TestNo++);
if( !DeviceIoControl( hWdm1, IOCTL_WDM1_ZERO_BUFFER,
NULL, 0,// Input
NULL, 0,// Output
&BytesReturned, NULL))
printf("XXX  Zero buffer failed %d\n",GetLastError());
else
printf("    Zero buffer succeeded\n");
if( !DeviceIoControl( hWdm1, IOCTL_WDM1_GET_BUFFER,
NULL, 0,// Input
Buffer, BufferSize,// Output
&BytesReturned, NULL))
printf("XXX  Could not get buffer\n");
else
printf("    First DWORD of buffer is %08X (%d bytes returned)\
n",*((DWORD*)Buffer),BytesReturned);


///////////////////////////////////////////////////////////////////
// Remove buffer
printf("\nTest %d\n",TestNo++);
if( !DeviceIoControl( hWdm1, IOCTL_WDM1_REMOVE_BUFFER,
NULL, 0,// Input
NULL, 0,// Output
&BytesReturned, NULL))
printf("XXX  Remove buffer failed %d\n",GetLastError());
else
printf("    Remove buffer succeeded\n");
if( !DeviceIoControl( hWdm1, IOCTL_WDM1_GET_BUFFER_SIZE,
NULL, 0,// Input
```

```
&BufferSize, sizeof(ULONG),// Output
&BytesReturned, NULL))
printf("XXX   Could not get buffer size\n");
else
printf("     Buffer size is %i (%d bytes returned)\n",BufferSize,BytesReturned);


/////////////////////////////////////////////////////////////////////////
// Unrecognised IOCTL
printf("\nTest %d\n",TestNo++);
if( !DeviceIoControl( hWdm1, IOCTL_WDM1_UNRECOGNISED,
NULL, 0,// Input
NULL, 0,// Output
&BytesReturned, NULL))
printf("     Unrecognised IOCTL correctly failed %d\n",GetLastError());
else
printf("XXX   Unrecognised IOCTL unexpectedly succeeded\n");


/////////////////////////////////////////////////////////////////////////
// Write 0xabcdef01 to start of buffer
printf("\nTest %d\n",TestNo++);
dwNewPtr = SetFilePointer( hWdm1, 0, NULL, FILE_BEGIN);
if( dwNewPtr==0xFFFFFFFF)
printf("XXX   SetFilePointer failed %d\n",GetLastError());
else
printf("     SetFilePointer worked\n");
Wvalue = 0xabcdef01;
if( !WriteFile( hWdm1, &Wvalue, 4, &TxdBytes, NULL))
printf("XXX   Could not write %X\n",Wvalue);
else if( TxdBytes==4)
printf("     Write 0xabcdef01 succeeded\n");
else
printf("XXX   Wrong number of bytes written: %d\n",TxdBytes);


/////////////////////////////////////////////////////////////////////////
// Close device
printf("\nTest %d\n",TestNo++);
if( !CloseHandle(hWdm1))
printf("XXX   CloseHandle failed %d\n",GetLastError());
else
printf("     CloseHandle worked\n");
```

```
///////////////////////////////////////////////////////////////////////
delete Buffer;
printf("\nPress enter please");
char line[80];
gets(line);
return 0;
}


///////////////////////////////////////////////////////////////////////
//GetDeviceViaInterface:Open a handle via a device interface

HANDLE GetDeviceViaInterface( GUID* pGuid, DWORD instance)
{
// Get handle to relevant device information set
HDEVINFO info = SetupDiGetClassDevs(pGuid, NULL, NULL, DIGCF_PRESENT | DIGCF_INTER-
FACEDEVICE);
if(info==INVALID_HANDLE_VALUE)
{
printf("No HDEVINFO available for this GUID\n");
return NULL;
}


// Get interface data for the requested instance
SP_INTERFACE_DEVICE_DATA ifdata;
ifdata.cbSize = sizeof(ifdata);
if(!SetupDiEnumDeviceInterfaces(info, NULL, pGuid, instance, &ifdata))
{
printf("No SP_INTERFACE_DEVICE_DATA available for this GUID instance\n");
SetupDiDestroyDeviceInfoList(info);
return NULL;
}


// Get size of symbolic link name
DWORD ReqLen;
SetupDiGetDeviceInterfaceDetail(info, &ifdata, NULL, 0, &ReqLen, NULL);
PSP_INTERFACE_DEVICE_DETAIL_DATA ifDetail = (PSP_INTERFACE_DEVICE_DETAIL_DATA)(new
char[ReqLen]);
if( ifDetail==NULL)
{
SetupDiDestroyDeviceInfoList(info);
return NULL;
}
```

```
// Get symbolic link name
ifDetail->cbSize = sizeof(SP_INTERFACE_DEVICE_DETAIL_DATA);
if( !SetupDiGetDeviceInterfaceDetail(info, &ifdata, ifDetail, ReqLen, NULL, NULL))
{
SetupDiDestroyDeviceInfoList(info);
delete ifDetail;
return NULL;
}

printf("Symbolic link is %s\n",ifDetail->DevicePath);
// Open file
HANDLE rv = CreateFile( ifDetail->DevicePath,
GENERIC_READ | GENERIC_WRITE,
FILE_SHARE_READ | FILE_SHARE_WRITE,
NULL, OPEN_EXISTING, FILE_ATTRIBUTE_NORMAL, NULL);

delete ifDetail;
SetupDiDestroyDeviceInfoList(info);
return rv;
}

////////////////////////////////////////////////////////////////////////////
```

6

Chapter 6

Testing and Debugging

Before looking at other aspects of WDM drivers, it is worth while looking at how to test and debug drivers. This chapter explains how to use the *DebugPrint* tool that comes with the book software. This lets a driver "print" debug event messages to a Win32 application, the *DebugPrint Monitor*.

A later chapter looks in detail at how the *DebugPrint* software works.

Keep your driver as simple as possible to make it easier to test and debug.

Oh, and needless to say — but I will say it anyway — comment your driver very well. Document it well. Do not forget to update your comments and documentation as the driver develops.

Test, Test, Test

Developers usually want to start debugging only when they suspect something has gone wrong with their software. However, it is especially important to start work early with device drivers. Even if you think a driver is doing its job, check it thoroughly. Do not get a first cut working and ship it to customers immediately.

Here is an example of a situation where I did not test code properly. When developing the *DebugPrint* software, I needed to implement a formatted print function. I could not use `sprintf`, as this function might use facilities that are not available to kernel mode drivers.

To make things simple, I developed the routines in a Win32 program and got them working. I copied the source over to the `DebugPrint` driver. A quick test seemed to indicate that it worked OK. In fact, I had changed something and so my handling of ellipsis arguments went wrong (passing a variable number of arguments to a function). By a complete fluke, one of the simple tests I inserted did get the correct information from the stack. It took a while to realize the error of my ways.

Driver Tests

This section describes some of the testing issues that are particularly important for device drivers. The following debugging section highlights some of the typical ways that drivers fail.

If you — and only you — are writing the user mode program that accesses your driver, you may be able to impose certain constraints on your driver. For example, you could dictate that only one request is ever issued at a time. This would make some of your routines slightly easier to implement. However, it is best to write your routines properly straightaway. It is quite possible for someone to later change the Win32 code without realizing that it might break your driver.

Test your driver

Test That All Functions Work

This may sounds obvious, but check that your driver works. This means writing a test program (or other driver) that exercises all of its functions.

As with any other program, check that all sorts of data work as expected. For example, check that invalid parameters are detected properly. Check that small and large amounts of data can be transferred. Check boundary conditions. For example, if you have a maximum request size of 1024 bytes, check that transfers of 0, 1, 1023, and 1024 bytes succeed and that transfers of 1025 bytes fail.

Check that all the features of the hardware you are driving are exercised. Do you have a strategy for coping with hardware updates? For example, leave some room for more function codes. Make sure that your new hardware firmware can be used by the old driver.

Check that a driver works in stress situations (e.g., in low memory computers or when being pummelled by lots of requests). Do this by repeatedly issuing I/O requests to your driver. Test with other processes doing the same job simultaneously. Another good test is to run your driver while copying a large file to a floppy disk; in Windows 98, this seems to stop most other activities dead in their tracks.

Check that a driver works when more than one of its devices are in use. Check that several user programs or threads can access the driver simultaneously.

Check that your driver works on a clean PC that has never seen your driver. Test on plain vanilla W2000 and W98 systems, without any of the development environment. Check that you installed any recent DLLs that are needed. For example, your user mode installation code may not run with an out-of-date version of SETUPAPI.DLL.

W2000 and W98

Do all your tests in Windows 2000 and Windows 98 if you are writing a driver that runs under both operating systems. If you provide a Dec Alpha W2000 version, check that it works as well.

In particular, check that a driver installs correctly in both operating systems. W2000 and W98 use different locations for some registry entries, though this should not effect most drivers.

Some facilities are only available in NT or W2000. For example, a driver can make calls to write events to the system log. In W98, these functions do nothing. Check that they work properly in NT and Windows 2000.

As long as you use the correct WDM.H header, you should not be making any kernel calls that work only in W2000. For example, I was going to use the ExSystemTimeToLocalTime kernel call in the DebugPrint driver. However, the documentation says that you must include NTDDK.H. This means that the call is not available in W98.

Multiprocessor

If possible, check that your driver works in a multiprocessor computer. Even if you do run it on such a computer, you may have difficulty deciding whether it has passed the test as the time-critical events may not happen very often. Stress testing the driver should force the issue.

Cancelling I/O

Win32 programs can issue asynchronous "overlapped" I/O requests, meaning that they can issue a request and get on with another job while the I/O is happening. Windows 98 programs cannot issue such requests for file I/O, but they can for device I/O.

Initially, there seems to be no difference as far as a device driver is concerned. Both synchronous "blocking" I/O requests and asynchronous I/O requests arrive at the driver in the same way, through the driver dispatch entry points.

However, asynchronous I/O requests can be cancelled. This can happen either when the CancelIo function[1] is called, or if the file handle is closed with the I/O still pending.

Synchronous I/O requests are also cancelled if a process crashes. The operating system tries to cancel any requests and close any open file handles.

How a driver copes with cancelling I/O is covered in a later chapter. Ensure that your test program checks these situations.

---

1.   CancelIo first became available in NT 4 and W98.

Debugging

Win32 user mode programs run within the protective arms of the operating system. For example, it stops them accessing memory that is not in a task's address space. Windows 98 programs have some of the kernel mapped into its upper addresses, so user mode programs could trample on the kernel. The protection offered by NT and Windows 2000 is better. If a user mode application runs into problems, it is usually possible to stop the program and carry on with other work.

A device driver has none of this protection because it is part of the kernel. Be especially careful to test your driver fully or you may lose data. Be prepared for the worst on your own development computer by making regular backups.

Although problems with a driver can be a real pain, it can eventually give you a greater understanding of what you are doing. You have to inspect each call or section of code carefully. Rather than just following the instructions given here, you may eventually really know what is going on. Or even, heaven forbid, why the system was designed to work the way it does. You may have suggestions for how drivers or the system might be improved.

How Do Things Go Wrong?

Windows 98 and Windows 2000 sometimes react differently to errors in drivers.

Crashes

A fatal error in NT or W2000 causes the "blue screen of death", properly called a "bugcheck". You must reset the computer to continue. The blue screen gives an indication of the error, and usually a list of the kernel modules and a stack trace. More details of how to decode this information are given later. Annoying though these bug checks are, they do stop further damage being done to the operating system, such as corrupting disks.

NT and W2000 usually log a "Save Dump" event for a bugcheck, listing the most pertinent information. An event may not be logged if the bugcheck occurred at boot time, before the event log service has started.

In W98, a similar blue screen appears for fatal errors, again giving a brief description of what went wrong. W98 can usually carry on, but it may be best to restart the PC.

The two most common causes of fatal errors are

- Accessing nonexistent memory
- Accessing paged memory at or above DISPATCH_LEVEL IRL

Core Dumps

If NT or W2000 has a bugcheck, you can also get it to produce core dumps in a file on disk, called memory.dmp by default. You have to enable this option in the Control Panel System applet Advanced tab Startup and Recovery section. You can use the *WinDbg* and *DumpExam* tools to inspect the dump. However, I found that using *WinDbg* in this way was not particularly productive.

In NT and W2000, you can include information in the core dump by calling KeInitializeCallbackRecord and KeRegisterBugCheckCallback. In the event of a bugcheck, your callback routine is called to store any state information in the core dump.

Driver Will Not Start

When you update a driver like Wdm1, the Device Manager may state that you need to restart the system before the driver will run. This means that the driver returned an error while loading.

There are two possible causes for such errors. Suppose your driver will start when the system reboots. This means that when it unloads it must be leaving something around that stops it from starting again. A common problem is forgetting to delete your device or the symbolic link to your device.

If your driver will not start when the system reboots, it means that you must have changed your driver so that the DriverEntry routine fails. Trace through the code to work out what is going wrong.

Just to complicate matters, in some circumstances, Windows needs to juggle the revised system resources when a driver is reloaded. In this case, Windows will not even attempt to start your driver if a reboot is necessary to satisfy the resource allocation process.

Hang Ups

A user thread can hang up if you never complete an IRP. The thread will never be able to complete. Have you queued the IRP somewhere and never processed it?

A driver can also hang up if it cannot acquire a resource that it needs. The different resources are described later. If a resource is unavailable, it may mean that just one IRP cannot be processed, or it could mean that all IRP processing grinds to halt.

A "deadly embrace" occurs if two different pieces of code are trying to access the same resources. For example, code A might hold resource X and want resource Y. If code B holds Y and wants X, then a deadly embrace occurs, hanging the whole system. The simplest solution is to always acquire resources in the same order.

You can also get hang ups if you try to acquire a resource more than once, or if your interrupt service routine never returns. Continual interrupts can seriously degrade the system performance.

Resource Leaks

Resource leaks are less dire, but still important to fix. For example, if you forget to free some memory allocated during each read, the system will eventually run out of memory.

Some resources are more important than others. For example, nonpaged memory should be used conservatively.

Time Dependencies

Possibly the worst type of problem to sort out is related to timing.

At the simplest level, a timing problem can simply mean that you have not filled a buffer quickly enough. For example, if your hardware needs to output data regularly, you may not have provided it with the data. Or did you fill it too quickly? Check that your driver works on different speed computers.

Another possible problem is that your driver may not be reading data quickly enough. For example, an isochronous device, such as a microphone, may generate a regular number of samples per second. Check that you can keep reading data at this speed, even in a stressed system. Alternatively, have a strategy for skipping samples (e.g., a call to your device to drop samples).

Debugging Techniques

Incremental Development

An incremental approach makes debugging far easier. Get each stage working and tested before moving onto the next stage.

An incremental approach still means that you should do proper design work. Think very carefully about how your driver is going to work before starting any development. Having decided upon a design, plan a development path that allows you to implement your design in stages.

Checked Version

Driver development environments are usually set up to produce two different builds: an optimized retail release version, called the *free* build, and an unoptimised test debug build, called the *checked* build.

Put any debug or test code so that it appears in the checked build. The simplest way to do this is to check the value of the preprocessor variable DBG, which is set to 1 in the checked build and 0 in free build. Here is an example of how to use this technique.

```
#if DBG
    DebugPrintInit("Wdm1 checked");
#else
    DebugPrintInit("Wdm1 free");
#endif
```

I shall soon look at the different types of debug code that you can put in your checked version.

Running under the checked build version of NT or W2000 can also pick up driver errors as the operating system makes more internal checks.

W2000 or W98

One decision you will need to make is whether to work primarily in Windows 2000 or Windows 98. You could set up the development environment in one operating system, and only do test installs in the other. As stated previously, you ought to be testing your driver in both operating systems.

My advice is to use Windows 2000 for development. A few useful tools are only available in W2000 (and NT). And, significantly, W2000 is better at detecting driver errors. For example, W2000 will usually cause a bugcheck when you try to access a bad address, such as writing to a NULL pointer. Windows 98 may not detect this sort of error and your driver will continue, oblivious to the problem. The downside to W2000 is that it takes a bit longer to restart.

It is useful to be able to dual boot a computer with Windows 2000 and Windows 98. If a device fails during system boot, you can use the other operating system to remove the offending driver temporarily. For this technique to work, you need the Windows 2000 system drive to be accessible to Windows 98 (i.e., not formatted in NTFS).

Debugging Tools

Apart from using your brain, there are two basic techniques for debugging. The first is to include "print" statements of some sort in your driver. The second is to do source code level debugging.

Windows 2000 Events

One way of reporting data is to generate NT and Windows 2000 event messages. For a commercial driver, you should use this facility to report any abnormal events to system administrators. However, it is somewhat complicated to set up a connection between a message number and the string it represents. You can include strings and data values as a part of the event.

The NT and Windows 2000 event viewer does not have the best display in the world, and it does not automatically update itself. Windows 98 has no event viewer.

Generating Windows Management Instrumentation (WMI) events is theoretically a way for a driver to report events. While this has the advantage that you should be able to view the data across a network, it is not a serious proposition for driver development, as it is even harder work to see events.

Tracing Tools

There are better tools available to display driver "print" trace statements. The next section discusses the *DebugPrint* software. This lets a driver use formatted print statements in the code. The output appears in a user mode monitor application.

As mentioned previously, two similar tools are available commercially: *Driver Monitor* in the Compuware Numega DriverWorks software, and *OSRTracer* in the Open Systems Resources *OSR DDK* software.

Driver Verifier

Windows 2000 includes *Driver Verify* for catching some common types of driver error. Use the Driver Verifier Manager (Start+Programs+Development Kits+Windows 2000 DDK+Driver Verifier) to change the driver verify settings.

Drivers can corrupt memory by accessing outside their allocated buffers. W2000 can use a special pool, with inaccessible memory at either side of a buffer. If you try to write to the inaccessible memory, a bugcheck occurs. The facility can be turned on for all your driver's memory requests. Alternatively, you can use it selectively by calling ExAllocatePoolWithTag with an appropriate Tag parameter.

Another common mistake is to access paged memory at DISPATCH_LEVEL IRQL or higher. A driver may "get away with it" if the paged memory is resident. *Driver Verify* can flush all paged memory to disk before your driver is called at elevated IRQL.

Drivers must also cope well if a memory request fails. Another *Driver Verify* feature can forcibly fail a random number of memory allocation requests.

Debuggers

The final and most powerful tool is a debugger. Standard user mode debuggers like Visual Studio do not work with drivers.

Instead, you must use a debugger designed for the job. The Microsoft DDKs come with the *WinDbg* debugger for NT and Windows 2000 systems. To use this, you must have two PCs — one running a checked build of NT/W2000 and another the free build — connected together using a serial cable. A normal network connection will also be useful. You can insert trace statements and debug at source-level. There are instructions for *WinDbg* in the DDKs. It will not be covered in this book.

NuMega Compuware at www.numega.com sells a debugger called SoftICE, which lets you to debug at source-level on a single PC.

DebugPrint

This book includes the *DebugPrint* software, which lets you use formatted print statements to trace the execution of a driver. Although not the ultimate debugging solution, it certainly gives a driver writer a useful tool for tracing what code a driver runs and what data is in variables. *DebugPrint* works in Windows 2000 and Windows 98, but not in earlier versions of these operating systems.

The source code of the DebugPrint driver and user mode monitor are included with the book. The DebugPrint driver serves to illustrate several driver development techniques. The *DebugPrint* source is described in Chapter 14.

DebugPrint is described in full on the web site www.phdcc.com\DebugPrint.

Trying out *DebugPrint*

To use *DebugPrint*, you must first install the DebugPrint driver. Do this in a similar way to the Wdm1 driver. This time, browse to the DebugPrint\sys directory and install the free build driver called "DebugPrint driver debugging tool".

You can now use the *DebugPrint Monitor* application to listen for trace print events from drivers that you are testing. You will probably find it convenient to set up a shortcut to this application at DebugPrint\exe\Release\DebugPrintMonitor.exe.

The Wdm1 driver includes various calls to the *DebugPrint* software. However, these calls only appear in the checked build of the driver. (You can opt to include the *DebugPrint* output in free builds.)

If you now reinstall the Wdm1 driver, selecting the checked build, you should see various events in the *DebugPrint Monitor* window, as shown in Figure 6.1. The displayed output is described in the next chapter.

Figure 6.1   Sample DebugPrint Monitor output

Using the *DebugPrint Monitor*

The *DebugPrint Monitor* program runs on the same computer as the drivers you are testing. It is a standard user mode MFC Win32 application that needs the shared MFC42.DLL library.

The *Monitor* is very easy to use. When started, it begins listening for *DebugPrint* trace events. If any events are buffered up, these are read and displayed first.

The *Monitor* displays one event per line. It has columns for the driver name, a timestamp, and the actual trace message. The latest event is always scrolled into view.

Use the *Edit+Delete events* menu to clear all the events from the display.

You can save the current list of events to a .dpm file. This is an ASCII text file with the columns separated by tab characters. You can reload .dpm files.

The *Monitor* remembers its position on the screen and the column widths. Printing events is not yet supported.

Using *DebugPrint* in Drivers

First, you must copy two standard source files into your driver project, DebugPrint.c and DebugPrint.h. Include DebugPrint.h in the driver's main header file. Amend the SOURCES file so that DebugPrint.c is built. These files are in the Wdm1 project source code.

Make calls to the *DebugPrint* functions in your driver code. You can only make these calls at DISPATCH_LEVEL IRQL or lower. This means that you can make *DebugPrint* calls in your DriverEntry, the main IRP dispatch routines, and in StartIo and Deferred Procedure Call (DPC) routines, but not in interrupt handling routines. The DebugPrintInit routine must be called at PASSIVE_LEVEL.

By default, the *DebugPrint* source only prints in the "checked" debug build. However, you can force it to work in the "free" build by setting the DEBUGPRINT preprocessor variable to 1 before you include DebugPrint.h, e.g.,

```
#define DEBUGPRINT 1
```

The *DebugPrint* functions will never overflow the internal output buffer that they allocate. All wide characters are simply cast to single-byte ANSI characters when they are put in an output string.

DebugPrint calls should only introduce minor delays into the execution of your driver. The main work of the *DebugPrint* calls takes place in a system thread that runs in the background at a low real-time priority.

DebugPrint Calls

Listing 6.1 illustrates a typical series of *DebugPrint* calls in three standard driver routines.

Listing 6.1   Typical DebugPrint calls

```
NTSTATUS DriverEntry(...)
{
#if DBG
    DebugPrintInit("Wdm1 checked");
#else
    DebugPrintInit("Wdm1 free");
#endif
    ...
    DebugPrint("RegistryPath is %T",RegistryPath);
    ...
}


NTSTATUS Read(...)
{
    ...
    DebugPrint("IRP %I.  Reading %u bytes from %x",
        Irp,
        IrpStack->Parameters.Read.Length,
        Irp->AssociatedIrp.SystemBuffer);
    ...
}


VOID Unload()
```

Listing 6.1   Typical DebugPrint calls (continued)

```
{
    ...
    DebugPrintClose();
}
```

Call `DebugPrintInit` to initialize the connection to the `DebugPrint` driver, passing the name of your driver as an ANSI NULL-terminated string. You typically use `DebugPrintInit` in your `DriverEntry` routine. Call `DebugPrintClose` to close the connection in your driver unload routine.

You can write simple NULL-terminated ANSI strings using `DebugPrintMsg`. For formatted print trace statements, use `DebugPrint` or `DebugPrint2`. `DebugPrint` uses an internal 100-byte buffer. `DebugPrint2` lets you specify the size of buffer to allocate.

Calls to `DebugPrint` or `DebugPrint2` must include a NULL-terminated ANSI format specification string, which may include one or more format specifier characters shown in Table 6.1 (e.g., `%c` for an ANSI character). You must provide an extra argument for each specifier in your format string. Failure to get this right could very easily result in an access violation. Be careful to use the correct uppercase or lowercase format specifier type character.

The available format specifiers cover a range of useful types. `%I` prints out the major and minor codes of an IRP. `%T` prints a `UNICODE_STRING`. Line feed and carriage return characters in the format string are ignored.

The `%1`, `%L`, `%s`, `%S`, and `%x` format specifier characters let you use one or more modifier characters to specify the maximum output width (e.g., `%*1` and `%n1`). The `*` modifier takes the maximum size from the next integer parameter to `DebugPrint` or `DebugPrint2`. The n modifier specifies the exact output width; where n is one or more characters between 1 and 9.

Table 6.1   *DebugPrint* format specifiers

	Format Specifier	Type
%c	ANSI character	char
%C	Wide character	wchar_t
%d, %i	Signed integer in decimal	int
%D	__int64 in decimal	__int64
%I	IRP major and minor codes	PIRP
%1	__int64 in hexadecimal	__int64
%L	LARGE_INTEGER in hexadecimal	LARGE_INTEGER
%s	NULL-terminated ANSI character string	char*
%S	NULL-terminated Wide character string	wchar_t*
%T	UNICODE_STRING	PUNICODE_STRING
%u	ULONG in decimal	ULONG
%x	ULONG in hexadecimal	ULONG

Debugging Notes

Here are some notes on how to deal with some common debugging problems.

Updating Drivers

Updating a driver is fairly straightforward. The makefile for the examples in this book copy the new driver to the Windows System32\drivers directory. However, the old version will still be running in memory.

To use the new version you have to update the drivers for your device(s). Use the Update/Change driver option in the Device Manager properties for a device. Opt to "display a list of drivers in a specific location, so you can select the driver you want".

For the **Wdm1** driver, select "Other Devices". You should see the list of book software drivers that you have installed. Select the one you want from this list. You do not have to select the Have Disk button (unless you want to install the checked version, say).

If asked, opt to use the new driver even if the Device Manager suggests that the current version may be newer. Windows 98 will force you to browse for the driver files. W2000 uses the driver files in the System32\drivers directory.

As noted above, if the new driver returns an error during its initialization then Windows will say that the system needs to reboot. Rebooting may not cure the problem.

NT Style Drivers

As mentioned in an earlier chapter you must update NT style drivers in a different way. As usual, you must get your driver into the Windows System32\drivers directory.

Run the Control Panel Devices applet. Find your driver, stop it, and start it again. You can also run the book software **Servicer** applet; type in the name of your driver, press Stop and then Start. If you run an NT style driver in Windows 98 you must reboot the computer to use an updated driver.

Driver Fails on Boot

If your driver fails catastrophically during system boot, the system will not load, which makes it tricky to delete or change the offending driver.

There are three solutions to this problem. In a dual boot system, simply reboot in the other operating system and delete or change the driver. Alternatively, both W2000 and W98 have "Safe mode" boot options that should allow you to start Windows, while only loading the most basic system drivers. Delete the offending driver and restart. A final option is to have a bootable floppy disk with a copy of Windows 98. Use this as a quick boot to delete the offending driver files.

Driver Dependencies

Watch out for one driver depending on another. The Wdm1 driver opens a handle to the Debug-Print driver during its initialization, and releases it only when the driver unloads. This means that the DebugPrint driver cannot be replaced while Wdm1 is running. To change the Debug-Print driver, Wdm1 has to be unloaded. Only when the new DebugPrint is safely running, can a Wdm1 device be installed again.

Uncanceled IRPs

If a driver queues IRPs but does not provide suitable cancel or cleanup routines, a Win32 program could hang up if it calls CancelIo or exits with file handles open. In this case, the Win32 program appears to have exited but it is still locked in memory, so you will not be able to update it.

The I/O Manager gives a driver five minutes to cancel IRPs. After this time, it displays a message to the user and any pending IRPs are dissociated from the terminating thread.

A reboot is usually necessary to clear this situation.

Bugcheck Codes

When Windows 2000 bugchecks, it displays information in the "blue screen of death" and stops dead. The information on this screen can help you to track down the source of the problem. You can cause a bugcheck deliberately using the KeBugCheck and KeBugCheckEx routines. A debug version of your driver might do this if it detects some unsolvable problem. Release drivers should never bugcheck voluntarily.

A problem in your driver may not directly cause a bugcheck. For example, you could overwrite another driver's memory and cause it to fail.

The top few lines of the bugcheck screen contain the most useful information. Down below might be a module list, a stack trace, and instructions. In NT, then it will say that is producing a physical dump of memory even if you have not enabled this option in the Control Panel System applet.

The most common bugcheck codes are listed in Table 6.2. The full list of stop codes is given in DDK bugcodes.h. The exception codes for "Unhandled Kernel exception" bugchecks are in the NTSTATUS.H header in the DDK.

For example, put this code in the Wdm1 driver.

```
char* NULLptr = NULL;
*NULLptr = 5;
```

This does not cause any errors in W98. However, in W2000, the following bugcheck occurs. (Interestingly, a read from address 0x0 does not seem to cause a bugcheck in W2000.)

```
*** STOP: 0x0000001E (0xC0000005,0xF2D7B875,0x00000001,0x00000000)
KMODE_EXCEPTION_NOT_HANDLED

*** Address 0xF2D7B875  base at 0xF2D7A000 Datestamp 362DF72F - Wdm1.sys
```

The bugcheck code is 0x0000001E, which translates as an "Unhandled Kernel exception". The four numbers in brackets are the four extra parameters that are passed to KeBugCheckEx. Table 6.2 shows the interpretations for the common bugcheck codes. For this bugcheck, the first parameter is the exception code. 0xC0000005 indicates an access exception. The fourth parameter indicates the memory address that you tried to access (0x0) and the second parameter gives the address of the instruction that caused the exception. The exception occurred at address 0xF2D7B875, a little way into the code of Wdm1, which is loaded at 0xF2D7A000.

A bug's life

Table 6.2   Common bugchecks

Code 0x0000000A IRQL_NOT_LESS_OR_EQUAL 1 Address referenced 2 IRQL (Not the correct IRQL) 3 0=read, 1=write 4 Address that referenced memory A driver tried to do something at an inappropriate IRQL (e.g., accessing paged memory at DISPATCH_LEVEL IRQL or higher)
Code 0x0000001E KMODE_EXCEPTION_NOT_HANDLED 1 Exception code: 0xC0000005 2 Address where exception occurred 3 4 Address referenced Access violation
Code 0x0000001E KMODE_EXCEPTION_NOT_HANDLED 1 Exception code: 0x80000003 2 Address where exception occurred Hard-coded breakpoint or ASSERT hit.
Code 0x000000BE Driver attempted to write to read-only memory
Code 0x000000C4 Driver Verifier detected exception. See its documentation for details.

Where Did the Bugcheck Happen?

How do you work out what code caused the bugcheck? By analyzing the linker map for a driver, you can work out which routine caused the problem. A source-level debugger is required if you are still having problems.

You build a linker map by adding a line like the following to your SOURCES file.

```
LINKER_FLAGS=-MAP:Wdm2.map
```

What routine caused the following access violation?

```
*** STOP: 0x0000001E (0xC0000005,0xF764C5F1,0x00000000,0x00000010)
KMODE_EXCEPTION_NOT_HANDLED


*** Address 0xF764C5F1  base at 0xF764A000 Datestamp 3653e5fb - Wdm2.sys
```

The first thing to note is that the access violation occurs at a very low address, 0x10. This suggests that the code had a NULL pointer to a structure and was trying to access a field at offset 0x10 in this structure. This turned out to be the case.

The problem seems to be in the Wdm2 driver. The offset into the executable image is 0xF764C5F1-0xF764A000 (i.e., 0x25F1).

Listing 6.2 shows part of the linker map for this build of the Wdm2 driver. The initial section shows that the load address is 0x00010000. The map then lists the segments that make up the executable image. However, the information of interest is buried in the next section, which lists each code and data object.

The entry for PnpDefaultHandler in Pnp.cpp has an Rva+Base of 0x00012512. If the load address is taken off, this shows that PnpDefaultHandler starts at offset 0x2512. The next line shows that the ForwardAndWait routine starts at 0x2611. Therefore, the access violation at offset 0x25F1 occurred towards the end of the PnpDefaultHandler routine.

Listing 6.2   Wdm2 linker map excerpt

```
Wdm2


Timestamp is 3653e5fb (Thu Nov 19 09:33:47 1998)


Preferred load address is 00010000


Start          Length      Name             Class
0001:00000000 0000152cH .text              CODE
0002:00000000 000000a0H .idata$5           DATA
0002:000000a0 00000632H .rdata             DATA
0003:00000000 00000119H .data              DATA
0003:00000120 00000042H .bss               DATA
0004:00000000 00000c1aH PAGE               CODE
0005:00000000 000000eeH INIT               CODE
0005:000000f0 00000028H .idata$2           CODE
0005:00000118 00000014H .idata$3           CODE
```

Listing 6.2   Wdm2 **linker map excerpt (continued)**

```
0005:0000012c 000000a0H .idata$4              CODE
0005:000001cc 00000344H .idata$6              CODE
0006:00000000 00000058H .rsrc$01              DATA
0006:00000060 00000338H .rsrc$02              DATA


 Address           Publics by Value          Rva+Base     Lib:Object


0001:00000012
   ?Wdm1Create@@YGJPAU_DEVICE_OBJECT@@PAU_IRP@@@Z 000102f2 f   dispatch.obj
0001:00000056
   ?Wdm1Close@@YGJPAU_DEVICE_OBJECT@@PAU_IRP@@@Z 00010336 f   dispatch.obj

...

0004:00000492
   ?PnpDefaultHandler@@YGJPAU_DEVICE_OBJECT@@PAU_IRP@@@Z 00012512 f   pnp.obj
0004:00000591
   ?ForwardAndWait@@YGJPAU_DEVICE_OBJECT@@PAU_IRP@@@Z 00012611 f   pnp.obj
```

Conclusion

Please test your driver well before releasing it. You must check that your driver works on a variety of computers. It must cope with user mode program aborts.

Debugging drivers is harder than user mode applications. In the worst case, a bugcheck occurs that tells you roughly what went wrong. Resource leaks and timing problems are more difficult to sort out.

The *DebugPrint* software lets you insert "print" statements into your driver code. You can do source-level debugging between two computers using *WinDbg*, or on a single computer using the NuMega **SoftICE** product.

The next chapter looks at the dispatch routines in the Wdm1 device driver.

Chapter 7

Dispatch Routines

This chapter looks at how to write driver dispatch routines that process I/O Request Packets (IRPs). Dispatch routines are used to handle requests from Win32 applications. It is crucial that you understand everything in this chapter clearly. If necessary, refer to Chapter 3 where IRPs are first introduced.

The dispatch routines for the Wdm1 driver are explained in full. These handle open, close, read, write, and IOCTL requests. The Wdm1 driver implements a global memory buffer that is shared by all Wdm1 devices.

This chapter takes a good hard look at I/O Request Packets (IRPs). It is worth reading this chapter carefully, as a good understanding of IRPs will ease your passage through the rest of the book.

Dispatch Routine IRPs

Table 7.1 lists the most common Win32 functions that are used to access devices. A Create-File call to your device ends up as a Create IRP, an I/O Request Packet with a major function code of IRP_MJ_CREATE. The driver routine to handle this IRP can have any name. However, I use a generic name for the Create IRP handler of Create. In your driver, you would usually put a short name or acronym in front of this base name. The Wdm1 device driver's Create IRP handler is called Wdm1Create.

Table 7.1 is not an exhaustive list of Win32 functions and their matching IRPs. For example, ReadFile has several variants, such as ReadFileEx, but they all end up as IRP_MJ_READ requests. IRPs can also be issued by Windows on behalf of the user program. For example, if an application terminates unexpectedly, the operating system will try to tidy any open files by issuing an IRP_MJ_CLOSE IRP to each file.

Table 7.1   Common dispatch routines

Win32 Function	IRP Major Code	Base Driver routine name
CreateFile	IRP_MJ_CREATE	Create
CloseHandle	IRP_MJ_CLOSE	Close
ReadFile	IRP_MJ_READ	Read
WriteFile	IRP_MJ_WRITE	Write
DeviceIoControl	IRP_MJ_DEVICE_CONTROL	DeviceControl

A driver need not handle all these IRPs, though handling Create and Close IRPs is an obvious minimum. Its DriverEntry routine sets up the entry points that are valid. If an entry point is not set, then the I/O Manager fails the Win32 request and GetLastError returns 1.

The following line in DriverEntry sets the Wdm1Read routine as the handler for Read IRPs.

```
DriverObject->MajorFunction[IRP_MJ_READ] = Wdm1Read;
```

I/O Request Packets

Dispatch Routine Handling

All dispatch routines have the same function prototype. The function is passed a pointer to your device object and the IRP. The function must return a suitable NTSTATUS value (e.g., STATUS_SUCCESS).

```
NTSTATUS Wdm1Read( IN PDEVICE_OBJECT fdo, IN PIRP Irp)
```

A dispatch routine is usually called at PASSIVE_LEVEL IRQL. (This is a bit unexpected, as you might expect dispatch routines to run at DISPATCH_LEVEL IRQL.) Running at PASSIVE_LEVEL means that a dispatch routine can very easily be interrupted by other parts of the kernel or even your own driver. However, a routine that runs at PASSIVE_LEVEL can issue most kernel calls, including writing to other files.

A dispatch routine can be called at DISPATCH_LEVEL if a higher level driver calls you at this IRQL level. This might happen if it calls you from a completion routine, as described in Chapter 9. If you think this might be happening, then assume that your driver dispatch routines are running at DISPATCH_LEVEL.

Dispatch routines run in an arbitrary thread context. As well as standard user mode threads, the kernel has its own threads that run only in kernel mode. Although your IRP will have originated in one of these threads, you cannot guarantee that you are running in its context. This means that a valid address in the originating thread may not be valid when your driver sees it. A later section in this chapter shows how to access user buffers correctly.

Reentrancy

A dispatch routine must be reentrant. This means that it may be called "simultaneously" to process two separate IRPs. In a multiprocessor Windows 2000 system, one processor could call Wdm1Read with one IRP, and a second process on another CPU could also call Wdm1Read simultaneously with another IRP. However, do not dismiss reentrancy as some arcane requirement that your driver will never encounter. In Windows 98, the single processor could start running Wdm1Read to handle a first IRP. Wdm1Read could issue a kernel request that blocks its operation[1]. Windows 98 might then schedule a different user application that issues another read resulting in another call to Wdm1Read.

The first technique to make your routine reentrant is to use local variables. Each separate call to Wdm1Read definitely has its own separate set of variables on the kernel stack.

Do not use global variables or variables in the device extension unless you protect access to them. However, when these variables are first set up (DriverEntry or AddDevice), you do not need to take special precautions.

The Wdm1 driver uses a *spin lock* to protect access to the shared memory buffer variables, as described later.

There are two other main techniques to achieve reentrancy in dispatch routines. The first is to use the services of the I/O Manager to create a device queue of IRPs. The IRPs in the device queue are passed one at a time to a StartIo routine in your driver.

The second technique is to use Critical section routines if your device interrupts the computer, as explained in full in Chapter 16. Calling KeSynchronizeExecution runs your Critical section routine safely. KeSynchronizeExecution raises the IRQL to the interrupt IRQL and acquires the interrupt spin lock. This technique ensures that your routine is run to completion without being interrupted by another part of the driver.

IRP Handling

There are three main techniques for handling IRPs.

- Handle immediately
- Put in a queue and process one by one
- Pass down to a lower-level driver

Only the simplest drivers, like Wdm1, can handle IRPs straightaway. Even so, you must still take precautions to ensure that your driver dispatch routines are reentrant.

Drivers that access real pieces of hardware usually want to serialize access to the hardware. As mentioned previously, the I/O Manager in the kernel provides a device queue that you can use. Dispatch routines call the kernel to put an IRP into the device queue. The I/O Manager calls your StartIo routine to process one IRP at a time. When your StartIo routine has completed an IRP, it should call the kernel to ensure that it is called again with the next available IRP.

Drivers might want to use more than one device queue. For example, a serial port driver will usually want to have two device queues; one for incoming read data and one for outgoing write data.

A driver can use a different queuing strategy. For example, the DebugPrint driver described in Chapter 14 lets only one read IRP be queued. Any further read IRPs are rejected while the first read IRP is outstanding.

The final main technique for handling IRPs is to pass them down for handling in a lower-level driver. This approach is common in WDM device drivers. Some drivers simply

---

1. Blocking in dispatch routines is not recommended as it defeats the purpose of overlapped operations.

pass the handling of some IRPs to the next lower driver and forget about them. If you take a more active interest, you can inspect the results after all the lower-level drivers have handled the IRP. These techniques are covered later.

A variant on this theme is for a driver to build a new IRP (or IRPs), send it down to lower drivers, and process the results afterwards.

IRP Completion

When a driver has finished working on an IRP, it must tell the I/O Manager. This is called IRP *completion*. As this code snippet shows, you must set a couple of fields in the IRP *IoStatus* field structure. IoStatus.Status is set to an NTSTATUS status code. The number of bytes transferred is usually stored in IoStatus.Information.

```
Irp->IoStatus.Status = STATUS_SUCCESS;
Irp->IoStatus.Information = BytesTxd;
IoCompleteRequest(Irp,IO_NO_INCREMENT);
```

Finally, IoCompleteRequest is called (at or below DISPATCH_LEVEL IRQL). As well as the IRP pointer, you must supply a PriorityBoost parameter to give a boost to the scheduling priority of the thread that originated the IRP. For example, keyboard drivers use the constant IO_KEYBOARD_INCREMENT. This is a high value of 6 because foreground threads should respond quickly to user input.

If a dispatch routine does not process an IRP immediately, it must mark the IRP as pending using IoMarkIrpPending and return STATUS_PENDING. When you eventually get round to completing the IRP, do it as described above.

If an IRP is queued and its associated process dies unexpectedly (or calls CancelIo to cancel overlapped IRPs), these pending IRPs must be cancelled. You must set an IRP's cancel routine to handle this circumstance, and to handle the Cleanup IRP. Chapter 16 describes these options in full. Wdm1 does not queue IRPs, so handling these options is not necessary.

IRP Structure

Figure 7.1 shows that an I/O Request Packet consists of a header IRP structure followed by a series of stack locations, each an IO_STACK_LOCATION structure. The information in the header and the current IRP stack location tell a driver what to do.

Figure 7.1   IRP overview

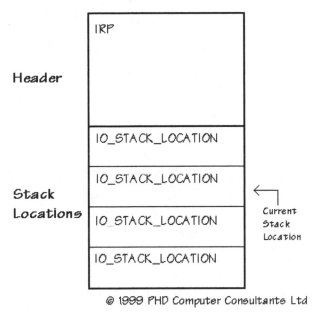

@ 1999 PHD Computer Consultants Ltd

Table 7.2 lists some of the fields in the IRP header structure that a driver can access, while Table 7.3 gives the general layout of the IO_STACK_LOCATION structure.

Table 7.2   Some IRP structure fields

Field	Description
IO_STATUS_BLOCK IoStatus	Completion status of IRP
PVOID AssociatedIrp.SystemBuffer	System space buffer (for Buffered I/O)
PMDL MdlAddress	Memory Description List (for Direct I/O)
BOOLEAN Cancel	Set if IRP has been cancelled
ULONG Flags	IRP Flags

Table 7.3   Some `IO_STACK_LOCATION` structure fields

```
typedef struct _IO_STACK_LOCATION
{
    UCHAR MajorFunction;
    UCHAR MinorFunction;
    // ...
    union
    {
        struct {  ...  } Create;
        struct {  ...  } Read;
        struct {  ...  } Write;
        struct {  ...  } DeviceIoControl;

        ...
    } Parameters;
    // ...
} IO_STACK_LOCATION, *PIO_STACK_LOCATION;
```

The fact that there is an IRP header and several associated I/O stack locations can be a source of confusion. However, the stack locations are a powerful tool for processing IRPs when several layered drivers have to access an IRP in turn.

The I/O stack location contains most of the important information about the IRP. The *MajorFunction* is the IRP code (e.g., IRP_MJ_READ for a Read IRP). Some IRPs, such as IRP_MJ_PNP, use the *MinorFunction* field to specify which particular Plug and Play function is being requested.

Each common IRP type has a struct within the *Parameters* union in the IO_STACK_LOCA-TION structure. For example, for Read IRPs, the `Parameters.Read.Length` field is the number of bytes to transfer. The parameters that are valid for each common IRP are described in the following text.

The key to understanding I/O stack locations is to realize that each driver needs to look at only one, the "current stack location". The information in the IRP header structure and the information in the current stack location are the parameters that a driver uses to process an IRP.

I/O Stack Locations

Let's go off track slightly to ask why Microsoft provides a set of I/O stack locations. If you have a stack of drivers that process an IRP, the highest might be a network protocol driver that can accept read requests of any length. This driver might know that the underlying transport driver can only cope with read transfers of up to 1024 bytes. Its job is to break up long transfers into a series of blocks, each with a maximum size of 1024 bytes. When the protocol driver calls the transport driver it, sets up the next I/O stack location with the transfer size set to 1024. When the transport driver processes the IRP, its current I/O stack location has this value, and it should proceed happily to process the IRP. When it has finished, the IRP is passed back to the protocol driver. This checks that the transfer worked and — assuming it did — sets up the next transfer and calls the transport driver again.

In this approach, the protocol driver sends the transport driver IRPs one by one. However, a more sophisticated protocol driver could allocate new IRPs, enough to move all the data. It could then issue them all to the transport driver. The protocol driver would have to check carefully that all the IRPs finished correctly. When all the data has been moved, one way or another, the protocol driver can finally complete its own original IRP.

This example reveals a problem. The field that contains the pointer to the data to be transferred is not in the I/O stack location, but in the IRP header. The transfer length is in the stack. Surely the data pointer ought to be in there as well. The fact that the original stack location is not changed by the call to the lower driver is good, as it makes it easier for a driver to remember how many bytes to transfer.

However, this reveals another common difficulty with IRPs — determining where to store a driver's own information about an IRP. Suppose the protocol driver wanted to remember something simple (e.g., how many bytes it had sent so far)[2].

The ideal place for some storage for drivers would be the I/O stack location. However, there is no space specifically reserved for drivers. Nonetheless, Read IRPs have a ULONG at Parameters.Read.Key in the current I/O stack location that can be used safely, although the DDK does not specifically say so. Write IRPs have a similar ULONG at Parameters.Write.Key.

There is some room in the IRP header that can be used by drivers — a PVOID DriverContext[4] in Tail.Overlay. Use the following code to access the first of these locations.

```
PVOID p = Irp->Tail.Overlay.DriverContext[0];
```

However, it is not safe to use these locations for storing context while an IRP is processed by lower drivers, for the simple reason that these other drivers may use this memory too.

The final point to note about I/O stack locations is that you can use different major function code when you call a lower driver. For example, you might implement a read request by sending the lower driver an IOCTL.

Common IRP Parameters

This section lists the parameters that are set for the common IRPs. In the following discussion, "the stack" means the current I/O stack location.

Create IRP, IRP_MJ_CREATE

The main parameter of interest to the Create IRP handler is the *FileObject* field in the stack. This is a pointer to a _FILE_OBJECT structure. The *FileName* field in here is a UNICODE_STRING with any characters after the basic device name. If you appended \file to the symbolic link name found in the GetDeviceViaInterface routine in *Wdm1Test*, \file would appear in *FileName*. If no characters are appended, *FileName* has a length of zero.

Other parameters to the Create IRP are given in the Parameters.Create structure in the stack, such as the *FileAttributes* and *ShareAccess*.

---

2. If the driver processes IRPs one by one, it could store this information in its device extension.

Close IRP, `IRP_MJ_CLOSE`

If need be, you can double-check that the *FileObject* in the IRP header matches the one you were sent in the create request.

If you have queued up Read or Write IRPs for this file, the I/O Manager will have cancelled them before the close request is received. It does this by calling an IRP's Cancel routine and issuing a Cleanup IRP, as described in Chapter 16.

Read IRP, `IRP_MJ_READ`

The `Parameters.Read` structure in the IRP stack has *Length* and *ByteOffset* fields that say how many bytes are requested and the file pointer. *ByteOffset* is a 64-bit integer stored in a `LARGE_INTEGER` structure. The Microsoft compiler can handle this type directly (i.e., `Parameters.Read.ByteOffset.QuadPart` is an __int64). If you need to specify 64-bit constant values in your code, append i64 to the constant (e.g., 100i64).

The user buffer can be specified in one of two ways, depending on whether your driver uses *Buffered I/O* or *Direct I/O*. See the following text for details of these terms. If using Buffered I/O, a pointer to the user buffer is in the IRP header at *AssociatedIrp.SystemBuffer*. For Direct I/O, a Memory Descriptor List (MDL) is in the IRP header in the *MdlAddress* field.

The *Key* field in the IRP stack `Parameters.Read` structure does not seem to be used for anything, and so could be used by a driver for any purpose.

Write IRP, `IRP_MJ_WRITE`

The parameters for Write IRPs are identical to Read IRPs, except that the relevant parameters are in the stack `Parameters.Write` structure.

IOCTL IRP, `IRP_MJ_DEVICE_CONTROL`

The `Parameters.DeviceIoControl` structure in the IRP stack has *IoControlCode*, *InputBufferLength*, and *OutputBufferLength* parameters.

The user buffer is specified using one or more of the *AssociatedIrp.SystemBuffer*, *MdlAddress*, or stack *Parameters.DeviceIoControl.Type3InputBuffer* fields. See the next section for details.

User Buffers

As a driver can run in the context of any thread, a plain pointer into the user's address space is not guaranteed to access the correct memory.

A driver can use two main methods to access the user's buffer properly, either Buffered I/O or Direct I/O. When you create a device, you must set the `DO_BUFFERED_IO` bit in the *Flags* field of the new device object to use Buffered I/O. For Direct I/O, set the `DO_DIRECT_IO` bit in *Flags*.

Buffered I/O

If you use Buffered I/O, the kernel makes the user's buffer available in some nonpaged memory and stores a suitable pointer for you in the *AssociatedIrp.SystemBuffer* field of the IRP header. Simply read or write this memory in your driver.

This technique is the easiest one for driver writers to use. However, it is slightly slower overall, as the operating system usually will have to copy the user buffer into or out of nonpaged memory.

Direct I/O

It is faster to use a Memory Descriptor List (MDL). However, this is only available to hardware that can perform Direct Memory Access (DMA). DMA and MDLs are not explained in this book, although Chapter 24 lists the changes in W2000 for those of you who have used DMA in NT 4 and earlier.

The MDL of the user's buffer is put in the *MdlAddress* field of the IRP header.

Neither

A final and uncommon technique for accessing a user's buffer is to use neither Buffered I/O nor Direct I/O. In this case, the user's buffer pointer is simply put in the *UserBuffer* field of the IRP header. If you are certain that your driver is the first driver to receive a request, the dispatch routine can directly access the buffer, as the driver will be operating in the context of the user's thread. Be very careful if you try to use this technique.

DeviceIoControl **Buffers**

DeviceIoControl requests can use a combination of these user buffer access techniques. Each IOCTL can use a different method, if need be. However, most drivers simply use Buffered I/O, as IOCTL buffers are usually fairly small. I shall show how to define IOCTLs shortly.

The TransferType portion of the actual IOCTL code indicates the buffer access technique. For Buffered I/O, specify METHOD_BUFFERED for TransferType. For Direct I/O, use either METHOD_IN_DIRECT or METHOD_OUT_DIRECT.

If you use METHOD_BUFFERED, *AssociatedIrp.SystemBuffer* is used for the input and output buffer. The buffer size is the maximum of the user's input and output buffer sizes. As the same memory is used for both input and output, make sure that you use (or copy) the input data before you start writing any output data.

For METHOD_IN_DIRECT, an MDL for the input buffer is put in the IRP header *MdlAddress* field. A buffered output buffer pointer is put in *AssociatedIrp.SystemBuffer*.

For METHOD_OUT_DIRECT, an MDL for the output buffer is put in the IRP header *MdlAddress* field. A buffered input buffer pointer is put in *AssociatedIrp.SystemBuffer*.

Finally, for METHOD_NEITHER, the input buffer user space pointer is put in *Parameters.DeviceIoControl.Type3InputBuffer* in the stack. The user space output buffer pointer is put in the IRP header *UserBuffer* field.

Wdm1 **Dispatch Routines**

The dispatch routines for the Wdm1 driver are in the file Dispatch.cpp, which is available on the book CD-ROM. These routines all run at PASSIVE_LEVEL IRQL, so they can be put in paged memory. All the basic dispatch routines complete the IRP straightaway.

The dispatch routines include various *DebugPrint* trace calls in the code. If you use the checked build version of Wdm1, you can view the trace output using the *DebugPrint Monitor* application. Listing 7.1 shows the (slightly edited) *DebugPrint* output on Windows 2000. The Wdm1 driver was started at around 12:00 and the *Wdm1Test* program was run at 12:10. You can follow the program execution as each test in *Wdm1Test* is run. The *DebugPrint* output would be different in Windows 98 because the SetFilePointer function does not work for device files.

Listing 7.1   *DebugPrint* output in Windows 2000

```
Monitor 12:03:04  Version 1.02 starting to listen under Windows 2000
(5.0 build 1877)
DebugPrint 11:59:09  Version 1.02 started
Wdm1    12:00:42  DebugPrint logging started
Wdm1    12:00:42  RegistryPath is
                  \REGISTRY\Machine\System\ControlSet002\SERVICES\Wdm1
Wdm1    12:00:42  DriverEntry completed
Wdm1    12:00:42  AddDevice
Wdm1    12:00:42  FDO is 80AAB020
Wdm1    12:00:42  Symbolic Link Name is \??\Root#UNKNOWN#0003#{c0cf0640...}
Wdm1    12:00:42  PnP IRP_MJ_PNP:IRP_MN_QUERY_CAPABILITIES
Wdm1    12:00:42  PnP IRP_MJ_PNP:IRP_MN_FILTER_RESOURCE_REQUIREMENTS
Wdm1    12:00:43  PnP IRP_MJ_PNP:IRP_MN_START_DEVICE
Wdm1    12:00:43  PnP IRP_MJ_PNP:IRP_MN_QUERY_CAPABILITIES
Wdm1    12:00:43  PnP IRP_MJ_PNP:IRP_MN_QUERY_PNP_DEVICE_STATE
Wdm1    12:00:43  PnP IRP_MJ_PNP:IRP_MN_QUERY_BUS_INFORMATION
Wdm1    12:00:43  PnP IRP_MJ_PNP:IRP_MN_QUERY_DEVICE_RELATIONS
Wdm1    12:10:14  Create File is
Wdm1    12:10:14  Read 4 bytes from file pointer 0
Wdm1    12:10:14  Read: 4 bytes returned
Wdm1    12:10:14  Write 4 bytes from file pointer 0
Wdm1    12:10:14  Write: 4 bytes written
Wdm1    12:10:14  Read 1 bytes from file pointer 3
Wdm1    12:10:14  Read: 1 bytes returned
Wdm1    12:10:14  Write 4 bytes from file pointer 3
Wdm1    12:10:14  Write: 4 bytes written
Wdm1    12:10:14  DeviceIoControl: Control code
                  0022200C InputLength 0 OutputLength 4
Wdm1    12:10:14  DeviceIoControl: 4 bytes written
Wdm1    12:10:14  DeviceIoControl: Control code
                  00222010 InputLength 0 OutputLength 7
```

Listing 7.1 *DebugPrint* output in Windows 2000 (continued)

```
Wdm1    12:10:14  DeviceIoControl: 7 bytes written
Wdm1    12:10:14  DeviceIoControl: Control code
                  00222010 InputLength 0 OutputLength 8
Wdm1    12:10:14  DeviceIoControl: 0 bytes written
Wdm1    12:10:14  DeviceIoControl: Control code
                  00222004 InputLength 0 OutputLength 0
Wdm1    12:10:14  DeviceIoControl: 0 bytes written
Wdm1    12:10:14  DeviceIoControl: Control code
                  00222010 InputLength 0 OutputLength 7
Wdm1    12:10:14  DeviceIoControl: 7 bytes written
Wdm1    12:10:14  DeviceIoControl: Control code
                  00222008 InputLength 0 OutputLength 0
Wdm1    12:10:14  DeviceIoControl: 0 bytes written
Wdm1    12:10:14  DeviceIoControl: Control code
                  0022200C InputLength 0 OutputLength 4
Wdm1    12:10:14  DeviceIoControl: 4 bytes written
Wdm1    12:10:14  DeviceIoControl: Control code
                  00222014 InputLength 0 OutputLength 0
Wdm1    12:10:14  DeviceIoControl: 0 bytes written
Wdm1    12:10:14  Write 4 bytes from file pointer 0
Wdm1    12:10:14  Write: 4 bytes written
Wdm1    12:10:14  Close
```

Create and Close

The Wdm1 create and close routines do nothing except complete the IRP successfully. A helper function, CompleteIrp, is used that sets the IRP header *IoStatus* fields to the given parameters and calls IoCompleteRequest.

The create routine shows how to access the current I/O stack location using IoGetCurrentIrpStackLocation. In the checked build version, it prints out the *FileName* field in the stack *FileObject*.

```
PIO_STACK_LOCATION IrpStack = IoGetCurrentIrpStackLocation(Irp);
DebugPrint( "Create File is %T", &(IrpStack ->FileObject->FileName));
```

Write

Things start to get interesting in the write dispatch routine, Wdm1Write, shown in Listing 7.2. It starts by getting the current stack location pointer and retrieving the current file pointer and the number of bytes to transfer. If the file pointer is less than zero (the kernel should ensure that it never is), it returns STATUS_INVALID_PARAMETER. It is possible to receive a transfer length of zero.

Listing 7.2   Wdm1 **write dispatch routine**

```
NTSTATUS Wdm1Write( IN PDEVICE_OBJECT fdo, IN PIRP Irp)
{
    PIO_STACK_LOCATION IrpStack = IoGetCurrentIrpStackLocation(Irp);
    NTSTATUS status = STATUS_SUCCESS;
    ULONG BytesTxd = 0;

    // Get call parameters
    LONGLONG FilePointer = IrpStack->Parameters.Write.ByteOffset.QuadPart;
    ULONG WriteLen = IrpStack->Parameters.Write.Length;
    DebugPrint("Write %d bytes from file pointer %d",
                (int)WriteLen,(int)FilePointer);

    if( FilePointer<0)
        status = STATUS_INVALID_PARAMETER;
    else
    {
        // Get access to the shared buffer
        KIRQL irql;
        KeAcquireSpinLock(&BufferLock,&irql);

        BytesTxd = WriteLen;

        // (Re)allocate buffer if necessary
        if( ((ULONG)FilePointer)+WriteLen>BufferSize)
        {
            ULONG NewBufferSize = ((ULONG)FilePointer)+WriteLen;
            PVOID NewBuffer =
                ExAllocatePool(NonPagedPool,NewBufferSize);
            if( NewBuffer==NULL)
            {
                BytesTxd = BufferSize - (ULONG)FilePointer;
                if( BytesTxd<0) BytesTxd = 0;
            }
            else
            {
                RtlZeroMemory(NewBuffer,NewBufferSize);
                if( Buffer!=NULL)
                {
                    RtlCopyMemory(NewBuffer,Buffer,BufferSize);
```

Listing 7.2   Wdm1 **write dispatch routine (continued)**

```
                    ExFreePool(Buffer);
            }
            Buffer = (PUCHAR)NewBuffer;
            BufferSize = NewBufferSize;
        }
    }


    // Write to shared memory
    if( BytesTxd>0 && Buffer!=NULL)
        RtlCopyMemory( Buffer+FilePointer,
            Irp->AssociatedIrp.SystemBuffer, BytesTxd);


    // Release shared buffer
    KeReleaseSpinLock(&BufferLock,irql);
    }

DebugPrint("Write: %d bytes written",(int)BytesTxd);


    // Complete IRP
    return CompleteIrp(Irp,status,BytesTxd);
}
```

The shared memory buffer is implemented using these three global variables.

```
KSPIN_LOCK BufferLock;
PUCHAR Buffer = NULL;
ULONG  BufferSize = 0;
```

If the buffer size is greater than zero, Buffer points to some nonpaged memory of this size. As mentioned earlier in this chapter, there must be some mechanism to protect access to such global variables in a multiprocessor environment (e.g., to prevent one dispatch routine from changing BufferSize while another, or even the same, routine tries to access or change it simultaneously).

Spin Locks

A kernel *spin lock* called BufferLock provides this protection. A spin lock can be used where code needs access to a resource of some sort for a short time.

The spin lock is initialized in the Wdm1 DriverEntry routine as follows.

```
KeInitializeSpinLock(&BufferLock);
```

Use the KeAcquireSpinLock function to acquire a spin lock and KeReleaseSpinLock to release it. Only one instance of a piece of code can acquire a spin lock at the same time. Other attempts to acquire the spin lock will "spin" until the resource becomes available. "Spinning" means that KeAcquireSpinLock keeps looking continuously. For this reason, make sure that you only hold a spin lock for a short time. The DDK recommends that you never hold a spin lock for more than 25 microseconds.

As shown in the code example, you must provide a pointer to a KIRQL variable in the call to KeAcquireSpinLock. This stores the original IRQL level before it is raised (if necessary) to DISPATCH_LEVEL. The call to KeReleaseSpinLock lowers the IRQL if necessary. If you are certain that your code is working at DISPATCH_LEVEL, you can use the KeAcquireSpinLockAtDpcLevel and KeReleaseSpinLockFromDpcLevel routines for better performance.

The Wdm1 driver acquires the BufferLock spin lock for the duration of any accesses to the Buffer and BufferSize variables. Do not access paged code or data while holding a spin lock, as the system will almost certainly crash. Definitely do not exit a main dispatch routine while holding a spin lock.

Write Algorithm

The write dispatch stores the write data in the shared memory buffer, starting from the given file pointer. It extends the buffer, if necessary.

If there is no buffer at all, or the buffer needs to be extended, ExAllocatePool is called to allocate some nonpaged memory. Notice that the algorithm checks for a NULL return value and copes as best as it can.

A new memory buffer is zeroed using RtlZeroMemory. If an old shorter buffer exists, it is copied to the start of the new buffer using RtlCopyMemory. RtlMoveMemory can be used if the source and destination pointers overlap. The old buffer is removed with ExFreePool.

Finally, Wdm1Write copies the data from the user buffer using RtlCopyMemory. As Wdm1 uses Buffered I/O, it can simply copy the data from Irp->AssociatedIrp.SystemBuffer.

This algorithm is fairly crude, because the buffer may have to be reallocated often. A much-enhanced version of Wdm1 could implement a RAM disk.

The driver unload routine frees any shared memory buffer.

```
if( Buffer!=NULL)
    ExFreePool(Buffer);
```

Read

The read dispatch routine for Wdm1, Wdm1Read, is simpler than the write handler. It acquires the spin lock while it accesses the global variables. The required number of bytes are copied to the user's buffer at Irp->AssociatedIrp.SystemBuffer. If the user requests more data than is in the buffer, the request is truncated.

IOCTL

The Wdm1DeviceControl dispatch routine handles the four IOCTLs defined for Wdm1 devices: Zero the buffer, Remove the Buffer, Get the buffer size, and Get the buffer.

All these IOCTLs use Buffered I/O, so any input and output data is found at Irp->AssociatedIrp.SystemBuffer. As usual, the routine acquires the shared buffer spin lock for the duration of the call. The actual implementation of each IOCTL is straightforward. The Get buffer size and Get buffer handlers check that the output buffer is large enough; if not, they return STATUS_INVALID_PARAMETER.

Defining IOCTLs

An IOCTL code is a 32-bit value formed using the CTL_CODE macro shown in Table 7.4. The Wdm1 example defines its IOCTL codes in Ioctl.h, as shown in this example.

```
#define IOCTL_WDM1_ZERO_BUFFER CTL_CODE(     \
        FILE_DEVICE_UNKNOWN,                 \
        0x801,                               \
        METHOD_BUFFERED,                     \
        FILE_ANY_ACCESS)
```

Table 7.4   CTL_CODE **macro parameters**

Parameter	Description
DeviceType	FILE_DEVICE_XXX value given to IoCreateDevice.
ControlCode	IOCTL Function Code 0x000-0x7FF Reserved for Microsoft 0x800-0xFFF Private codes
TransferType	METHOD_BUFFERED METHOD_IN_DIRECT METHOD_OUT_DIRECT METHOD_NEITHER
RequiredAccess	FILE_ANY_ACCESS FILE_READ_DATA FILE_WRITE_DATA FILE_READ_DATA|FILE_WRITE_DATA

Ioctl.h is also included in the *Wdm1Test* project. It includes the standard winioctl.h header file first to get the definition of CTL_CODE.

System Control

Dispatch.cpp also contains a handler for the Windows Management Instrumentation IRP, IRP_MJ_SYSTEM_CONTROL. Wdm1SystemControl simply passes the IRP down to the next driver in the stack. A full explanation of this process is given later.

```
IoSkipCurrentIrpStackLocation(Irp);
PWDM1_DEVICE_EXTENSION dx = (PWDM1_DEVICE_EXTENSION)fdo->DeviceExtension;
return IoCallDriver( dx->NextStackDevice, Irp);
```

Conclusion

This chapter has looked in detail at I/O Request Packets (IRPs) and how to write dispatch routines to handle common IRPs. In addition, it has covered how to define IOCTL codes, and how to use spin locks. The Wdm1 dispatch routines implement a shared memory buffer, protected by a spin lock.

The next chapter looks at Plug and Play in detail, enhancing Wdm1 to support PnP correctly.

Chapter 8

Plug and Play and Device Stacks

This chapter looks at the design of the Plug and Play (PnP) system and how PnP drivers fit into a device stack. A device stack is the means by which several layers of drivers can work together to process user requests. The next chapter looks at how to implement Plug and Play in detail, and describes the Wdm2 example. As well as PnP, Wdm2 supports Power Management, as described in Chapter 10.

Plug and Play — in a different form — is available in Windows 95 for virtual device drivers (VxDs) NT 4 does not support Plug and Play. Therefore, the Windows 98 and Windows 2000 support for Plug and Play in the Windows Driver Model is new for both environments. If you have any old NT 4 kernel mode device drivers, Microsoft recommends that you update them to support Plug and Play and Power Management. However, supporting Plug and Play requires a major change to driver code, so you will have to judge whether it is worthwhile.

Although Plug and Play initially looks complicated, it is possible to use the code the Pnp.cpp module described in the next chapter for most drivers. Careful use of the *LockDevice* and *UnlockDevice* routines is all that is needed elsewhere.

Design Overview

From a user's point of view, Plug and Play is straightforward. You just plug in some new equipment and Windows finds the device and prompts you for the correct drivers. Then you play with your new device.

The main benefit for users is that there should be no fiddling with DIP switches or jumpers to configure the device. Instead, Windows does all the necessary setting up. This is where your device driver comes in.

1. A standard bus driver detects when a device is added.
2. The device identifiers are used to find your driver.
3. Your driver is loaded and told that a device has been added.
4. A further message tells you what hardware resources to use.
5. Your driver then talks to your device, possibly using the services of a standard driver.

When a piece of equipment is unplugged, Windows detects this and tells your driver that the device has gone.

This chapter first looks at the overall Plug and Play design. It then covers the PnP messages that a driver receives. The bulk of the chapter is spent looking at how drivers work together in device stacks to find devices and process user requests.

Design Goals

The Plug and Play system is designed to satisfy these two goals.

- Cope with new devices when they are added to the system. Devices that are configurable in software must be told what resources or addresses to use.
- Make it easier for drivers to access complicated devices by providing standard drivers. For a relatively complicated bus like the Universal Serial Bus, it makes sense to provide a standard driver that other drivers must use if they want to access the bus.

The Plug and Play system brings both these design problems together. The standard drivers detect when a device is added or removed from the system.

Although Microsoft's solution satisfies both the original design goals, it still has some drawbacks. First, it makes a typical driver more complicated. Second, the device stack structure is complicated to understand, although it is not too difficult to use.

Plug and Play System

The PnP Manager controls these four main elements of the design.

- Standard bus drivers detect when devices are added or removed.
- Device identifiers are used to determine which driver or drivers to load.
- The following messages are sent to drivers when device related events occur:
     "A device has been added"
     "Your device has been removed"
     "Here are your device's resource assignments"
     "Please stop your device while its resources are reassigned"
- Device stacks are constructed so user requests can be processed in stages.

The rest of this chapter discusses this design in detail. The following chapter shows how to put this design into practice. Chapter 11 describes the installation process: how drivers are located and the format of installation INF files.

Detecting Devices

The first aim of the PnP system is to cope with new devices as they added to the system. Table 8.1 summarizes the requirements and the Microsoft solution, together with its one main drawback.

Standard bus drivers detect devices, both at power up and afterwards when devices are added or removed. Each bus driver retrieves one or more identifiers from its devices. These identifiers are used to find an appropriate device driver by looking through all the available installation files. If a suitable driver cannot be found, the user is prompted for a driver disk.

The driver is loaded and its AddDevice entry point is called to tell it that a new device has been found.

The bus driver or the installation file details the hardware resources that a device needs. Arbiters are used to decide what resources to give to each device. A *Start Device* PnP message tells the driver its device's resource assignments. It can then start to talk to its device properly.

In some cases, the bus driver must perform additional arbitration steps. For example, in the Universal Serial Bus, a device initially responds at a default address. The USB bus driver must assign the device a free USB address before it can use the full bus protocol. In addition, the USB bus driver must reserve some of the bus bandwidth for some types of device. If there is not enough bandwidth available then the device cannot be used.

The main drawback of this design is that resource assignments can be taken away from a device temporarily when it is up and running. Ideally, any I/O operations that are in progress should only pause briefly during this process. This is a complete pain because any I/O operations in progress at the time have to be halted, and any new I/O requests should be held in a queue until the device is restarted. Halting and queuing requests is not a trivial task for device drivers.

Suppose an existing PnP device currently uses IRQ7 and I/O ports 0x378 to 0x37A. The device can also be configured in software to use IRQ5 and ports 0x278 to 0x27A. Suppose a new device is now added that can only use IRQ7 and I/O ports 0x300 to 0x30F. It is possible to accommodate both these devices if the first one has its interrupt line changed from IRQ7 to IRQ5. To make this happen, the first driver must be halted and restarted straightaway with the new resource assignments[1]. If this works, the second device can then be given its resource assignments.

---

1. The first driver must disable its device interrupts, reprogram its hardware, and then enable interrupts again.

Table 8.1   Plug and Play device detection

Aim	Detect all devices at power up. Load the most appropriate driver. Give each driver their resource assignments. For hot pluggable devices, cope as devices are added and removed
Solution	Provide enumerator bus drivers to find new devices and detect when devices are added or removed. Each device provides identifiers which are used to find the most appropriate driver. The bus driver or INF file provides the device's resource requirements. Provide arbiters to decide which resources to allocate to which device. Send messages to indicate device events.
Drawback	Devices which are running may have to stopped so that their resources can be reassigned.

Driver Layers

The second aim of the Plug and Play system is to make standard drivers available for other drivers to use.

At the lowest level, a bus or class driver talks directly to the hardware. One or more client drivers are layered on top of such system drivers. These client drivers work at the functional level. For example, a USB printer client driver uses the USB bus driver to send its messages. However, the messages are only understood by the printer firmware. The client driver therefore controls the printer functions, letting the USB system driver handle all the messy low-level communications details.

Each bus driver has one device object that represents the whole bus. The bus driver creates a new device object, called the Physical Device Object (PDO), for each device that it finds on its bus. It provides various device identifiers to the PnP Manager. The PnP Manager finds and loads the most appropriate driver, and calls its AddDevice routine.

The newly loaded driver is called a **function** driver, as it should know how to control its device's function. The function driver must create its own new device object, called a Functional Device Object (FDO) which stores its information about the device[2]. The FDO is created in a function driver's AddDevice routine. The AddDevice routine then goes on to layer itself above the bus driver PDO.

In this simple case, two device objects refer to the same physical device. The PDO is the bus driver's representation of the device, while the FDO stores the function driver information about the device.

Figure 8.1 shows the situation when two devices are attached to a single bus. The bus driver has two 'child' PDOs, one for each device. A suitable function driver has been found for each device. Each function driver has created an FDO to store information about its device.

The bus driver controls three different device objects. Each child PDO receives requests from the function driver layered about it. In addition, the bus driver has an FDO of its own which it uses to store information about the whole bus. A bus driver might handle child PDO requests by sending them to its FDO for processing.

---

2.   Both FDOs and PDOs use exactly the same DEVICE_OBJECT structure.

Just to reiterate, each **function** driver knows how to control the functions in a device. However, a function driver usually uses the facilities of its underlying **bus** driver to talk to its device.

Even a virtual device driver like **Wdm1** has a device stack. **Wdm1** is a function driver and so its device object is an FDO. There is a bus driver underneath **Wdm1**, the system Unknown driver. The Unknown bus driver makes a PDO to represent each Wdm1 device. The Unknown driver does some important jobs, even though it does not correspond to a hardware bus. The following chapter describes the jobs that even a minimal bus driver like Unknown must do.

Figure 8.1   A bus driver, two devices, and their function drivers

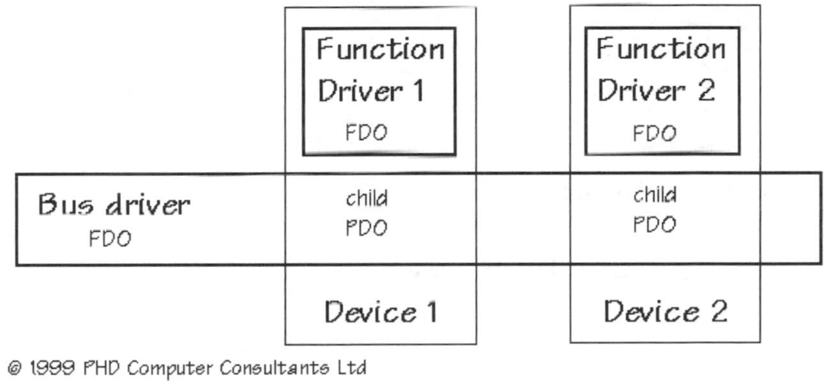

@ 1999 PHD Computer Consultants Ltd

Device Stacks

When one driver layers its device over another, it forms a *device stack*. Each device in the stack usually plays some part in processing user requests.

Figure 8.2 shows a generic device stack. A bus driver is always at the bottom of the stack. One or more function drivers do the main device control functions.

Filter drivers are used to modify the behavior of standard device drivers. Rather than rewrite a whole bus or class driver, a filter driver modifies its actions in the area of interest. As the figure shows, there are various types of filter driver. A bus filter driver acts on all devices that attach to a particular bus driver. Class filter drivers are installed in the stack for each device of the specified class. Finally, a device filter driver is installed for one particular device. *Lower-level* filter drivers are below the function driver, while *upper-level* drivers are above the function driver in the stack.

Filter drivers are only installed when a device is first installed, so they cannot be inserted once a device stack has been built.

"You said you wanted a device stack."

Figure 8.2 Bus, function, and filter drivers

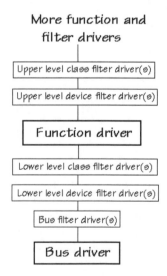

@ 1998 PHD Computer Consultants Ltd

Monolithic and Layered Drivers

When a standard keyboard is found, Windows loads the appropriate driver. This talks directly to the keyboard through its controller electronics. The keyboard driver is described as *monolithic* because it takes over all the processing required to handle the keyboard (i.e., it does not use other drivers to do its job).

In contrast, USB drivers use a layered approach. To use a USB keyboard, the keyboard driver uses the services of the USB class drivers.

As shown in Chapter 21, the USB class drivers expose an upper edge that drivers higher than it must use. Such drivers can issue USB Request Blocks (URBs), which transmit or receive information from a USB device. The keyboard driver does not care how the USB class drivers do their job.

The situation for the USB class drivers themselves is probably slightly different[3]. They are told where the USB bus controller electronics live, and they make use of the services of the PCI bus driver. However, I suspect that the USB class drivers talk directly to their own hardware. Asking the PCI bus driver to write or read memory would be too slow.

A different approach would have been to let each USB driver talk directly to the USB hardware in a monolithic driver. This would lead to two problems. The first is that each USB driver would need a large amount of code. Duplicating this code would waste memory and writing large drivers is decidedly error prone. The second problem is almost of more significance: how would all these different drivers coordinate their activities? The answer is, of course, with difficulty. Using different layers of drivers solves both these problems.

Plug and Play Messages

This section looks at the messages that a PnP driver receives. Its AddDevice routine is called when a device is added. After that, messages are sent using the PnP IRP. The PnP messages have the same major function code IRP_MJ_PNP, but each message uses a different minor function code. Each message has different parameters in the IRP.

A PnP driver must implement an AddDevice routine and handle various PnP IRPs. Table 8.2 is a list of all the minor function codes. The first eight of these minor codes are the most important. The other minor function codes are usually just handled by bus drivers.

Table 8.2   Plug and Play minor function codes

Common PnP IRPs	
IRP_MN_START_DEVICE	Assign resources and start a device
IRP_MN_QUERY_REMOVE_DEVICE	Ask if a device can be removed
IRP_MN_CANCEL_REMOVE_DEVICE	Cancel a query remove request
IRP_MN_REMOVE_DEVICE	Device has been unplugged or uninstalled Deallocate resources and remove a device
IRP_MN_SURPRISE_REMOVAL (W2000 only)	A user has unexpectedly unplugged a device
IRP_MN_QUERY_STOP_DEVICE	Ask if a device can be stopped
IRP_MN_CANCEL_STOP_DEVICE	Cancel a query stop request
IRP_MN_STOP_DEVICE	Stop a device for resource reallocation

---

3. I don't know for sure because I don't know how they work. And I don't care!

Table 8.2   Plug and Play minor function codes (continued)

Unusual PnP IRPs	
IRP_MN_QUERY_DEVICE_RELATIONS	Ask for PDOs that have certain characteristics
IRP_MN_QUERY_INTERFACE	Let a driver export a direct-call interface
IRP_MN_QUERY_CAPABILITIES	Ask about device capabilities (e.g., whether it can be locked or ejected)
IRP_MN_QUERY_RESOURCES	Get a device's boot configuration resources
IRP_MN_QUERY_RESOURCE_REQUIREMENTS	Ask what resources a device requires
IRP_MN_QUERY_DEVICE_TEXT	Get a device's description or location string
IRP_MN_FILTER_RESOURCE_REQUIREMENTS	Let filter and function drivers filter a device's resource requirements
IRP_MN_READ_CONFIG	Read configuration space information
IRP_MN_WRITE_CONFIG	Set configuration space information
IRP_MN_EJECT	Eject the device from its slot
IRP_MN_SET_LOCK	Set device locking state
IRP_MN_QUERY_ID	Get hardware, compatible, and instance IDs for a device
IRP_MN_QUERY_PNP_DEVICE_STATE	Set bits in a device state bitmap
IRP_MN_QUERY_BUS_INFORMATION	Get type and instance number of parent bus
IRP_MN_DEVICE_USAGE_NOTIFICATION	Notify whether a device is in the path of a paging, hibernation, or crash dump file.
IRP_MN_QUERY_LEGACY_BUS_INFORMATION	Returns legacy bus information (W2000 only)

Figure 8.3 shows the PnP states for a device and the messages that are sent to change state. The minor function code names are shortened in the figure (e.g., IRP_MN_START_DEVICE is shown as START_DEVICE). In this book, a PnP IRP with an IRP_MN_START_DEVICE minor function code is called a *Start Device* message. The other PnP function codes are given similar names.

When a device is added to the system, Windows finds the correct driver and calls its DriverEntry routine. Chapter 11 explains how Windows finds the correct drivers.

The PnP Manager then calls the driver's AddDevice routine to tell it that a device has been added. It is at this point that the driver makes its own device object, the Functional Device Object (FDO). However, the driver should not try to access its device hardware yet.

In due course, the driver receives an IRP_MN_START_DEVICE IRP that includes information about the resources the device has been assigned. It can then start talking properly to the device hardware.

If a device is about to be unplugged, Windows asks the driver if it is all right for the device to be removed using an IRP_MN_QUERY_REMOVE_DEVICE IRP. If the driver agrees, an IRP_MN_REMOVE_DEVICE IRP is sent to remove the device. If the driver does not want its device removed (e.g., if it is in the middle of a long transfer) it fails the remove request. It is then sent an IRP_MN_CANCEL_REMOVE_DEVICE IRP to put it back in the started state.

If a user unexpectedly pulls out a device, the driver is sent an `IRP_MN_REMOVE_DEVICE` IRP in Windows 98 or an `IRP_MN_SURPRISE_REMOVAL` IRP in Windows 2000. You have to cope with interrupted transfers as well as you can.

The other main state change occurs when the PnP Manager wishes to reallocate some of the driver's resources. This might happen if a new device of some sort is plugged in, meaning that the resource assignments need to be juggled about. The PnP Manager asks to stop the driver temporarily while its resources are reassigned. Similarly, in response to remove requests, an `IRP_MN_QUERY_STOP_DEVICE` IRP asks if it is OK to stop your device. If it is, an `IRP_MN_STOP_DEVICE` IRP is issued to take the device into the stopped state. If not, an `IRP_MN_CANCEL_STOP_DEVICE` IRP moves the device back into the started state. While stopped, the driver should not access its device. An `IRP_MN_START_DEVICE` IRP informs the driver of its new resources and starts the device again.

Figure 8.3   Plug and Play device states and messages

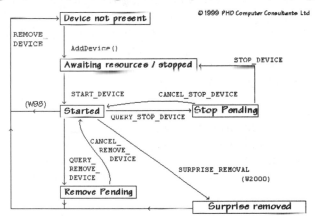

Note that a driver might receive a remove device message while in the stopped or awaiting resources states. These state changes are not shown in the diagram for the sake of clarity.

The next chapter looks in detail at how to handle these state changes. For the moment, I will look at device enumeration and then how layers of PnP drivers work together to form a device stack.

Device Enumeration

The enumeration process finds all the devices on a user's PC.

Fixed and Configurable Devices

As you probably know, most of the hardware devices on a PC motherboard are found at fixed locations. For example, on most PCs, the keyboard controller is something that looks like an 8042 processor that can be accessed at I/O ports 0x60 and 0x64. The keyboard controller interrupts on line IRQ1.

This approach works satisfactorily for the one keyboard controller that every motherboard has. However, with serial ports and the old ISA bus, things soon started to get too complicated for most users. If an old ISA bus card just uses fixed addresses and interrupts, it could easily conflict with another card. An interim solution was to provide jumpers and switches on each card to make it configurable. However, configuring this hardware was too much for most mortals to cope with.

The solution to this problem is to have devices that are configurable by software. This means that a bus driver tells each card or device where it should be located, possibly according to assignments given by the PnP Manager. All the newer buses, such as PCI, USB, and IEEE 1394 are software-configurable. Some PnP ISA devices are configurable in software, as well.

Usually, this works as follows. When a card or device powers up, it is detected by its bus driver. The bus driver uses the slot or port number to interrogate the device. Some configuration information or a device descriptor tells the bus driver what sort of resources the card needs. The bus driver or the PnP Manager then allocates the resources appropriately and tells the card this information, either by sending it a command or by writing to its registers. The card then configures itself to respond at the addresses it has been given. The card or device can then start operating normally.

The PnP Configuration Manager sorts out the basic system resources for drivers: I/O ports, Memory addresses, DMA channels, and Interrupts. However, the bus resources required in a particular bus environment are usually controlled by the bus driver. For example, only the USB drivers know about bandwidth allocation on the USB bus. When a USB client driver tries to configure its device, the USB bus driver decides if its bandwidth requirements can be satisfied.

Some types of device can be reconfigured after they have powered up. Configurable *hot-pluggable* devices can be plugged in or unplugged while the computer is switched on. The appropriate bus driver detects these changes and allocates or deallocates the card or device's resources. However, subsidiary buses are usually designed so that resources do not need to be reassigned in mid-flow.

Enumeration

When Windows is started, it does not know which devices are attached to the computer. It can work out some basic information for itself, such as how much memory it has, but how does it find out about the rest?

Windows uses drivers to *enumerate* the available hardware. Enumeration means finding and listing any available devices. *Arbitrators* are then used to juggle all their resource requirements. An appropriate driver (or drivers) is found for each device. The drivers are then told which resources to use, and off they go.

Figure 8.4 shows how enumeration works. Enumeration starts at the lowest level. The *root device* in the PC finds the basic chips that are on the motherboard[4]. It finds any simple devices, such as built-in serial ports and the keyboard. It also finds the PCI adapter, the electronics that control the PCI bus.

The PCI bus driver then enumerates and configures any hardware it finds. First, it finds a bridge to the ISA bus. The driver for this then finds a PnP ISA sound card, and the sound card drivers are loaded.

---

4. For ACPI systems, the ACPI bus driver implements the root device.

The PCI bus driver also finds a USB bus controller. The USB drivers enumerate the USB bus and find a keyboard and printer attached. These are configured and the appropriate drivers are loaded.

Figure 8.4   Hardware enumeration

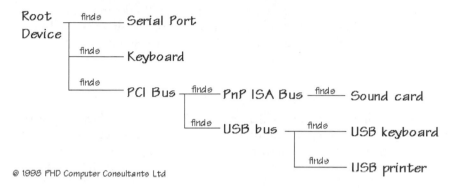

@ 1998 PHD Computer Consultants Ltd

Device Tree

Figure 8.5 shows a device tree for part of my PC. A device tree is usually shown growing upwards from a root device at the bottom. The lower-level drivers are the ones that interact with the hardware.

As can be seen, the main keyboard driver can get information from two sources, either the legacy keyboard or a USB keyboard. The USB keyboard driver is layered above the HID and USB class drivers and the PCI bus driver. Whether the keyboard is legacy or USB, Win32 gets the same response: the drivers hide the hardware.

Figure 8.5   Device tree

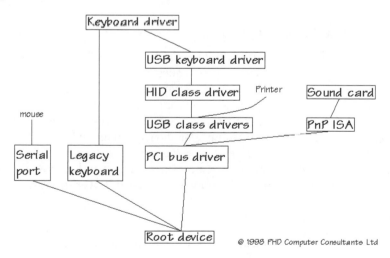

@ 1998 PHD Computer Consultants Ltd

The advantage of this approach is that each driver builds upon the work undertaken in lower layers, making it much easier to write most drivers. Indeed, it is the only way to write some types of drivers. If you are writing a USB driver, you must access your device through the USB class drivers. This means that you have to learn the specification of the relevant class driver. However, this is a far easier task than writing a huge monolithic driver that works with other similar drivers.

One possible drawback is that all this layering of drivers will take more processing time. While this is certainly a valid criticism, it is likely that a monolithic driver would use similar layering internally, so I/O should not take too much longer. More importantly, drivers that are easier to write are more likely to be stable.

Device Stacks

It does not make sense to have huge monolithic drivers. Wherever possible, a series of driver layers are built up, each performing an appropriate task. The advantage of this layering is that it breaks up I/O into a series of manageable tasks. If each layer has a standard specification, it means that a whole layer can be replaced and the higher layers will not know the difference. The lower layers hide the implementation details.

Microsoft has helped this layering process by providing standard *bus* and *class* drivers that implement one driver layer for a whole class of devices. For example, the Human Input Device (HID) class driver provides all the common functionality that these devices must implement. A driver that wants to talk to a specific type of HID device will layer itself above the Windows HID class driver. It will make I/O requests to this HID class driver. The HID driver layers itself above other drivers that let it talk to the real world.

As an another example, the USB class drivers are layered internally. Two types of electronics can be used to interface a PC to a physical USB bus: the Open Host Controller Interface (OpenHCI) and the Universal Host Controller Interface (UHCI). Windows chooses either the OpenHCI.sys driver or the UHCD.sys driver as the layer to interface to the electronics. Each driver implements the same upper-edge functionality. Other internal layers of the USB class drivers are built upon this common base.

PnP Support and the Device Stack

The PnP system has been designed to work with configurable devices and layers of drivers. A device stack represents the layers of drivers that process requests.

When a bus driver enumerates its bus, it gets the PnP Manager to call each new driver's AddDevice routine for each device it finds. The resources are assigned using the *Start Device* message. PnP messages are used to stop other drivers while the resources are being rejigged. Appropriate PnP messages are issued when a hot-pluggable device is removed.

The PnP system works with layers of drivers. I/O requests can be passed down to lower-level drivers for processing. A driver can then inspect the results from lower drivers and act accordingly. A driver can also generate new I/O requests to send to lower drivers.

Device Objects

Chapter 5 mentioned briefly that a PnP driver, such as Wdm1, has to deal with several different types of device object.

Each driver layer must create a device object to hold its information about a device. These device objects are arranged in a device stack, as shown in Figure 8.6. Note that it is the device objects that are directly connected together, not the drivers themselves (though each device object obviously knows with which driver it is associated). The arrangement is called a device stack, even though it does not correspond with the usual programmer definition of a stack.

The important point is that layers of device objects are built up. I/O requests are sent to the top of the stack and are gradually sent down the driver layers for processing. The results are sent back up the stack for post-processing.

In many cases, however, IRPs do not simply flow down to the bottom of the device stack and rise back up again. If one driver rejects a request, it can fail the IRP immediately and send it back up straightaway. Another common scenario is as follows. Suppose one driver receives a read request. To process the read, it might have to issue two read requests to its underlying drivers. The driver issues these two requests in turn and then waits for the results before finally completing its own request (i.e. sending it back up the stack).

Figure 8.6   Device objects

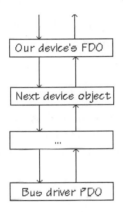

© 1998 PHD Computer Consultants Ltd

As Figure 8.6 shows, the various device objects have different names. Each device object is, in fact, the same DEVICE_OBJECT structure. Different names are given to each type of device object simply to remind us what type of driver is involved.

The device object at the bottom of the stack is called the Physical Device Object (PDO). A PDO is created and serviced by the appropriate *bus driver*. For example, the USB bus driver provides the PDO for the USB keyboard and USB printer devices.

The main driver that services a device is called a *function driver*. An installation INF file may specify that more than one Function driver is put into the device stack.

Each function driver creates its own device object called a Functional Device Object (FDO). As Chapter 5 demonstrated, IoCreateDevice is used to create an FDO. After an FDO is created, it is usually attached to the device stack using IoAttachDeviceToDeviceStack. This returns a pointer to the next device down the stack that is stored in a field called *NextStack-Device* in the device extension. This typical device driver does not know where it is in the stack, so it passes any requests to *NextStackDevice*. In a simple case, the next device object might be the same as the PDO.

Some DDK examples refer to the next device in the stack as *TopOfStack*. I thought that this was slightly misleading, as the next device is not the top of all the device stack. Some other drivers call this field *LowerDeviceObject*.

A final category of device object is called a Filter Device Object (Filter DO). *Filter device drivers* are slipped into the driver stack as the stack is built to modify the behavior of other drivers.

The AddDevice routine in each function or filter driver is called whenever a new device stack is built. Each AddDevice routine is passed a pointer to the same bus driver PDO. AddDevice then makes an FDO that is then attached to the device stack. The order in which the driver AddDevice routines are called determines the order of drivers in the stack. In this way, the device stack is built from the bottom up. Similarly, when a device is removed, the device stack is deconstructed by removing the highest drivers first. The PDO serves as the anchor point for the whole device stack, as each driver in the stack is given the same PDO pointer.

Upper Edges

This section uses a full example to illustrate two points. First, it shows how I/O requests are handled by a real driver stack. Second, it shows how drivers can have an upper edge that is different from that provided by lower layers.

USB Keyboard Example

Figure 8.7 shows how a USB keyboard might be used. A USB keyboard must be accessed via the HID class driver. The figure shows how two possible HID clients talk to the keyboard.

The items in this figure are not discussed in detail. The chapters on USB and HID will explain all. The major information flows are of concern just now.

The USB keyboard driver is a kernel mode HID client that sends requests to the HID class driver in the form of standard IRP_MJ_READ and IRP_MJ_WRITE IRPs. These must be in the right format to be recognized by the HID class driver. Alternatively, a user mode HID application can access the HID keyboard directly (rather than by waiting for standard Windows character messages). It does this using the Win32 ReadFile and WriteFile routines. These calls appear to the HID class driver as IRP_MJ_READ and IRP_MJ_WRITE IRPs. Again, these requests must be in the correct HID format.

Internally, the HID class driver uses one of its minidrivers to talk to the lower drivers. In this case, it is using its USB minidriver to talk to the top of the USB stack. Although it is not shown on the diagram, the main HID class driver actually uses Internal IOCTLs to request I/O from a minidriver.

The HID USB minidriver generates Internal IOCTLs to use the services of the USB class drivers. The most common Internal IOCTL has a control code of IOCTL_INTERNAL_USB_SUBMIT_URB. This submits a USB Request Block (URB) to the USB class driver. To get some input data, the minidriver will almost certainly use the URB_FUNCTION_BULK_OR_INTERRUPT_TRANSFER function code within a URB.

As mentioned before, the USB class drivers are layered internally. In this case, the diagram shows that the OpenHCI USB host controller driver is in use. The OpenHCI driver will make some use of its lower driver, the PCI bus driver. However, it will do most of its work by talking directly to the OpenHCI USB Controller hardware. It will read and write memory and controller registers. It will almost certainly handle interrupts and use DMA to transfer information.

The USB controller hardware itself is responsible for the very last stage of the keyboard I/O request handling. It generates the appropriate signals on the USB bus, according to the instructions given to it by the USB class drivers. The USB keyboard understands these signals and responds according to its specification.

When a key is pressed on the keyboard, information percolates its way back up through the chain of drivers, ending up at one of the client drivers as a keypress input report.

Figure 8.7   HID USB Keyboard I/O Request handling

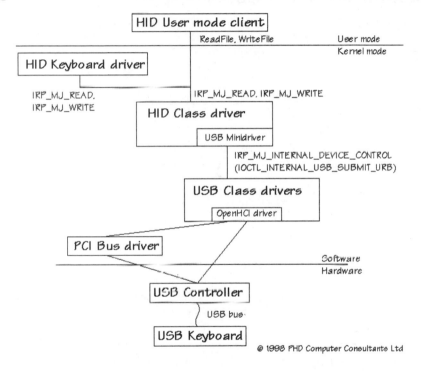

@ 1998 PHD Computer Consultants Ltd

Functional and Physical Device Objects

Just to complicate matters, the previous example does not involve just one device stack. Instead, there are actually four, as shown in Table 8.3.

A device stack must always end at a Physical Device Object (PDO). Each bus driver creates a PDO for each device it finds.

1. Starting from the HID Keyboard driver's FDO, going down the stack, there is an FDO created by the HID class driver and a PDO created by the USB bus driver. These three device objects are in the first device stack.

2. The DDK documentation says that the USB class drivers are layered internally, with a hub driver at the top and a host controller driver at the bottom (OpenHCI, in this case).

3. The USB host controller was originally found by the PCI bus driver. Therefore, the host controller itself has an FDO that is layered on top of a PDO created by the PCI bus driver.

4. The PCI bus was originally found by the Windows root bus driver. The final device stack therefore consists of the PCI bus FDO layered on top of the root bus PDO.

This arrangement looks complicated. However, if you are trying to use only the USB class drivers, you do not really care how they process your requests. All the details of the other device stacks are hidden from you.

Table 8.3   USB Keyboard device stacks

Driver	Device stack
HID Keyboard driver	HID Keyboard device FDO HID class device FDO PDO (created by USB hub bus driver)
USB Hub	USB Hub device FDO PDO (created by USB host controller bus driver)
USB Host controller	USB Host Controller device FDO PDO (created by PCI bus driver)
PCI Adapter	PCI device FDO PDO (created by root bus driver)

Upper Edge Definitions

This example shows how several standard Windows system drivers have been used in the device stack for a USB keyboard. Each driver writer usually needs to know only the specification of the next layer down[5]. The USB Keyboard driver writer needs to understand only the HID class driver specification. Knowledge of the different USB controllers and how to use them is definitely not required.

The *upper edge* of a driver is the specification of how to use it. The upper edge of the HID class driver responds to standard read and write IRPs. However, the upper edge of the USB class drivers only responds to Internal IOCTL IRPs. This is a sensible option for the USB class drivers, as it is not appropriate for user mode drivers to call them directly. User mode programs could not adhere to the timing requirements of the USB class drivers.

The example also shows that the upper edge presented by a driver does not determine its lower edge. The HID class drivers accept standard read and write IRPs. However, it implements these by sending Internal IOCTLs to its own minidriver. The USB minidriver implements its upper edge by sending URBs to the USB class drivers in its own Internal IOCTLs.

---

5.   However, note that the USB Open HCI driver accesses hardware directly, once it has its configuration information from the PCI bus driver.

Thus, you have to look at some of the earlier example figures with care. Figure 8.5 shows function, filter, and bus drivers in a smooth hierarchy. In real life, the chain of events is much more complicated, as I/O requests are processed by several drivers.

A filter driver must have the same upper and lower edge because it must slip into the stack without affecting other drivers. A filter can modify or inspect the requests in which it is interested. However, all other requests must be passed down the stack unmodified. It is possible for a filter driver to perform some extra checking, which will mean that some IRPs are never passed down. However, it is vital that user mode applications or higher-level drivers can continue to function normally.

Conclusion

This chapter has presented the background for Plug and Play device drivers, the messages they handle, and how they fit into the WDM device stack. The next chapter looks at how to implement Plug and Play in practice and illustrates it in the Wdm2 device driver.

Chapter 9

Plug and Play Implementation

The last chapter went through the theory of Plug and Play (PnP) and the device stack. This chapter looks at how to implement Plug and Play in practice, with a discussion of the advanced topics of Plug and Play Notification and Bus drivers.

The Wdm2 example includes full PnP and Power Management facilities. Power Management is described in the next chapter.

The *Wdm2Test* user mode application tests some of the Wdm2 PnP functionality. The *Wdm2Notify* Win32 program displays PnP Notification device change events for the Wdm2 device interface.

Implementing Plug and Play

Supporting Plug and Play primarily means implementing an AddDevice routine and an IRP_MJ_PNP handler. This PnP IRP has eight minor function codes that most WDM drivers need to support.

- IRP_MN_START_DEVICE (*Start Device*)
- IRP_MN_QUERY_REMOVE_DEVICE (*Query Remove*)
- IRP_MN_REMOVE_DEVICE (*Remove Device*)
- IRP_MN_CANCEL_REMOVE_DEVICE (*Cancel Remove*)
- IRP_MN_STOP_DEVICE (*Stop Device*)
- IRP_MN_QUERY_STOP_DEVICE (*Query Stop*)

- IRP_MN_CANCEL_STOP_DEVICE (*Stop Device*)
- IRP_MN_SURPRISE_REMOVAL (*Surprise Removal*)

Looking at this list, it might not seem too complicated to handle Plug and Play in a driver. In fact, there are many things to get right. At a basic level this means:

- Coping with adding and removing devices
- Getting resource assignments
- Handling *Query Stop* and *Query Remove* messages
- Handling *Stop* messages
- Handling *Surprise Removal* messages

However, as will be shown, it soon becomes apparent that you must also do the following tasks:

- Allow only I/O requests while the device is started
- Not allow a device to be removed while there are any open handles
- Queue I/O requests while the device is not started
- Wait until any I/O requests have completed before handling remove requests
- Process *Start Device* messages after lower devices have started
- Pass unsupported IRPs down the stack

The rest of this chapter will look at the theory behind all the important PnP messages. The Wdm2 example driver shows how to implement most of the related tasks listed above. However, the most complicated — queuing I/O requests — is left until Chapter 16.

Refer to the previous chapter if you need to remind yourself of the difference between function and filter drivers, and the different types of device objects that the Wdm2 driver deals with.

Adding and Removing Devices

When a PnP driver handles an AddDevice message, it means that a new device has be found. Either a bus driver has found the device at power on or when it was inserted, or the user has added it by hand from the Control Panel. As Wdm2 devices are virtual, these have to be added by hand.

The PnP Manager will have worked out which drivers are going to be in the stack. The Physical Device Object (PDO) is created by the bus driver first. Then, going up the stack, each driver's AddDevice routine is called in turn. Chapter 11 describes how the PnP Manager works out which drivers to put in the stack.

If a driver's AddDevice routine fails, any drivers below it (whose AddDevice succeeded) will be sent a *Remove Device* message. Be prepared to accept a *Remove Device* message straightaway after your AddDevice routine has completed.

The job of an AddDevice routine is to create and initialize a device object for the current driver to use, and to connect it to the device stack. For function drivers, the device object is called a Functional Device Object (FDO). Filter drivers make a Filter Device Object (Filter DO). As mentioned in the last chapter, PDOs, FDOs, and Filter DOs all use the same DEVICE_OBJECT structure. Microsoft recommends that we use different names to help us remember what sort of driver owns the device object.

When a `Wdm2` device is added, the Unknown bus driver makes a PDO for it. The PnP Manager passes the PDO to `Wdm2`. The `Wdm2AddDevice` routine, shown later, calls `IoCreateDevice` to create the `Wdm2` Functional Device Object and eventually calls `IoAttachDeviceToDeviceStack` to attach it to the device stack. In between, the FDO and its device extension are initialized and a device interface for it is set up.

The eventual job of the *Remove Device* message handler is to stop the device and delete the FDO. In `Wdm2`, `PnpRemoveDeviceHandler`, shown later, eventually calls `IoDetachDevice` to detach the device object from the stack and calls `IoDeleteDevice` to delete the device object and its device extension memory. Processing *Remove Device* PnP messages safely is, in fact, a bit more complicated than this, as will be shown. Note that a *Remove Device* request will be the last IRP a driver receives before it is unloaded.

Basic PnP Handlers

The `Wdm1` driver handles adding a device in its `AddDevice` routine and its `Wdm1Pnp` routine handles the remove device minor code `IRP_MN_REMOVE_DEVICE`.

The `AddDevice` and `IRP_MJ_PNP` handlers are called at `PASSIVE_LEVEL` IRQL in the context of a system thread. Calls to these routines may be issued while other IRPs are running in the same driver. Even in a uniprocessor computer, processing of a Read IRP could have stalled for some reason. A PnP call could then be issued.

AddDevice

In the `Wdm2` driver, the code for adding and removing devices is substantially the same as `Wdm1`. The `Wdm2AddDevice` routine in Listing 9.1 is exactly the same, apart from new lines initializing some extra fields in the device extension. These all relate to handling PnP and Power IRPs correctly and are explained in due course.

Most of the PnP handling code is in the `Wdm2` `Pnp.cpp` module. Some changes have been made to the `Dispatch.cpp` code from the `Wdm1` version. A new module `DeviceIo.cpp` handles device starting and stopping.

Listing 9.1   Wdm2 Wdm2AddDevice

```
NTSTATUS Wdm2AddDevice( IN PDRIVER_OBJECT DriverObject, IN PDEVICE_OBJECT pdo)
{
    DebugPrint("AddDevice");
    NTSTATUS status;
    PDEVICE_OBJECT fdo;

    // Create our Functional Device Object in fdo
    status = IoCreateDevice( DriverObject, sizeof(WDM2_DEVICE_EXTENSION), NULL,
                            FILE_DEVICE_UNKNOWN, 0, FALSE, &fdo);
    if( NT_ERROR(status))
        return status;

    // Initialise device extension
    PWDM2_DEVICE_EXTENSION dx = (PWDM2_DEVICE_EXTENSION)fdo->DeviceExtension;
```

Listing 9.1 Wdm2 Wdm2AddDevice **(continued)**

```
dx->fdo = fdo;
dx->pdo = pdo;
dx->UsageCount = 1;
KeInitializeEvent( &dx->StoppingEvent, NotificationEvent, FALSE);
dx->OpenHandleCount = 0;
dx->GotResources = false;
dx->Paused = false;
dx->IODisabled = true;
dx->Stopping = false;
dx->PowerState = PowerDeviceD3;
dx->PowerIdleCounter = NULL;
DebugPrint("FDO is %x",fdo);

// Initialise device power state
POWER_STATE NewState;
NewState.DeviceState = dx->PowerState;
PoSetPowerState( fdo, DevicePowerState, NewState);

// Register and enable our device interface
status =
    IoRegisterDeviceInterface( pdo, &WDM2_GUID, NULL, &dx->ifSymLinkName);
if( NT_ERROR(status))
{
    IoDeleteDevice(fdo);
    return status;
}
IoSetDeviceInterfaceState( &dx->ifSymLinkName, TRUE);
DebugPrint("Symbolic Link Name is %T",&dx->ifSymLinkName);

// Attach to the driver stack below us
dx->NextStackDevice = IoAttachDeviceToDeviceStack(fdo,pdo);

// Set fdo flags appropriately
fdo->Flags &= ~DO_DEVICE_INITIALIZING;
fdo->Flags |= DO_BUFFERED_IO;

dx->PowerIdleCounter =
    PoRegisterDeviceForIdleDetection( pdo, 30, 60, PowerDeviceD3);

return STATUS_SUCCESS;
}
```

Remove Device handler

The code that handles removing devices has been moved to PnpRemoveDeviceHandler, as shown in Listing 9.2. This is the same as before, apart from calling the PnpStopDevice routine, which is explained in the following text.

Listing 9.2   Wdm2 PnpRemoveDeviceHandler

```
NTSTATUS PnpRemoveDeviceHandler( IN PDEVICE_OBJECT fdo, IN PIRP Irp)
{
    PWDM2_DEVICE_EXTENSION dx=(PWDM2_DEVICE_EXTENSION)fdo->DeviceExtension;
    DebugPrintMsg("PnpRemoveDeviceHandler");

    // Wait for I/O to complete and stop device
    PnpStopDevice(dx);

    // Pass down stack and carry on immediately
    NTSTATUS status = PnpDefaultHandler(fdo, Irp);

    // disable device interface
    IoSetDeviceInterfaceState( &dx->ifSymLinkName, FALSE);
    RtlFreeUnicodeString(&dx->ifSymLinkName);

    // unattach from stack
    if (dx->NextStackDevice)
        IoDetachDevice(dx->NextStackDevice);

    // delete our fdo
    IoDeleteDevice(fdo);

    return status;
}
```

Main PnP IRP Handler

The main IRP_MJ_PNP dispatch routine, Wdm2Pnp, has changed considerably from Wdm1, as shown in Listing 9.3. The bulk of the code is a switch statement based on the PnP minor function code. Most of the interesting minor function handling is delegated to subsidiary routines, but some are handled inline. All other minor function codes are handled by the PnpDefaultHandler routine. The PnpQueryCapabilitiesHandler routine is described in the chapter on Power Management.

Listing 9.3   Wdm2 Wdm2Pnp

```
NTSTATUS Wdm2Pnp( IN PDEVICE_OBJECT fdo, IN PIRP Irp)
{
    PWDM2_DEVICE_EXTENSION dx=(PWDM2_DEVICE_EXTENSION)fdo->DeviceExtension;
    DebugPrint("PnP %I",Irp);
```

Listing 9.3 Wdm2 Wdm2Pnp **(continued)**

```
if (!LockDevice(dx))
    return CompleteIrp(Irp, STATUS_DELETE_PENDING, 0);

// Remember minor function
PIO_STACK_LOCATION IrpStack = IoGetCurrentIrpStackLocation(Irp);
ULONG MinorFunction = IrpStack->MinorFunction;

NTSTATUS status = STATUS_SUCCESS;
switch( MinorFunction)
{
case IRP_MN_START_DEVICE:
    status = PnpStartDeviceHandler(fdo,Irp);
    break;
case IRP_MN_QUERY_REMOVE_DEVICE:
    status = PnpQueryRemoveDeviceHandler(fdo,Irp);
    break;
case IRP_MN_SURPRISE_REMOVAL:
    status = PnpSurpriseRemovalHandler(fdo,Irp);
    break;
case IRP_MN_REMOVE_DEVICE:
    status = PnpRemoveDeviceHandler(fdo,Irp);
    return status;
case IRP_MN_QUERY_STOP_DEVICE:
    dx->Paused = true;
    dx-> IODisabled = true;
    status = PnpDefaultHandler(fdo,Irp);
    break;
case IRP_MN_STOP_DEVICE:
    status = PnpStopDeviceHandler(fdo,Irp);
    break;
case IRP_MN_QUERY_CAPABILITIES:
    status = PnpQueryCapabilitiesHandler(fdo,Irp);
    break;
case IRP_MN_CANCEL_REMOVE_DEVICE:     // fall thru
case IRP_MN_CANCEL_STOP_DEVICE:
    dx->Paused = false;
    dx-> IODisabled = false;
    status = PnpDefaultHandler(fdo,Irp);
    break;
default:
```

Listing 9.3   Wdm2 Wdm2Pnp **(continued)**

```
        status = PnpDefaultHandler(fdo,Irp);
    }


    UnlockDevice(dx);
#if DBG
    if( status!=STATUS_SUCCESS)
        DebugPrint("PnP completed %x",status);
#endif
    return status;
}
```

Passing Unsupported IRPs Down the Stack

PnpDefaultHandler (Listing 9.4) passes the PnP IRP down the device stack for processing by all the lower device drivers. Call IoCallDriver whenever you want a driver to process an IRP. If you have created an IRP from scratch, you must set up all the IRP and IRP stack fields correctly. The different ways of allocating and sending your own IRPs are discussed in Chapter 21.

For PnpDefaultHandler, all the IRP structure fields are already set up correctly in the existing PnP IRP that is to be passed onto the next driver. What about the IRP stack?

One IRP stack location is reserved for each possible device in a device stack. When an IRP is passed to the next driver, the next stack location must be set up for it. However, in this case, the Wdm2 driver is never going to need to look at the IRP or its stack location again. IoSkipCurrentIrpStackLocation does not copy the current stack location to the next one. In fact, it simply sets up the IRP internally so that the next driver's call to IoGetCurrentIrpStackLocation returns the same stack location as Wdm2 saw.

You need to set up the next IRP stack location properly if you are going to inspect the IRP processing results or even simply wait for the IRP to complete. Use the routine IoCopyCurrentIrpStackLocationToNext if you simply want to copy the current stack location without changing any of the information. An example of this function is given in Listing 9.4 when Wdm2 waits for an IRP to be processed by all the lower drivers.

IoCallDriver, IoSkipCurrentIrpStackLocation, IoCopyCurrentIrpStackLocationToNext, and the other relevant function IoSetCompletionRoutine must be called at DISPATCH_LEVEL IRQL or lower.

PnpDefaultHandler therefore simply sets up the stack location and passes the IRP to the next lower driver in the device stack. It does not wait for the IRP to complete. It returns the status code that IoCallDriver returns.

Listing 9.4   Wdm2 PnpDefaultHandler

```
NTSTATUS PnpDefaultHandler( IN PDEVICE_OBJECT fdo, IN PIRP Irp)
{
    DebugPrintMsg("PnpDefaultHandler");
```

Listing 9.4 Wdm2 PnpDefaultHandler (continued)

```
    PWDM2_DEVICE_EXTENSION dx=(PWDM2_DEVICE_EXTENSION)fdo->DeviceExtension;
    IoSkipCurrentIrpStackLocation(Irp);
    return IoCallDriver( dx->NextStackDevice, Irp);
}
```

PnP States and Messages

Before continuing with the rest of the PnP implementation, it is worth looking again at the Plug and Play message and state diagram that was shown in the last chapter.

Figure 9.1 shows the main theoretical device states that a device can be in and the messages that are sent to change between these states. As mentioned before, a message such as START_DEVICE in this diagram corresponds to a PnP IRP with a minor code of IRP_MN_START_DEVICE.

Note that I said "theoretical" device states. There are no visible flags in the kernel device structure that say which state a device is in. Wdm2 has to maintain its own state variables.

Another important point to note is that your code should be prepared to accept more-or-less any message from any state. The DDK documentation says in at least two places that an unexpected message may occasionally be sent when in one particular state.

Figure 9.1 Plug and Play device states and messages

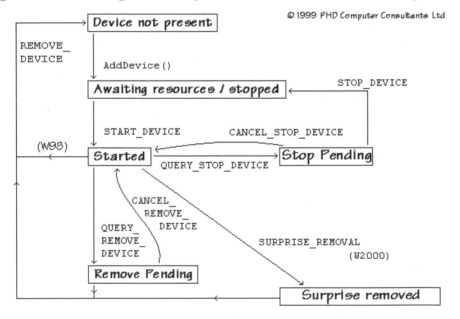

State Flags

The device extension for Wdm2 device objects is shown in Listing 9.5. Four state flags are used to ensure that I/O requests are only begun when the device is in the *Started* state.

Paused	Device has a remove pending or stop pending
GotResources	Device running normally or paused (i.e., not stopped)
IODisabled	Paused or stopped
Stopping	Device is in process of being removed or stopped

The *GotResources* flag is set when the device has retrieved and allocated any hardware resources that it needs. In Figure 9.1, the *GotResources* flag is set when the device is in the *Started*, *Stop pending,* and *Remove Pending* states.

The *Paused* flag is set when the device is in the *Stop Pending* or *Remove Pending* States.

The *IODisabled* flag is set when *GotResources* is false or *Paused* is true (i.e., in the *Stop pending, Remove Pending, Stopped,* and *Surprise removed* states).

The *Stopping* flag is used during the processing of *Remove Device* and *Stop Device* messages, as described in the following text.

In the *Device not present* state, a device object and its flags simply do not exist.

Some devices may want to have an *InterruptsEnabled* flag, as well, to indicate when device interrupts are enabled.

Listing 9.5   Wdm2 **device extension**

```
typedef struct _WDM2_DEVICE_EXTENSION
{
    PDEVICE_OBJECT    fdo;
    PDEVICE_OBJECT    NextStackDevice;
    UNICODE_STRING    ifSymLinkName;

    bool GotResources; // Not stopped
    bool Paused;       // Stop or remove pending
    bool IODisabled;   // Paused or stopped

    LONG OpenHandleCount; // Count of open handles

    LONG UsageCount;      // Pending I/O Count
    bool Stopping;        // In process of stopping
    KEVENT StoppingEvent; // Set when all pending I/O complete

    DEVICE_POWER_STATE PowerState;  // Our device power state
    PULONG PowerIdleCounter;        // Device idle counter

    // Resource allocations
    bool GotPortOrMemory;
    bool PortInIOSpace;
```

Listing 9.5   Wdm2 **device extension (continued)**

```
    bool PortNeedsMapping;
    PUCHAR PortBase;
    PHYSICAL_ADDRESS PortStartAddress;
    ULONG PortLength;

    bool GotInterrupt;
    ULONG Vector;
    KIRQL Irql;
    KINTERRUPT_MODE Mode;
    KAFFINITY Affinity;
    PKINTERRUPT InterruptObject;


} WDM2_DEVICE_EXTENSION, *PWDM2_DEVICE_EXTENSION;
```

The *GotResources* flag indicates that Wdm2 has been assigned its resources. Wdm2 does not use any hardware resources. However, it still needs this flag to indicate the state it is in.

The *Paused* flag has no hardware interpretation. It is simply a way of stopping IRPs from starting when in the *Stop Pending* or *Remove Pending* states. In a full PnP implementation, IRPs must be queued while the *Paused* flag is set. Wdm2 should never get in the *Stop Pending* state, as it has no resources to reallocate. It should only be in the *Remove Pending* state briefly. For simplicity sake, Wdm2 does not queue IRPs. Chapter 16 gives this subject the full airing it deserves.

The *IODisabled* flag is provided so that the dispatch routines have to check only one flag quickly, not both *GotResources* and *Paused*. Each normal IRP dispatch routine, therefore, has the following code at the top that fails the IRP straightaway with an appropriate error code if the device is disabled.

```
if( dx->IODisabled)
    return CompleteIrp( Irp, STATUS_DEVICE_NOT_CONNECTED, 0);
```

To recap, three flags in the Wdm2 device extension are used to ensure that normal I/O requests are permitted only when the Wdm2 device is fully started. In fact, there will be very few times in the life of a Wdm2 device when I/O requests are not permitted. (Strictly speaking, I could have combined the *Paused* and *IODisabled* flags, but it is clearer to have them separate.)

Holding IRPs

When a device is not fully started, a full PnP implementation ought to queue I/O request IRPs.

Consider the various PnP stop messages. These occur when some new hardware has been added to the system or a new device has been plugged in. The PnP Manager may decide that it can only accommodate the new device by reassigning the resources that an existing device currently uses. It does this by issuing a *Query Stop* message first. If all devices in the stack agree that a stop can take place, then the PnP Manager issues a *Stop Device* request. If any of the drivers in the stack do not want the device stopped, the PnP Manager issues a *Cancel Stop* message. It then probably informs the user that there are not enough resources available currently and so a restart is necessary. When the resources have been reassigned, it sends a *Start Device* message with the new resource allocations.

During this entire process, it seems reasonable that I/O requests on existing devices carry on as normal. In practice, this means not starting any new requests and holding them in a queue for processing when the device is started again. A user might therefore notice a slight pause in proceedings while resources are reassigned, but I/O requests should not fail.

The DDK documentation recommends that drivers do not start any new I/O requests after a *Query Stop* or *Query Remove* message has been received. This lets any following stop or remove request proceed quickly.

The end result is that IRPs ought to be queued when in the *Stop Pending*, *Remove Pending*, and *Stopped* states. In the Wdm2 driver, this means that IRPs ought to be queued when the `IODisabled` flag is `true`. However, as stated before, Wdm2 does not hold IRPs, as this is a complicated subject to be covered in Chapter 16.

As shown in Listing 9.3, Wdm2 handles *Query Stop* message by setting the *Paused* and `IODisabled` flags before passing the IRP down the stack. The *Query Remove* request does the same job. The *Cancel Stop* and *Cancel Remove* messages undo these actions by clearing *Paused* and `IODisabled`.

Open Handles

What happens if the user asks to remove the Wdm2 device while there are open handles to it?

The PnP Manager sends a *Query Remove* request to the driver. The simplest approach is to refuse to let a device be removed while there are any open handles. The DDK documentation says that the *Query Remove* must be failed if "there are open handles that cannot be closed".

Wdm2 uses the simple approach. It keeps a count of open handles to a device in the *OpenHandleCount* variable in the device extension. *OpenHandleCount* is initialized to zero when the device is created in Wdm2AddDevice. The `InterlockedIncrement` and `InterlockedDecrement` routines are used in the Wdm2Create and Wdm2Close dispatch routines, respectively, to maintain this count safely.

```
InterlockedIncrement(&dx->OpenHandleCount);
```

Listing 9.6 shows the complete *Query Remove* handler for Wdm2. If *OpenHandleCount* is greater than zero, PnpQueryRemoveDeviceHandler simply fails the IRP straightaway. Notice that it does not need to pass the PnP IRP down the stack as it is failing it. Instead, it just completes the IRP with the STATUS_UNSUCCESSFUL status code.

If there are no open handles, Wdm2 sets its *Paused* and *IODisabled* flags, as discussed before. However, in this case, PnpQueryRemoveDeviceHandler must pass the IRP down the stack in PnpDefaultHandler to give lower devices a chance to reject the *Query Remove* IRP.

Listing 9.6   Wdm2 PnpQueryRemoveDeviceHandler

```
NTSTATUS PnpQueryRemoveDeviceHandler( IN PDEVICE_OBJECT fdo, IN PIRP Irp)
{
    PWDM2_DEVICE_EXTENSION dx=(PWDM2_DEVICE_EXTENSION)fdo->DeviceExtension;
    DebugPrintMsg("PnpQueryRemoveDeviceHandler");
    if( dx->OpenHandleCount>0)
    {
        DebugPrint("PnpQueryRemoveDeviceHandler: %d handles still open",
                    dx->OpenHandleCount);
        return CompleteIrp( Irp, STATUS_UNSUCCESSFUL, 0);
    }
    dx->Paused = true;
    dx->IODisabled = true;
    return PnpDefaultHandler(fdo,Irp);
}
```

When to Process PnP IRPs

You must be careful to process PnP IRPs at the right time. When a *Start Device* message is received for a USB device, for example, the USB drivers must enable the device at the bus level first. It is only then that the function drivers above can access the device. In fact, when processing the *Start Device* message, the drivers must process the message in order, going up the device stack.

Similar considerations apply when processing *Stop*, *Remove,* and *Surprise Removal* messages. In these cases, all the drivers in the stack must process the IRP first, in order going down the stack. Each driver must do whatever it needs to do to stop its device before the lower drivers pull the rug out from under its feet.

Handle the *Cancel Remove* and *Cancel Stop* requests on the way up the stack so that all the lower devices have restarted. However, the Wdm2 driver enables requests straightaway and then passes the IRP down the stack for processing.

The other PnP messages are usually processed by drivers on the way down the device stack. However, in a few circumstances, you may wish to see what results the lower drivers have produced.

WDM drivers can process IRPs in both these orders. So far, I have only shown how to process IRPs in order going down the stack. I shall now look at how to process IRPs in the other order.

IRP Completion Routines

As mentioned before, the IoCallDriver routine is used to call another driver. It is important to realize that IoCallDriver may return before the IRP has been completely processed. In this case, IoCallDriver returns STATUS_PENDING.

If a driver wants to process an IRP when all lower drivers have completed processing it, the driver must set a completion routine for the IRP. The completion routine is called when all the lower drivers have finished processing the IRP. The completion routine is called in an arbitrary context. I now show how a completion routine signals that it has been run using a kernel event.

Listing 9.7 shows how the ForwardIrpAndWait routine forwards an IRP to the lower drivers and waits for its completion. As ForwardIrpAndWait waits for the completion routine event to become signalled, it must run at PASSIVE_LEVEL IRQL. Waiting for dispatcher objects, such as events, is covered in full in Chapter 14. PnP IRPs are always called at PASSIVE_LEVEL, so it is safe to call ForwardIrpAndWait.

To set a completion routine, the next IRP stack location must be set up correctly. As described previously, IoCopyCurrentIrpStackLocationToNext copies all the current stack location parameters. Having done this, IoSetCompletionRoutine is used to set the completion routine to the ForwardedIrpCompletionRoutine function. ForwardIrpAndWait is then ready to call the next driver using IoCallDriver.

The last three BOOLEAN parameters of IoSetCompletionRoutine specify the circumstances in which you want the completion routine called. If the first BOOLEAN, InvokeOnSuccess, is TRUE, the completion routine is called if the IRP completes successfully. The other two BOOLEAN parameters, InvokeOnError and InvokeOnCancel, state whether the completion routine should be called if an error is returned or the IRP is cancelled. In ForwardIrpAndWait, I want the completion routine called in all circumstances, so all these parameters are set to TRUE.

ForwardIrpAndWait now has two tasks to perform. The completion routine has to signal when it has run, and the main code must wait for this signal. The signalling mechanism is a kernel event, which is basically the same as its Win32 equivalent. Chapter 14 discusses kernel events in full. The event is initialized to the nonsignalled state using KeInitializeEvent. When the completion routine runs, it simply calls KeSetEvent to set the event into the signalled state. ForwardIrpAndWait uses KeWaitForSingleObject to wait for the event to become signalled.

A completion routine has a standard prototype, passing the device object, the IRP, and a context pointer. In this case, ForwardIrpAndWait uses a pointer to the event as the context pointer. The context pointer is set in the IoSetCompletionRoutine call. When the IRP has been processed by all the lower drivers, ForwardedIrpCompletionRoutine is called. It simply sets the event.

ForwardedIrpCompletionRoutine returns a status of STATUS_MORE_PROCESSING_REQUIRED. This means that the IRP has not been completed by this driver and some other part of the driver will complete it. The only other alternative is to return STATUS_SUCCESS, in which case the IRP continues its journey back up the device stack.

These are good layers

IoCallDriver returns STATUS_PENDING if the IRP has not completed its processing by the lower drivers. If this value is returned, ForwardIrpAndWait must wait for the completion routine to run. Once the event has been set, the call to KeWaitForSingleObject returns. ForwardIrpAndWait retrieves the status returned by the lower drivers from the IRP's *IoStatus.Status* field.

If IoCallDriver returns any value apart from STATUS_PENDING, this means that the IRP has been processed by all the lower drivers. The completion routine has been run and the event set. However, ForwardIrpAndWait does not need to wait for the event, as it already knows that the IRP has been processed in full.

There is no way to know in advance if IoCallDriver will return a pending status. Therefore, if you want to use an IRP after calling IoCallDriver you must set a completion routine.

Completion routines may run at DISPATCH_LEVEL IRQL or lower in an arbitrary thread context. To be safe, you should assume that it is running at DISPATCH_LEVEL IRQL. This means that the completion routine itself and the context pointer must not be in paged memory. The ForwardIrpAndWait event variable is in the kernel stack. This is normally in nonpaged memory. Apparently, the kernel stack can be pageable if a user mode wait is issued. However, as a kernel mode wait is used here, the event memory should be safely nonpaged.

Listing 9.7 Wdm2 **Forwarding IRPs and waiting for completion**

```
NTSTATUS ForwardIrpAndWait( IN PDEVICE_OBJECT fdo, IN PIRP Irp)
{
    DebugPrintMsg("ForwardIrpAndWait");
    PWDM2_DEVICE_EXTENSION dx=(PWDM2_DEVICE_EXTENSION)fdo->DeviceExtension;

    KEVENT event;
    KeInitializeEvent( &event, NotificationEvent, FALSE);
```

Listing 9.7   Wdm2 **Forwarding IRPs and waiting for completion (continued)**

```
    IoCopyCurrentIrpStackLocationToNext(Irp);
    IoSetCompletionRoutine( Irp,
        (PIO_COMPLETION_ROUTINE)ForwardedIrpCompletionRoutine,
        (PVOID)&event, TRUE, TRUE, TRUE);

    NTSTATUS status = IoCallDriver( dx->NextStackDevice, Irp);
    if( status==STATUS_PENDING)
    {
        DebugPrintMsg("ForwardIrpAndWait: waiting for completion");
        KeWaitForSingleObject( &event, Executive, KernelMode, FALSE, NULL);
        status = Irp->IoStatus.Status;
    }
#if DBG
    if( status!=STATUS_SUCCESS)
        DebugPrint("ForwardIrpAndWait: completed %x",status);
#endif
    return status;
}


NTSTATUS ForwardedIrpCompletionRoutine( IN PDEVICE_OBJECT fdo, IN PIRP Irp,
    IN PKEVENT ev)
{
    KeSetEvent( ev, 0, FALSE);
    return STATUS_MORE_PROCESSING_REQUIRED;
}
```

Some completion routines include code to check the IRP *PendingReturned* field. If pending was initially returned as the status for this IRP, then *PendingReturned* is set TRUE. Some higher level drivers need to know as soon as possible that the IRP is still pending, so call IoMarkIrp-Pending.

```
if (Irp->PendingReturned)
    IoMarkIrpPending(Irp);
```

PnP Start Device Handler

The Wdm2 driver handles the PnP *Start Device* message in its PnpStartDeviceHandler routine shown in Listing 9.8. The state diagram in Figure 9.1 shows that the *Start Device* message is received either from the *Awaiting resources* state or the *Stopped* state.

PnpStartDeviceHandler calls ForwardIrpAndWait to have the *Start Device* request processed by all the drivers below it. PnpStartDeviceHandler completes the IRP with the error status code if it has failed during processing by these lower drivers.

PnpStartDeviceHandler can then do whatever it needs to do to start its device. Wdm2 delegates this to the routine StartDevice, which will be covered later. StartDevice is passed a pointer to the allocated resources, in the IRP stack Parameters.StartDevice.AllocatedResourcesTranslated field.

If StartDevice succeeds, it will have set the *GotResources* flag. PnpStartDeviceHandler also then clears the *Paused* and *IODisabled* flags.

The drivers above Wdm2 in the device stack could fail the *Start Device* IRP. If this happens, Wdm2 and the other lower drivers will be sent a *Remove Device* request.

Listing 9.8   Wdm2 **PnP start device handler**

```
NTSTATUS PnpStartDeviceHandler( IN PDEVICE_OBJECT fdo, IN PIRP Irp)
{
    DebugPrintMsg("PnpStartDeviceHandler");
    PWDM2_DEVICE_EXTENSION dx=(PWDM2_DEVICE_EXTENSION)fdo->DeviceExtension;
    PIO_STACK_LOCATION IrpStack = IoGetCurrentIrpStackLocation(Irp);

    NTSTATUS status = ForwardIrpAndWait( fdo, Irp);
    if( !NT_SUCCESS(status))
        return CompleteIrp( Irp, status, Irp->IoStatus.Information);

    DebugPrint("PnpStartDeviceHandler: post-processing");
    status = StartDevice( dx,
        IrpStack->Parameters.StartDevice.AllocatedResourcesTranslated);
    if( NT_SUCCESS(status))
    {
        dx->Paused = false;
        dx->IODisabled = false;
    }

    return CompleteIrp( Irp, status, 0);
}
```

Device Locking

When a user wants to remove a Wdm2 device, the PnP Manager always asks if it is all right to remove it using a *Query Remove* request. As described previously, Wdm2 agrees to a removal request only if there are no open handles to the device. Therefore, for the main *Remove Device* request, the Wdm2 driver can be certain that there is no I/O in progress on the device.

However, some devices are hot-pluggable (i.e., a user can ruthlessly pull out the plug, or it could be bashed out by mistake). In this case, a PnP driver does not receive a *Query Remove* message. Instead, in Windows 98, it simply gets a *Remove Device* message. Windows 2000 sends a *Surprise Removal* message first and then sends a *Remove Device* request when all the open handles are closed.

A driver needs to cope with surprise removals in the best way possible. A *Surprise Removal* IRP must succeed. Obviously, the driver needs to stop any further I/O as soon as possible. However, one or more I/O requests may be in progress or queued up. It is fairly straightforward to cancel all the IRPs in a device queue; see Chapter 16 for details. However, the best method to handle I/O IRPs that are in progress is more complicated.

If you cannot somehow fail any I/O requests in progress, the recommended solution is to wait for them to complete. The main IRP processing routines are more than likely to be using the device or device extension, so these structures cannot be deleted until it is certain that they will not be used again. The main IRP processing routines almost certainly encounter some sort of problem. Make sure that they can handle a device being removed. A routine might be expecting an interrupt. Chapter 17 later shows how two different types of timers can be used to provide a time-out for I/O requests.

What is best way to keep track of how many I/O requests are in progress? The answer is to laboriously keep track of the number of open I/O requests in a *UsageCount* field in the device extension. A call to LockDevice is made at the beginning of each IRP request to increment *UsageCount*. UnlockDevice is called when each IRP is completed to decrement *Usage-Count*.

UsageCount is set to one when the device is created. To remove or stop a device, an extra call to UnlockDevice is made. If there are no other IRPs in progress, this call should have decremented *UsageCount* to zero. If there are IRPs in progress, *UsageCount* will at this stage have a value greater than zero. However, when these IRPs finish, UnlockDevice will be called and so *UsageCount* will in due course become zero.

The *Remove Device* or *Stop Device* PnP IRP needs to know when UnlockDevice decrements *UsageCount* to zero. Another kernel event, *StoppingEvent*, is used for this purpose. UnlockDevice sets *StoppingEvent* into the signalled state when *UsageCount* drops to zero.

Listing 9.9 shows the PnpStopDevice routine that the Wdm2 driver uses to stop a device. It is called by the *Remove Device, Surprise Removal,* and *Stop Device* PnP IRP handlers. Pnp-StopDevice sets the *IODisabled* flag straightaway to stop any new requests from starting. The device is already stopped if *GotResources* is false, so no more processing is required.

PnpStopDevice resets *StoppingEvent* and then calls UnlockDevice twice. The first undoes the call to LockDevice at the start of the main PnP IRP handler. The second will reduce *UsageCount* to zero, either straightaway or when all the IRPs in progress complete. PnpStop-Device then waits for *StoppingEvent* to be set by UnlockDevice.

PnpStopDevice then calls the StopDevice routine, described later. StopDevice resets the *GotResources* flag. PnpStopDevice's last task is to call LockDevice again to increment *Usage-Count* again. The main PnP IRP handler calls UnlockDevice in due course to get *UsageCount* down to its correct value of zero.

Listing 9.9   PnpStopDevice **routine**

```
void PnpStopDevice( IN PWDM2_DEVICE_EXTENSION dx)
{
    // Stop I/O ASAP
    dx->IODisabled = true;

    // Do nothing if we're already stopped
    if( !dx->GotResources)
        return;
```

Listing 9.9   `PnpStopDevice` **routine (continued)**

```
    // Wait for any pending I/O operations to complete
    dx->Stopping = true;
    KeResetEvent( &dx->StoppingEvent );
    UnlockDevice(dx);
    UnlockDevice(dx);
    KeWaitForSingleObject( &dx->StoppingEvent, Executive, KernelMode,
        FALSE, NULL);
    DebugPrint("PnpStopDevice: All pending I/O completed");
    dx->Stopping = false;


    // Stop our device before passing down
    StopDevice(dx);


    // Bump usage count back up again
    LockDevice(dx);
    LockDevice(dx);
}
```

Listing 9.10 shows how a typical I/O IRP dispatch routine fits into the Wdm2 device Plug and Play handling. Initially, it checks the *IODisabled* flag and then it tries to lock the device using LockDevice. If LockDevice fails, the correct response is to return STATUS_DELETE_PEND-ING. The Wdm2 PnpStopDevice routine always sets the *IODisabled* flag first, so the dispatch routine will never return STATUS_DELETE_PENDING. This could be fixed by adding another suitable flag to the device extension.

The dispatch routine ends by completing the IRP and calling UnlockDevice. If the IRP is not completed straightaway, do not call UnlockDevice straightaway. Instead, wait until the IRP is completed or cancelled before calling UnlockDevice.

All dispatch routines should include calls to LockDevice when an IRP arrives and Unlock-Device when an IRP is completed or passed onto another driver. PnP and WMI IRP handlers should also call these routines.

Listing 9.10 **Dispatch routine entry and exit code**

```
NTSTATUS Wdm2Read( IN PDEVICE_OBJECT fdo, IN PIRP Irp)
{
    PWDM2_DEVICE_EXTENSION dx = (PWDM2_DEVICE_EXTENSION)fdo->DeviceExtension;
    if( dx->IODisabled)
        return CompleteIrp( Irp, STATUS_DEVICE_NOT_CONNECTED, 0);
    if (!LockDevice(dx))
        return CompleteIrp( Irp, STATUS_DELETE_PENDING, 0);


    // ...
```

Listing 9.10 Dispatch routine entry and exit code (continued)

```
    // Complete IRP
    CompleteIrp(Irp,status,BytesTxd);
    UnlockDevice(dx);
    return status;
}
```

PnpStopDevice also sets another flag in the device extension, *Stopping*, to true while it is waiting for all pending I/O to complete. Although not strictly necessary in Wdm2, it is another way of forcing an IRP routine to see if it is OK to start during the call to LockDevice.

Listing 9.11 shows the LockDevice and UnlockDevice routines. They use InterlockedIncrement and InterlockedDecrement calls to ensure that *UsageCount* is maintained safely in multiprocessor systems. If UnlockDevice finds that *UsageCount* has decremented to zero, it sets the *StoppingEvent* flag. LockDevice also checks the *Stopping* flag. If it is set, a PnP IRP is trying to stop the device and so the caller should not continue. In this case, *UsageCount* is decremented and *StoppingEvent* is set, if appropriate, before false is returned.

Listing 9.11 LockDevice **and** UnlockDevice **routines**

```
bool LockDevice( IN PWDM2_DEVICE_EXTENSION dx)
{
    InterlockedIncrement(&dx->UsageCount);

    if( dx->Stopping)
    {
        if( InterlockedDecrement(&dx->UsageCount)==0)
            KeSetEvent( &dx->StoppingEvent, 0, FALSE);
        return false;
    }
    return true;
}


void UnlockDevice( IN PWDM2_DEVICE_EXTENSION dx)
{
    LONG UsageCount = InterlockedDecrement(&dx->UsageCount);
    if( UsageCount==0)
    {
        DebugPrintMsg("UnlockDevice: setting StoppingEvent flag");
        KeSetEvent( &dx->StoppingEvent, 0, FALSE);
    }
}
```

To wrap up the last loose ends, the handlers for the *Stop Device* and *Surprise Removal* messages are exactly the same, as shown in Listing 9.12. They both simply call PnpStopDevice and call PnpDefaultHandler to pass the IRP down the stack.

Listing 9.12 Stop device handler

```
NTSTATUS PnpStopDeviceHandler( IN PDEVICE_OBJECT fdo, IN PIRP Irp)
{
    DebugPrintMsg("PnpStopDeviceHandler");
    PWDM2_DEVICE_EXTENSION dx=(PWDM2_DEVICE_EXTENSION)fdo->DeviceExtension;

    // Wait for I/O to complete and stop device
    PnpStopDevice(dx);

    return PnpDefaultHandler( fdo, Irp);
}
```

W2000 Device Locking

W2000 provides standard routines to replace our LockDevice and UnlockDevice routines, and the associated variables. You must provide an IO_REMOVE_LOCK field in your device extension. Initialise this field using IoInitializeRemoveLock function in your AddDevice routine. Replace all calls to LockDevice with IoAcquireRemoveLock and calls to UnlockDevice with IoReleaseRemoveLock. In the PnpStopDevice routine above, replace the code that waits for all pending I/O to complete with a call to IoReleaseRemoveLockAndWait.

The DDK recommended that you call IoAcquireRemoveLock whenever you pass out a reference to your code, e.g., a timer, DPC, or any other call-back routine. Call IoReleaseRemoveLock when these call-backs are disabled, i.e., usually in your *Remove Device* handler.

Getting Resource Assignments

A *Start Device* PnP IRP passes a list of the resources that have been assigned to the device. Although the Wdm2 driver never gets any resources, the code in DeviceIo.cpp goes through the steps to find these out. In addition, there is a lot of commented out code that shows how to use the resources. The WdmIo driver described in Chapters 15–17 show how hardware resources are actually used.

DeviceIo.cpp primarily contains the StartDevice and StopDevice routines. StartDevice must get the list of assigned resources, check that the resources are suitable, allocate them, and do any hardware-related initialization for the driver. If that all goes well, it should set the *GotResources* flag to true. StopDevice must release any hardware resources and reset the *GotResources* flag.

StartDevice calls RetrieveResources to find out which resources have been assigned to it by the PnP Configuration Manager. The *Start Device* PnP message passes two fields in the IRP stack Parameters.StartDevice structure, *AllocatedResources*, and *AllocatedResources-Translated*. *AllocatedResources* lists the resources in "raw form". This is how the device itself will see addresses, etc. Use *AllocatedResources* to program the device itself. Usually the "translated form" of the resource list, *AllocatedResourcesTranslated*, is of more interest to the driver. Use the *AllocatedResourcesTranslated* to connect to interrupt vectors, map I/O space, and memory.

If there are no resources assigned, W2000 sets these fields to NULL. In W98, a resource list is allocated but the "partial resource" count is zero.

Table 9.1 shows the raw and translated resource assignments for an I/O Port and two different interrupts in Windows 98 and Windows 2000.

Table 9.1   Resource assignments

I/O Port	Address			
AllocatedResources	378			
AllocatedResourcesTranslated	378			
Interrupt (W98)	**Vector**	**IRQL**	**Affinity**	**Mode**
AllocatedResources	7	7	1	Latched
AllocatedResourcesTranslated	37	20	1	Latched
Interrupt (W2000)	**Vector**	**IRQL**	**Affinity**	**Mode**
AllocatedResources	3	3	-1	Latched
AllocatedResourcesTranslated	33	24	1	Latched

Partial Resource Descriptors

A resource list is an array of *full resource descriptors*. A full resource descriptor has a *partial resource list*, an array of *partial resource descriptors*. The relevant structures (CM_RESOURCE_ LIST, CM_FULL_RESOURCE_DESCRIPTOR, CM_PARTIAL_RESOURCE_LIST, and CM_PARTIAL_ RESOURCE_DESCRIPTOR) are also covered in Chapter 18.

WDM drivers have only one full resource descriptor[1]. Therefore, inspect the array of partial resource descriptors to see what resources have been assigned. The resource descriptors are given in no particular order.

Table 9.2 shows the different types of resource information that can be found in a partial resource descriptor. The *Type* field specifies what resource is described. There is also a *Share- Disposition* field that specifies how the resource can be shared. One of the following values can be found: CmResourceShareDeviceExclusive, CmResourceShareDriverExclusive, or CmResourceShareShared.

Table 9.2   Partial resource descriptors

I/O Port	(Type==CmResourceTypePort)
PHYSICAL_ADDRESS u.Port.Start	Port Bus specific start address
ULONG u.Port.Length	Number of addresses
USHORT Flags	CM_RESOURCE_PORT_IO or CM_RESOURCE_PORT_MEMORY if you need to map the port into memory with MmMapIoSpace

---

1. Non-WDM NT style drivers may use more than one full resource descriptor.

Memory	**(Type==CmResourceTypeMemory)**
PHYSICAL_ADDRESS u.Memory.Start	Bus specific start address
ULONG u.Memory.Length	Number of addresses
USHORT Flags	CM_RESOURCE_MEMORY_READ_WRITE CM_RESOURCE_MEMORY_READ_ONLY CM_RESOURCE_MEMORY_WRITE_ONLY

Interrupt	**(Type==CmResourceTypeInterrupt)**
ULONG u.Interrupt.Level	The interrupt IRQL
ULONG u.Interrupt.Vector	The interrupt vector
ULONG u.Interrupt.Affinity	The set of processors to which the interrupt is dispatched.
USHORT Flags	CM_RESOURCE_INTERRUPT_LEVEL_SENSITIVE CM_RESOURCE_INTERRUPT_LATCHED

DMA	**(Type==CmResourceTypeDma)**
ULONG u.Dma.Channel	System DMA controller channel
ULONG u.Dma.Port	MCA type device port
USHORT Flags	See the DDK

Other resource types	
UCHAR Type	CmResourceTypeDeviceSpecific CmResourceTypeBusNumber CmResourceTypeDevicePrivate CmResourceTypePcCardConfig

The RetrieveResources code shown in Listing 9.13 extracts and checks the resources that have been assigned for this device. It first checks whether any resources have been assigned. As Wdm2 does not need any resources, it simply returns STATUS_SUCCESS if there are no resources.

RetrieveResources then goes through the partial resource descriptor list checking for I/O Port, Memory, and Interrupt resource types. The assigned information is stored in the device extension and printed out using DebugPrint. Other resource types result in an error.

Listing 9.13 RetrieveResources **routine**

```
NTSTATUS RetrieveResources( IN PWDM2_DEVICE_EXTENSION dx,
                    IN PCM_RESOURCE_LIST AllocatedResourcesTranslated)
{
    if( AllocatedResourcesTranslated==NULL ||
        AllocatedResourcesTranslated->Count==0)
    {
        DebugPrintMsg("RetrieveResources: No allocated translated resources");
        return STATUS_SUCCESS;    // or whatever
    }


    // Get to actual resources
```

Listing 9.13 RetrieveResources **routine (continued)**

```
PCM_PARTIAL_RESOURCE_LIST list =
            &AllocatedResourcesTranslated->List[0].PartialResourceList;
PCM_PARTIAL_RESOURCE_DESCRIPTOR resource = list->PartialDescriptors;
ULONG NumResources = list->Count;

DebugPrint("RetrieveResources: %d resource lists %d resources",
            AllocatedResourcesTranslated->Count, NumResources);

bool GotError = false;

// Clear dx
dx->GotInterrupt = false;
dx->GotPortOrMemory = false;
dx->PortInIOSpace = false;
dx->PortNeedsMapping = false;

// Go through each allocated resource
for( ULONG i=0; i<NumResources; i++,resource++)
{
    switch( resource->Type)
    {
    case CmResourceTypePort:
        if( dx->GotPortOrMemory) { GotError = true; break; }
        dx->GotPortOrMemory = true;
        dx->PortStartAddress = resource->u.Port.Start;
        dx->PortLength = resource->u.Port.Length;
        dx->PortNeedsMapping = (resource->Flags & CM_RESOURCE_PORT_IO)==0;
        dx->PortInIOSpace = !dx->PortNeedsMapping;
        DebugPrint("RetrieveResources: Port %x%x Length %d NeedsMapping %d",
            dx->PortStartAddress.HighPart, dx->PortStartAddress.LowPart,
            dx->PortLength, dx->PortNeedsMapping);
        break;

    case CmResourceTypeInterrupt:
        dx->GotInterrupt = true;
        dx->Irql = (KIRQL)resource->u.Interrupt.Level;
        dx->Vector = resource->u.Interrupt.Vector;
        dx->Affinity = resource->u.Interrupt.Affinity;
        dx->Mode = (resource->Flags == CM_RESOURCE_INTERRUPT_LATCHED)
                    ? Latched : LevelSensitive;
```

Listing 9.13 RetrieveResources **routine (continued)**

```
        DebugPrint("RetrieveResources: Interrupt vector
            %x IRQL %d Affinity %d Mode %d",
            dx->Vector, dx->Irql, dx->Affinity, dx->Mode);
        break;

    case CmResourceTypeMemory:
        if( dx->GotPortOrMemory) { GotError = true; break; }
        dx->GotPortOrMemory = true;
        dx->PortStartAddress = resource->u.Memory.Start;
        dx->PortLength = resource->u.Memory.Length;
        dx->PortNeedsMapping = true;
        DebugPrint("RetrieveResources: Memory %x%x Length %d",
            dx->PortStartAddress.HighPart, dx->PortStartAddress.LowPart,
            dx->PortLength);
        break;

    case CmResourceTypeDma:
    case CmResourceTypeDeviceSpecific:
    case CmResourceTypeBusNumber:
    default:
        DebugPrint("RetrieveResources: Unrecognised resource type %d",
            resource->Type);
        GotError = true;
        break;
    }
}

// Check we've got the resources we need
if( GotError /*|| !GotPortOrMemory || !GotInterrupt*/)
    return STATUS_DEVICE_CONFIGURATION_ERROR;

return STATUS_SUCCESS;
}
```

Allocating Resources

The StartDevice routine now allocates the system resources it requires. The Wdm2 driver does not need or expect any resources, so this code is commented out. The WdmIo driver covered in Chapters 15–17 does use system resources. These chapters cover Critical section routines and Interrupts in detail. However, I will briefly introduce these topics now.

For memory mapped I/O ports (with the CM_RESOURCE_PORT_MEMORY flag bit set) you must call MmMapIoSpace to get a pointer that can be used by a driver. Do not forget to call MmUnmap-IoSpace when the device is stopped.

```
dx->PortBase =
    (PUCHAR)MmMapIoSpace( dx->PortStartAddress, dx->PortLength, MmNonCached);
```

For ordinary I/O ports and memory, simply use the low 32 bits of the *PortStartAddress*[2].

```
dx->PortBase = (PUCHAR)dx->PortStartAddress.LowPart;
```

For interrupts, IoConnectInterrupt is called to install an interrupt handler. You must be ready to handle an interrupt straightaway, so make sure that everything is set up correctly. It is common to be able to disable interrupts by writing some value to a device register. Write a DisableDeviceInterrupts routine to disable interrupts before calling IoConnectInterrupt and call your EnableDeviceInterrupts routine when you are ready to receive interrupts.

Any code that tries to access some real hardware must synchronize its activities with the interrupt handler. It is no good having an interrupt during a complicated device access procedure. Critical section routines solve this problem. You call KeSynchronizeExecution passing the name of the function you want called. KeSynchronizeExecution raises the IRQL to the correct interrupt level and calls your routine. When it has completed, the IRQL is lowered again. Critical section routines obviously need to be in nonpaged memory and cannot access paged memory.

EnableDeviceInterrupts and DisableDeviceInterrupts are usually Critical section routines and will usually need to be called via KeSynchronizeExecution.

Finally, StartDevice powers up its device, as described in the next chapter.

Port and Memory I/O

There are several standard kernel routines to access I/O ports and memory, as this code snippet shows.

```
void WriteByte( IN PWDM2_DEVICE_EXTENSION dx, IN ULONG offset, IN UCHAR byte)
{
    if (dx->PortInIOSpace)
        WRITE_PORT_UCHAR(dx->PortBase+offset, byte);
    else
        WRITE_REGISTER_UCHAR(dx->PortBase+offset, byte);
}
```

Read data using the routines with READ in their name and write data with WRITE routines. PORT routines access I/O registers in I/O port space, while REGISTER routines access registers in memory space. Each type has UCHAR, USHORT, and ULONG variants. Finally, BUFFER variants transfer more than one data value. For example, use the following code to read a set of ULONG values from I/O port space into a buffer.

```
READ_PORT_BUFFER_ULONG( PortBase, Buffer, Count);
```

---

2. Obviously, this may well change in 64-bit systems.

As mentioned previously, synchronize all your hardware accesses with any interrupt routines using Critical section routines and KeSynchronizeExecution. Chapter 16 covers Critical section routines.

Device Access

Most WDM drivers do not, in fact, need to access I/O ports and memory or handle interrupts. Instead, they access their devices using the facilities provided by class drivers. For example, a USB client driver uses the oft-mentioned USB Request Blocks (URBs) to access its device.

When a USB client driver handles a PnP *Start Device* message, it first waits for the IRP to be processed by lower level drivers. It then typically issues one or more URBs to its device before completing its IRP.

For *Stop Device* messages, a USB client driver might well want to issue one or more URBs to its device before the IRP is sent down the stack.

Testing Wdm2

The *Wdm2Test* Win32 console application in the Wdm2\exe subdirectory of the book software tests the Wdm2 driver. *Wdm2Test* is the same as *Wdm1Test* with only one change, apart from referencing the Wdm2 driver. *Wdm2Test* halts halfway through, waiting for the user to press a key. While it is waiting, it has a handle to the first Wdm2 device still open.

First, install a Wdm2 device using one of the installation INF files in the book software Wdm2\sys directory. It does not matter whether you use the free or checked build, although the checked build produces *DebugPrint* trace output.

Wdm2Test tests to see if a *Query Remove* request is rejected while there are any open handles to a Wdm2 device. Run *Wdm2Test* but do not press a key when it stops half way through. Now try to remove the Wdm2 device or reinstall its driver. The request should be rejected. Windows will state that the system must be restarted for the operation to complete.

Check that the Wdm2 driver can be removed or reinstalled when *Wdm2Test* has completed.

It is very difficult to test the other new aspects of Plug and Play support in the Wdm2 driver. First, *Stop Device* and *Surprise Removal* requests should never be issued for virtual devices that have no resources. Second, it is not possible to suddenly remove a Wdm2 device in such a way that the Wdm2 driver has to wait for pending I/O to complete. Do appropriate tests for drivers that have resources and can be suddenly removed.

If the test for open handles is not made, then W2000 in fact will still not allow the device to be removed, as it must contain its own internal reference count for the device. However, W98 would let the device be removed. Any I/O requests on open handles would then simply fail.

Actual Plug and Play Messages

The *DebugPrint* output from Wdm2 shows exactly which Plug and Plug messages are sent by Windows during add device and remove device operations.

Adding a Device

The following PnP calls are made when a Wdm2 device is successfully added, or when the driver for a device is reinstalled. Two of the messages are issued only by Windows 2000.

```
AddDevice
(W2000) IRP_MN_QUERY_LEGACY_BUS_INFORMATION
IRP_MN_FILTER_RESOURCE_REQUIREMENTS
IRP_MN_START_DEVICE
IRP_MN_QUERY_CAPABILITIES
IRP_MN_QUERY_PNP_DEVICE_STATE
IRP_MN_QUERY_DEVICE_RELATIONS BusRelations
IRP_MN_QUERY_DEVICE_RELATIONS BusRelations (W2000)
```

Removing a Device

The following PnP messages are sent when a Wdm2 device is successfully removed, or when the driver for a device is reinstalled.

```
IRP_MN_QUERY_DEVICE_RELATIONS RemoveRelations
IRP_MN_QUERY_REMOVE_DEVICE
IRP_MN_REMOVE_DEVICE
```

Unknown Status Returns

It is interesting to note the IRP status values are returned by the lower Unknown driver when Wdm2 sends IRPs down the stack. Windows is supposed to set the IRP status return value to STATUS_NOT_SUPPORTED before it is issued to the top of the device stack. If Wdm2 sees this value on return from its PnpDefaultHandler routine, it means that the lower drivers have not processed the IRP or have deliberately not returned STATUS_NOT_SUPPORTED.

For Wdm2, W98 succeeds all the PnP IRPs that it receives.

For Wdm2, W2000 does not process IRP_MN_FILTER_RESOURCE_REQUIREMENTS, IRP_MN_QUERY_PNP_DEVICE_STATE, IRP_MN_QUERY_BUS_INFORMATION, and IRP_MN_QUERY_DEVICE_RELATIONS IRPs.

Other PnP IRPs

This section briefly describes the Plug and Plug minor function code IRPs that have not been described in full before. These IRPs are handled by PnP bus drivers. The notes for each function code indicate if it is possible for a function driver to intercept the IRP.

Most function drivers ignore all these PnP IRPs. However, a function driver that performs Power Management may well want to handle IRP_MN_QUERY_CAPABILITIES. The following chapter describes how this is done.

IRP_MN_DEVICE_USAGE_NOTIFICATION

This message tells a driver if its device is in the path of a paging, hibernation, or crash dump file. Do not allow a device to be removed until you are notified that no critical file is on its path.

IRP_MN_FILTER_RESOURCE_REQUIREMENTS

The Windows 2000 PnP Manager sends this IRP to a device stack so filter and function drivers can adjust the resources required by the device, if appropriate. Function, filter, and bus drivers can handle this request.

IRP_MN_QUERY_BUS_INFORMATION

The PnP Manager uses this IRP to request the type and instance number of a device's parent bus.

Bus drivers should handle this request for their child devices (PDOs). Function and filter drivers do not handle this IRP.

IRP_MN_QUERY_CAPABILITIES

The PnP Manager sends this IRP to get the capabilities of a device, such as whether the device can be locked or ejected, and various Power Management features. Function and filter drivers can handle this request if they alter the capabilities supported by the bus driver. Bus drivers must handle this request for their child devices.

This IRP is sent twice, both before and after function drivers are loaded and started.

A driver can send one of these IRPs down the stack to see what the bus driver capabilities are.

IRP_MN_QUERY_DEVICE_RELATIONS

This IRP asks how this device relates to other devices and comes in five different forms. All forms return an array of pointers to the relevant PDOs.

A BusRelations query asks for the PDOs of all the devices physically present on the bus. EjectionRelations asks which devices are also ejected if this device is ejected. PowerRelations asks which devices are also powered down when this device is powered down. RemovalRelations asks which devices must be removed when this device is removed. TargetDeviceRelation calls ObReferenceObject for the device PDO and returns the PDO.

IRP_MN_QUERY_DEVICE_TEXT

The PnP Manager uses this IRP to get a device's description or location information. Bus drivers must handle this request for their child devices if the bus supports this information. Function and filter drivers do not handle this IRP.

Parameters.QueryDeviceText.DeviceTextType is either DeviceTextDescription or DeviceTextLocationInformation. Parameters.QueryDeviceText.LocaleId is an LCID specifying the locale for the requested text.

IRP_MN_QUERY_ID

This IRP gets device, hardware, compatible, or instance IDs for a device, depending on whether `Parameters.QueryId.IdType` is `BusQueryDeviceID`, `BusQueryHardwareIDs`, `BusQuery-CompatibleIDs`, or `BusQueryInstanceID`.

IRP_MN_QUERY_INTERFACE

The `IRP_MN_QUERY_INTERFACE` request enables a driver to export a direct-call interface to other drivers.

IRP_MN_QUERY_PNP_DEVICE_STATE

The query asks the drivers in the device stack to set any of the state bits shown in Table 9.3. Be careful not to overwrite any bits that are set by other drivers.

Table 9.3   Query device state bits

PNP_DEVICE_DISABLED	The device is physically present but is disabled in hardware.
PNP_DEVICE_DONT_DISPLAY_IN_UI	Don't display the device in the user interface. The device is physically present but not usable in the current configuration.
PNP_DEVICE_FAILED	The device is present but not functioning correctly. When both this flag and PNP_DEVICE_RESOURCE_REQUIREMENTS_CHANGED are set, the device must be stopped before the PnP Manager assigns new hardware resources.
PNP_DEVICE_REMOVED	The device has been physically removed.
PNP_DEVICE_RESOURCE_REQUIREMENTS_CHANGED	The resource requirements for the device have changed.
PNP_DEVICE_NOT_DISABLEABLE	The device cannot be disabled.

If any of the state characteristics change after the initial query, a driver notifies the PnP Manager by calling `IoInvalidateDeviceState`. In response to a call to `IoInvalidateDevice-State`, the PnP Manager queries the device's `PNP_DEVICE_STATE` again.

IRP_MN_QUERY_RESOURCE_REQUIREMENTS

The PnP Manager uses this IRP to get a device's resource requirements list. Bus drivers must handle this request for their child devices that require hardware resources. Function and filter drivers do not handle this IRP.

IRP_MN_QUERY_RESOURCES

The PnP Manager uses this IRP to get a device's boot configuration resources. Bus drivers must handle this request for their child devices that require hardware resources. Function and filter drivers do not handle this IRP.

IRP_MN_READ_CONFIG

Bus drivers for buses with configuration space must return the relevant information for their child devices. Filter and function drivers do not handle this request.

IRP_MN_SET_LOCK

This request is used to lock or unlock a device. Bus drivers must handle this IRP for a child device that supports device locking. Function and filter drivers do not handle this request.

IRP_MN_WRITE_CONFIG

Bus drivers must write the given information into the configuration space of the child device. Function and filter drivers do not handle this request.

Plug and Play Notification

Plug and Play Notification informs Win32 programs and device drivers of device change events that interest them, such as device arrival and removal. They can also refuse requests to remove a device. PnP Notification uses a device interface GUID or a file handle to identify which devices are of interest.

Win32 applications and device drivers register to receive PnP notifications so that they can cope if a device is about to be removed, or if they want to use a device that has just been added. For example, if a program is in the middle of a long transfer, it could refuse permission for a device to be removed. In addition, device drivers can be informed of all devices that expose the required device interface.

Win32 PnP Notification

A Win32 program calls RegisterDeviceNotification to register that it wants to receive PnP Notification device change messages. You must pass either a window handle or (in W2000 only) a service status handle. The NotificationFilter parameter points a structure that says what type of events you want to receive.

To receive events about devices with a particular device interface, pass a DEV_BROADCAST_ DEVICEINTERFACE structure pointer to RegisterDeviceNotification. Set the *dbcc_classguid* field to the device interface GUID of interest. Set the *dbcc_size* and *dbcc_devicetype* fields appropriately.

I have not tried it, but passing a DEV_BROADCAST_HANDLE structure pointer should let you receive events from one open file handle, including custom events.

Do not forget to call UnregisterDeviceNotification when you do not wish to receive any further PnP Notification events.

Device Change Message

Each PnP Notification event is sent to a Win32 program as a WM_DEVICECHANGE message. In MFC applications, this appears in an OnDeviceChange handler.

Table 9.4 lists the main event types returned in wParam. lParam may point to an appropriate structure. For device interface change events, it is a DEV_BROADCAST_DEVICEINTERFACE structure whose *dbcc_name* field contains the device filename.

For a WM_DEVICECHANGE message that asks permission to remove a device, the handler must return TRUE to agree or BROADCAST_QUERY_DENY[3] to deny the requests. Returning FALSE seems to have the same effect as returning TRUE.

Table 9.4   WM_DEVICECHANGE **event types**

DBT_CONFIGCHANGECANCELED	A request to change the current configuration (dock or undock) has been cancelled
DBT_CONFIGCHANGED	The current configuration has changed, due to a dock or undock
DBT_CUSTOMEVENT	A custom event has occurred
DBT_DEVICEARRIVAL	A device has been inserted and is now available
DBT_DEVICEQUERYREMOVE	Permission is requested to remove a device. Any application can deny this request and cancel the removal
DBT_DEVICEQUERYREMOVEFAILED	A request to remove a device has been cancelled
DBT_DEVICEREMOVEPENDING	A device is about to be removed. Cannot be denied
DBT_DEVICEREMOVECOMPLETE	A device has been removed
DBT_DEVICETYPESPECIFIC	A device-specific event has occurred
DBT_QUERYCHANGECONFIG	Permission is requested to change the current configuration (dock or undock)
DBT_DEVNODES_CHANGED	Device tree has changed
DBT_USERDEFINED	The meaning of this message is user-defined

Wdm2Notify **application**

Wdm2Notify is a Win32 MFC dialog application that displays any device change events for the Wdm2 device interface. It can be found in the Wdm2\Notify directory of the book software.

I had a little difficulty setting up the project so that it would compile and link correctly. As I was using VC++ 5, some of the header files and libraries were out of date. I therefore had to change the project settings to use the Platform SDK include directories first, before the Visual C++ directories. My Platform SDK was installed at D:\MSSDK, so you may need to change the project settings if you want to recompile *Wdm2Notify*. I also had to link to the version of user32.lib in the Platform SDK. If you are running a later version of Visual Studio, you may be able to undo these project changes.

---

3. BROADCAST_QUERY_DENY has a value of 0x424D5144, ASCII "BMQD", which I presume stands for Broadcast Message Query Deny!

Some of the device change structures are defined in dbt.h, so I included this header file in the *Wdm2Notify* main MFC header, stdafx.h. I also had to include the following line in the stdafx.h to ensure that the correct functions were declared.

```
#define WINVER 0x0500
```

Listing 9.14 shows the main *Wdm2Notify* PnP Notification routines in file Wdm2NotifyDlg.cpp. RegisterDeviceNotification is called in the dialog OnInitDialog routine and unregistered in DestroyWindow. Device change messages are handled in OnDeviceChange. The rest of the source code for *Wdm2Notify* can be found on the book software disk.

Listing 9.14 Wdm2Notify **device change routines**

```
BOOL CWdm2NotifyDlg::OnInitDialog()
{
    CDialog::OnInitDialog();

    // ...

    // Register for Wdm2 device interface changes
    DEV_BROADCAST_DEVICEINTERFACE dbch;
    dbch.dbcc_size = sizeof(dbch);
    dbch.dbcc_devicetype = DBT_DEVTYP_DEVICEINTERFACE;
    dbch.dbcc_classguid = WDM2_GUID;
    dbch.dbcc_name[0] = '\0';
    WdmNotificationHandle = RegisterDeviceNotification(
            GetSafeHwnd(), &dbch, DEVICE_NOTIFY_WINDOW_HANDLE);
    if( WdmNotificationHandle==NULL)
        GetDlgItem(IDC_STATUS)->SetWindowText(
            "Cannot register for Wdm2 class device notification");

    return TRUE;
}


BOOL CWdm2NotifyDlg::DestroyWindow()
{
    if( WdmNotificationHandle!=NULL)
    {
        UnregisterDeviceNotification(WdmNotificationHandle);
        WdmNotificationHandle = NULL;
    }
    return CDialog::DestroyWindow();
}


BOOL CWdm2NotifyDlg::OnDeviceChange( UINT nEventType, DWORD dwData )
```

Listing 9.14 `Wdm2Notify` **device change routines (continued)**

```
{
    CString Msg = "duh";
    switch( nEventType)
    {
    case DBT_CONFIGCHANGECANCELED:
        Msg.Format("DBT_CONFIGCHANGECANCELED"); break;
    case DBT_CONFIGCHANGED:
        Msg.Format("DBT_CONFIGCHANGED"); break;
    case DBT_CUSTOMEVENT:
        Msg.Format("DBT_CUSTOMEVENT"); break;
    case DBT_DEVICEARRIVAL:
        Msg.Format("DBT_DEVICEARRIVAL"); break;
    case DBT_DEVICEQUERYREMOVE:
        Msg.Format("DBT_DEVICEQUERYREMOVE"); break;
    case DBT_DEVICEQUERYREMOVEFAILED:
        Msg.Format("DBT_DEVICEQUERYREMOVEFAILED"); break;
    case DBT_DEVICEREMOVEPENDING:
        Msg.Format("DBT_DEVICEREMOVEPENDING"); break;
    case DBT_DEVICEREMOVECOMPLETE:
        Msg.Format("DBT_DEVICEREMOVECOMPLETE"); break;
    case DBT_DEVICETYPESPECIFIC:
        Msg.Format("DBT_DEVICETYPESPECIFIC"); break;
    case DBT_QUERYCHANGECONFIG:
        Msg.Format("DBT_QUERYCHANGECONFIG"); break;
    case DBT_DEVNODES_CHANGED:
        Msg.Format("DBT_DEVNODES_CHANGED"); break;
    case DBT_USERDEFINED:
        Msg.Format("DBT_USERDEFINED"); break;
    default:
        Msg.Format("Event type %d",nEventType);
    }


    PDEV_BROADCAST_DEVICEINTERFACE pdbch =
        (PDEV_BROADCAST_DEVICEINTERFACE)dwData;
    if( pdbch!=NULL && pdbch->dbcc_devicetype==DBT_DEVTYP_DEVICEINTERFACE)
    {
        CString Msg2;
        Msg2.Format("%s: %s",Msg,pdbch->dbcc_name);
        Msg = Msg2;
    }
```

Listing 9.14 Wdm2Notify **device change routines (continued)**

```
    CListBox* EventList = (CListBox*)GetDlgItem(IDC_EVENT_LIST);
    EventList->AddString(Msg);
    return TRUE;    // or BROADCAST_QUERY_DENY to deny a query remove
}
```

Running Wdm2Notify **in Windows 98**

Figure 9.2 shows some typical output in the *Wdm2Notify* window in Windows 98 as the Wdm2 driver is updated for an installed Wdm2 device. First the PnP Manager asks if it is OK for the existing Wdm2 device to be removed. Then it issues remove pending and remove complete messages. When the driver has been updated, the Wdm2 device is added again resulting in a device arrival message. Finally, a DBT_DEVNODES_CHANGED event indicates that the device tree has changed.

I am not sure why there are so many duplicate device change messages.

If *Wdm2Notify* returns BROADCAST_QUERY_DENY to the *Query Remove* message, the remove is correctly refused and a DBT_DEVICEQUERYREMOVEFAILED message is then issued.

Figure 9.2   Wdm2Notify **output in Windows 98**

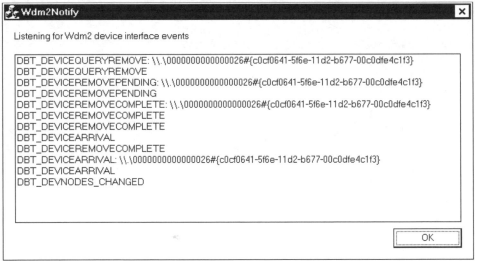

@ 1999 PHD Computer Consultants Ltd

Running *Wdm2Notify* **in Windows 2000**

The Beta 2 version of Windows 2000 produces output very different from *Wdm2Notify*. It issues only a DBT_DEVICEREMOVECOMPLETE message and then a DBT_DEVICEARRIVAL message. A *Query Remove* request is not issued and there are no repeat messages. I hope that the final version of W2000 will issue more PnP Notification device change messages to Win32 applications.

Device Driver PnP Notification

A driver uses PnP Notification to find devices with a particular device interface. A driver does this so that it can issue calls to the devices that it finds. It may wish to layer its own device on top of each found device. This technique is commonly used when finding Human Input Devices, and a full example of this is given in Chapter 23.

PnP Notification in device drivers works in a similar way to Win32 applications. Remember that this is one device driver asking for device change notifications about devices controlled by another driver.

The driver must call `IoRegisterPlugPlayNotification` to indicate which events it is interested in receiving, and eventually call `IoUnregisterPlugPlayNotification` when done. You must pass the name of a callback routine to `IoRegisterPlugPlayNotification`, along with a context pointer.

In device drivers, you can ask for three different categories of PnP Notification events. If you ask for `EventCategoryDeviceInterfaceChange` events your callback routine is passed a `DEVICE_INTERFACE_CHANGE_NOTIFICATION` structure that notifies you only of *device removal* and *device arrival* events, along with the appropriate symbolic link. Specify the relevant GUID in the `EventCategoryData` parameter of `IoRegisterPlugPlayNotification`. If you specify `PNPNOTIFY_DEVICE_INTERFACE_INCLUDE_EXISTING_INTERFACES` for the `EventCategoryFlags` parameter, *arrival* events are also sent straightaway for any existing devices that have a matching device interface GUID.

If you register for `EventCategoryHardwareProfileChange` events, you receive *query change*, *change complete*, and *change cancelled* events.

Finally, if you ask for `EventCategoryTargetDeviceChange` events, you receive *query remove*, *removal completes*, and *remove cancelled* messages for one specific device. You must supply a pointer to the relevant file object as the `EventCategoryData` parameter of `IoRegisterPlugPlayNotification`. I assume that you can reject *query remove* requests.

Driver PnP Notification is particularly important for client drivers that use device interfaces exposed by the system class drivers. For example, kernel mode Human Input Device (HID) client drivers can use PnP Notification to identify all installed HID devices. See Chapter 23 for a full example of PnP Notification in device drivers.

As mentioned earlier, device stacks cannot be changed once they are built. PnP Notification lets you find existing devices and effectively layer on top of them.

Notification Request Driver Interactions

As might expect, Win32 programs and device drivers that register for PnP Notification events will effect the operation of a driver. *Query remove* requests are processed by PnP Notification applications and drivers before the main driver receives an `IRP_MN_QUERY_REMOVE_DEVICE` IRP.

Cancel remove notification messages are sent to applications and other drivers after the main driver has processed its `IRP_MN_CANCEL_REMOVE_DEVICE` message. *Remove pending* messages are sent before a device is removed (or surprise removed) and a *remove complete* notification message is sent afterwards.

Advanced Plug and Play

This section looks at some advanced Plug and Play topics.

Bus Drivers

This section looks briefly at the job a bus driver must do. Please consult the DDK documentation for full details. The W2000 DDK source code `kernel\serenum` example shows how a bus driver is coded.

Bus drivers must manage their device objects carefully. A bus driver has a Functional Device Object (FDO) for each instance of a bus. A bus driver detects any devices on its bus and creates child device objects for each device. These are actually the Physical Device Objects (PDOs) to which higher driver layers attach.

The IRP dispatch routines in a bus driver handle requests for both FDOs and child PDOs. Use a flag in a common portion of the FDO and PDO device extensions to indicate whether the device is an FDO or a child PDO.

If a bus driver detects a Plug and Play device arriving or removing, it must call `IoInvalidateDeviceRelations`. The PnP Manager then sends an `IRP_MN_QUERY_DEVICE_RELATIONS` PnP IRP to get the new details. Non-Plug and Play devices can be reported in W2000 using `IoReportDetectedDevice`.

Creating Child Devices

A bus driver must enumerate its own bus to find any attached devices. It must also keep checking to see if any devices are removed or added. A bus driver might use a system worker thread, described in chapter 14, to schedule its device checks.

A bus driver must call `IoInvalidateDeviceRelations` if it detects any new devices or if one of its child devices has been removed. This call forces the PnP Manager to issue an IRP_MN_QUERY_DEVICE_RELATIONS `BusRelations` IRP to the bus driver FDO. The bus driver then enumerates its bus fully. It calls `IoCreateDevice` to make the child PDO for each new child device.[4] The bus driver returns a list of all valid child PDOs. The PnP Manager works out what PDOs have been added or removed.

If a new device has arrived, the PnP Manager starts to configure this device and build a new device stack. It issues various PnP IRPs to this child PDO. The IRP_MN_QUERY_ID IRP requests the hardware and compatible ids for the device. The PnP Manager also issues IRP_MN_QUERY_CAPABILITIES, IRP_MN_QUERY_DEVICE_TEXT, IRP_MN_QUERY_RESOURCES and IRP_MN_QUERY_RESOURCE_REQUIREMENTS requests to get further information about the device. Chapter 11 shows how the PnP Manager uses all this information and INF files to determine which drivers should be loaded on top of the child PDO. The stack drivers' `AddDevice` routines are run and *Start Device* IRPs issued as usual.

If a function driver decides that a device is no longer present, it can call `IoInvalidateDeviceState`. The PnP Manager issues an `IRP_MN_QUERY_PNP_DEVICE_STATE` request. If it finds the `PNP_DEVICE_FAILED` state bit set, then it issues a (surprise) remove PnP message.

---

4. I think PDOs must be created with FILE_AUTOGENERATED_DEVICE_NAME as the Device-Characteristics parameter to `IoCreateDevice`.

FDO IRP Handling

Normally, a bus driver FDO will receive only AddDevice and Plug and Play IRPs. As mentioned previously, it should enumerate its bus when it receives an IRP_MN_QUERY_DEVICE_ RELATIONS BusRelations PnP IRP. It should return some meaningful information for an IRP_ MN_QUERY_BUS_INFORMATION request.

The bus driver FDO should also process all the standard Plug and Play IRPs in the usual way. Make sure that it can cope if it receives a *Remove Device* IRP while child devices are still attached to the bus. The DDK documentation says that *Query Remove* requests are sent to the child devices before being sent to the main FDO.

Child PDO PnP Handling

The child PDO receives IRPs that have been passed down the device stack. Note that it will not receive an AddDevice call. AddDevice is called to create a new device. The bus driver has already done this job, so there is no need for an AddDevice call for a child PDO.

The child PDO handles all the PnP IRPs that come down the device stack. A bus driver handles some of these IRPs differently from function and filter drivers. In particular, it must always complete the PnP IRP so that the IRP can begin its journey up the stack. However, if it does not handle a particular minor function code, it should not change the value in the IoStatus.status field as the PnP Manager sets this field to STATUS_NOT_SUPPORTED in advance.

A bus driver should consider handling these IRPs correctly for the child PDO: IRP_MN_ QUERY_CAPABILITIES, IRP_MN_QUERY_DEVICE_TEXT, IRP_MN_QUERY_ID, and IRP_MN_QUERY_ DEVICE_RELATIONS.

Child PDO I/O IRP Handling

The child PDO also receives any other IRPs that come down the stack. These IRPs will be the create, close, read, write, or IOCTL I/O requests that do a useful job.

The child PDO handler for these requests could interact with the bus directly. However, it is possible that all bus operations have to be performed in a controlled manner, rather than have all child PDOs trying to access it at once. One option, therefore, is to pass any incoming I/O IRPs to the bus driver FDO. The bus driver FDO handler for these IRPs can do whatever is necessary (e.g., serialize all I/O requests).

If you use this technique, you will obviously need to store a pointer to the parent bus driver FDO in the child PDO device extension. Use IoCallDriver as usual to pass on any IRPs.

There is one thing to remember if using this technique. The number of IRP stack locations for the child PDOs will have to be increased as the IRPs will be passed down through more driver layers. Simply add the FDO *Stacksize* field to the child PDO *Stacksize* after IoCreat- eDevice has returned.

Sending PnP IRPs

Most PnP IRPs are sent by the PnP Manager, but some can be sent by drivers (e.g., IRP_MN_ QUERY_INTERFACE). A driver must send a PnP IRP to the driver at the top of the device stack. Call IoGetAttachedDeviceReference to get a pointer to the device object for the driver at the top of the device stack.

Device Properties

In AddDevice (or elsewhere) you can ask for various properties of the PDO using IoGetDevi-ceProperty. The property information is stored in the registry but is best accessed using IoGetDeviceProperty. The DeviceProperty parameter has the information you want. You also supply a buffer that is filled with the appropriate information.

Among other things, you can ask for the device's hardware ID and the GUID of the bus driver and the device's setup class. Sometimes you may need to call IoGetDeviceProperty twice. The first call informs you of the size of buffer required, while the second call actually gets the data.

Conclusion

This chapter has looked in detail at a practical Plug and Play function driver implementation for Wdm2 devices. The Wdm2 driver now handles all its PnP operations safely. The *Wdm2Test* Win32 test program checked that the Wdm2 driver could not be removed or replaced while its devices had open handles.

We also looked at how Plug and Play Notification can inform Win32 and device drivers of device change events. The *Wdm2Notify* Win32 test program displays any device change events for the Wdm2 device interface. Finally, bus drivers were briefly described.

The next chapter completes the discussion of the Wdm2 driver by describing its Power Management features.

10

Chapter 10

Power Management

This chapter investigates how to support Power Management in WDM device drivers. NT style drivers in Windows 2000 and Windows 98 can also support Power Management. A device can change its power usage in response to system power state changes, and can reduce its own power usage when it has been idle for a while. A sleeping device can wake the system up, such as when a modem receives an incoming call.

The Power Management aspects of the example Wdm2 driver are explained. The Win32 application *Wdm2Power* can be used to test Power Management facilities.

The Power Picture

Device drivers play a crucial role in helping Windows make the most of the available power resources. This means reducing power consumption in low power situations, such as when running on batteries. Drivers that support power management can also help to minimize startup and shutdown times. Finally, turning off some devices reduces system noise and turning off the display can save a monitor screen.

Most desk-based computers only run on AC power. However, they may have one or more Uninterruptible Power Supplies (UPS) with batteries to provide power in case of emergencies when main power is lost. Most portable computers are able to run off both battery (DC) and AC power.

The system should detect when the computer is running from battery power and reduce power consumption wherever possible. If not in use, it should *suspend* itself into a *standby* sleep mode. When running from batteries, the system should shut itself down in an orderly fashion when the battery condition becomes critical. When AC power is available, batteries should be recharged, if possible. When devices are idle for a certain amount of time, they

203

should turn themselves off to reduce noise and power consumption, even when running from AC power.

OnNow is a term for a PC that appears to be off but is always on, so it can respond immediately to user or other requests. When a computer is **sleeping** like this, it must shut down all inessential devices. Make sure that your drivers cooperate in this task to save energy. An alternative for energy-conscious speed freaks in Windows 2000 is to make their computers *hibernate*, a state in which a memory image is stored on disk to make startup quicker.

ACPI

Obviously, some of these desired features are implemented in hardware, such as smooth switching between AC and DC power. However, Windows' Power Manager can take part in the decision making process if the hardware meets the Advanced Configuration and Power Interface (ACPI) specification, found at `www.teleport.com/~acpi`[1].

Most drivers do not need to interact with ACPI. Instead, the Power Manager and Plug and Play Manager use the ACPI facilities to manage system power and enumerate hardware devices. Non-ACPI systems use a BIOS enumerator to find all PnP devices.

For devices described by an ACPI BIOS, the Power Manager inserts an ACPI filter between the bus driver and higher-level drivers. This ACPI filter driver powers devices on and off according to its power policy. The ACPI filter is transparent to other drivers and is not present on non-ACPI machines. In addition, an ACPI bus driver implements the Plug and Play root device for the whole computer, as described in Chapter 8.

Win32 Power Management

The Control Panel Power Management applet is where users change their power options. The settings here determine the time delay before Windows goes into standby, turns off hard disks, etc. In W2000, you can enable its Hibernate option.

User mode applications might be interested in receiving power events. If an application does background processing, it should stop this work if the system goes to sleep. Alternatively, it can insist that the system does not sleep.

Wdm2Power Application

The *Wdm2Power* application in the `Wdm2\Power` directory of the book software shows how to handle the following aspects of Power Management in Win32.

- Capturing `WM_POWERBROADCAST` messages
- Attempting to suspend or hibernate the system with `SetSystemPowerState`
- Getting the system power status using `GetSystemPowerStatus`
- Getting the hard disk status using `GetDevicePowerState`

You may find *Wdm2Power* useful when testing the Power Management capabilities of your driver. Figure 10.1 shows how *Wdm2Power* looks in Windows 98 on a desktop system.

I found that it was necessary to alter some VC++ 5 settings to be able to compile *Wdm2Power*. I needed to use the Platform SDK standard include files and libraries. This

---

1. Note that there are various optional elements to the ACPI, but the PC99 specification makes several of these mandatory.

meant altering the Directories in the Tools+Options menu. I had to put the Platform SDK include directory at the head of the "Include files" directories listing (i.e., `D:\MSSDK\INCLUDE` on my PC). Similarly, I had to put its library directory at the head of the "Library files" directories listing (e.g., `D:\MSSDK\LIB`). This may effect your other projects, so change these settings back afterwards.

Figure 10.1   Example *Wdm2Power* output

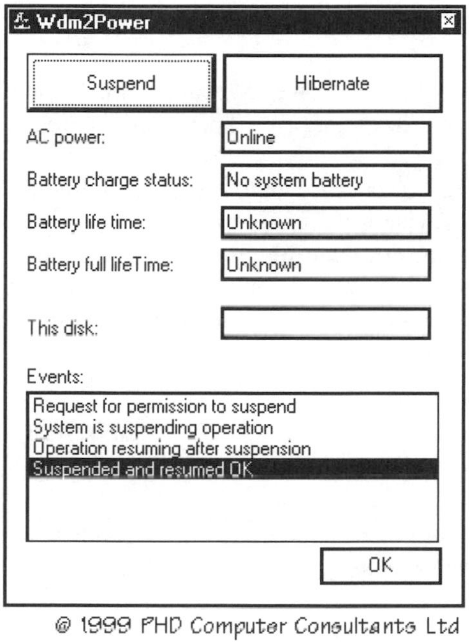

@ 1999 PHD Computer Consultants Ltd

The system broadcasts `WM_POWERBROADCAST` messages to indicate various power management events or requests. In the ***Wdm2Power*** MFC dialog box application these are handled in the `OnPowerBroadcast` method in `Wdm2PowerDlg.cpp`. It simply adds the event information to the window list box.

Both the Suspend and Hibernate options eventually call the `SetSystemPowerState` function in the `SuspendOrHibernate` dialog method. In W2000 you have to enable the shutdown privilege for this process. The Hibernate option is only available in Windows 2000.

Wdm2Power sets a timer that calls its `GetPowerStatus` method every second. This calls the `GetSystemPowerStatus` function to get the system power state and `GetDevicePowerState` to determine if *Wdm2Power*'s disk has been spun down. On my system, `GetDevicePowerState` always returned `FALSE`, which I presume means that the file system does not handle the IOCTL command `IOCTL_GET_DEVICE_POWER_STATUS`.

Win32 applications can also ask that the system not be put to sleep using the `SetThreadExecutionState` function. Alternatively, they can request that the lowest latency sleep be used using `RegisterWakeupLatency`. Finally, they can request that the system wake up at a certain time or in response to a device event, such as a modem ring.

Battery Miniclass Drivers

The Power Manager uses battery miniclass drivers to provide an interface to batteries. These miniclass drivers must fit in with the specification given in the DDK documentation. For example, it must register its various callback entry points using `BatteryClassInitializeDevice`.

Battery miniclass drivers must include routines to support PnP and to support battery management and monitoring.

System Power Policies

The rest of this chapter looks at how device drivers handle Power Management. Some classes of device ought to have certain Power Management characteristics. These details are found in the Device Class Power Management specification.

System power states indicate the overall energy usage of a whole system, while a *device* power state says how much energy an individual device is using. Even though the system may be fully powered up, a device can power itself down. For example, if a battery-operated computer is fully on but the hard disk is not being used, the disk driver may reasonably decide to power the disk down to save energy. Conversely, a sleeping computer may keep its modem powered up if it is waiting for incoming faxes.

System and Device States

Windows defines six *system power states* and four *device power states*. Please be very clear about whether you are referring to system or device power states.

A `POWER_STATE` variable represents either a system or device power state. The `POWER_STATE` typedef is a union of the two enum typedefs, `SYSTEM_POWER_STATE` and `DEVICE_POWER_STATE`.

```
typedef union _POWER_STATE {
    SYSTEM_POWER_STATE SystemState;
    DEVICE_POWER_STATE DeviceState;
} POWER_STATE, *PPOWER_STATE;
```

System Power States

Table 10.1 shows the system power states along with the `SYSTEM_POWER_STATE` enum names and values. Sleeping state S1 has the lowest latency so the system can return to the fully on state in the quickest time possible. States S2 and S3 gradually increase power up latency and decrease power consumption.

Table 10.1   System power states

ACPI State	Description	Enum **name**
S0	Working/Fully on	PowerSystemWorking (1)
S1	Sleeping	PowerSystemSleeping1 (2)
S2	Sleeping	PowerSystemSleeping2 (3)
S3	Sleeping	PowerSystemSleeping3 (4)
S4	Hibernating: Off, except for trickle current to the power button and similar devices.	PowerSystemHibernate (5)
S5	Shutdown/Off	PowerSystemShutdown (6)

Windows moves to and from only S0. There are never changes between the states S1 to S5 (e.g., from S4 to S2). Windows always assumes that it can power up to S0, so it does not have to ask drivers for permission (using the IRP_MN_QUERY_POWER request).

The IRP stack *Parameters.Power.ShutdownType* value gives extra information about system state changes. It is particularly useful for transitions to S5. PowerActionShutdownReset indicates a reboot while PowerActionShutdownOff means the computer is being switched off.

Device Power States

Table 10.2 shows the available device power states along with the DEVICE_POWER_STATE enum names and values. Again, state D1 has lower latency than D2.

Table 10.2   Device power states

State	Description	Enum **name**
D0	Fully working	PowerDeviceD0 (1)
D1	Sleeping	PowerDeviceD1 (2)
D2	Sleeping	PowerDeviceD2 (3)
D3	Off	PowerDeviceD3 (4)

Both the state values increase from 1 the sleepier they are.

A device need not support all the device states. D0 for on and D3 for off are a basic minimum. A device is not limited in its state transitions, so a change from D1 to D3 is possible.

Later, I describe how each device (or its bus driver) provides a table for device states for each system state.

Power IRPs

A driver's Power Management routines revolve around the Power IRP, IRP_MJ_POWER, by both handling it and generating it when necessary. There are four minor function code variants of the Power IRP shown in Table 10.3.

Table 10.3   `IRP_MN_POWER` **minor function codes**

Minor function code	Description
IRP_MN_SET_POWER	Set system or device power state
IRP_MN_QUERY_POWER	Ask if a system or device state change is OK
IRP_MN_WAIT_WAKE	Wake computer in response to an external event
IRP_MN_POWER_SEQUENCE	Send this IRP to determine whether your device actually entered a specific power state

The Power Manager maintains a separate internal queue of Power IRPs. This ensures that there is only one Set System Power IRP being processed in the system. It also ensures that there is only one Set Device Power IRP running for each device. These rules ensure that power transitions are handled smoothly. As each device is powered up, it may require an inrush of energy. Handling Power IRPs one at a time ensures that the peak power required is kept as low as possible.

As the Power Manager has its own queue of IRPs, do not use the standard I/O Manager routines when handling IRPs. Use `PoCallDriver` instead of `IoCallDriver` to call the next driver down the stack.

You must tell the Power Manager when you have finished processing a power IRP so that it can start the next. If you are simply passing an IRP down the stack (with no completion routine) then you should call the `PoStartNextPowerIrp` function before you skip or copy the current IRP stack location. If you use a completion routine, it must also usually call `PoStartNextPowerIrp`. Full examples of these calls are given later.

Processing Power IRPs

A Query Power IRP request is sent to see if a state change is acceptable to all the drivers in the device stack. A Set Power IRP is then issued to actually make the change.

You can send yourself a Set Device Power State IRP at any time. However, only the Power Manager can send Set System Power State IRPs. To process a Set System Power State IRP, you must send yourself a Set Device Power State IRP. That's right — you must ask for a Power IRP to be sent to the top of your device stack.

Handling Device Power IRPs

Handling a Set Device Power State is relatively straightforward. You must power down your device before all the lower drivers. Conversely, power your device up after all the lower drivers have powered up. This means setting a completion routine and doing the power up there.

Handling System Power IRPs

Handling a Set System Power State is a different kettle of fish. If you receive a Set System Power IRP, you must first determine the equivalent device power state. The `Wdm2` driver is put in the fully on device state `D0` for the fully on system state `S0`. For all other system states, the `Wdm2` device state is fully off, `D3`.

If the current device state is not the same as the required device state, you must act to change it. As before, powering down requires that you change the device state before you pass the Set System Power State down to the lower drivers. Power up after all the lower drivers have had their say.

You must change device power state by issuing a Set Device Power State IRP to yourself. You must then wait for this IRP to complete. You can then continue processing the Set system state IRP.

Same System Power

Suppose you receive a Set System Power State IRP. You must work out the corresponding device power state. If your device is already at this power level, then all you need to do is pass the IRP to the lower drivers.

System Powering Down

If you decide that your device needs to power down, then you must send yourself a Set Device Power IRP before you pass down the system IRP. Figure 10.2 illustrates this scenario. It assumes that the first Query Set System Power State IRP has completed OK, at ①.

In the figure, the Set System State IRP handler decides that it must power its device down at ②. It sends a Set Device State IRP to itself, and sets a completion routine so that it knows when this second IRP has been completed.

In due course, your driver will receive the Set Device State IRP (that you sent yourself) at ③. Your driver will decide again that it must power down. It does this at ④ and then sends the IRP down to the lower drivers at ⑤.

When the Set Device State IRP has been processed by all the lower drivers, the completion routine is called at ⑥. This signals to the original IRP handler that it can continue, at ② again. All this has to do is pass the IRP down to the lower drivers at ⑦.

Figure 10.2   Power down system processing

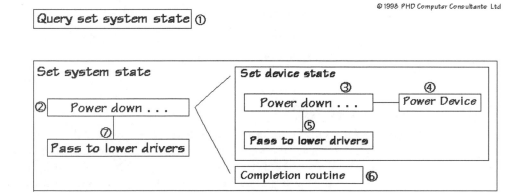

© 1998 PHD Computer Consultants Ltd

System Powering Up

In contrast, Figure 10.3 shows what might happen when your device has to power up to get to the required new system power state. First, a Query Power IRP is sent at ① asking if a new system state is OK. Then a Set Power IRP is issued for the system power state. The driver determines that it must power up its device. It sets a completion routine and passes the IRP to the lower drivers at ②.

The completion routine for the Set System State sends itself a Set Device State Power IRP at ③, setting a completion routine. When this IRP gets to your driver, it again works out that it must power up. It passes the IRP to the lower drivers at ④ and sets another completion routine. When this completion routine runs at ⑤, the driver can finally power its device up in whatever way is appropriate at ⑥.

The completion routine at ⑦ runs. The original Set System State Power IRP completion routine at ③ eventually continues its run. It completes the first IRP.

This example appears complicated, as there are three different IRP completion routines involved. However, the full example given later should make the process clear. The original Set System State IRP needs a completion routine at ③ so that the driver knows when the IRP has been processed by the rest of the stack. This completion routine must send a Set Device State IRP to itself. It needs a completion routine at ⑦ so that it knows when this IRP has completed its rites of passage. Finally, when the driver processes the Set Device State IRP, it needs a completion routine at ⑤ so that it knows when the lower drivers have finished processing this IRP.

Figure 10.3   Power up system processing

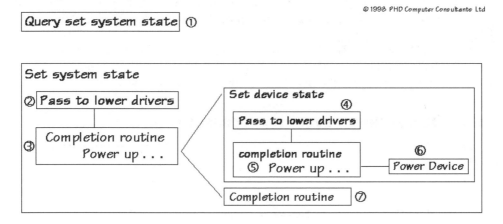

© 1998 PHD Computer Consultants Ltd

Not Processing Power IRPs

Some drivers like Wdm1 (and in fact, Wdm2) can happily ignore all Power IRPs. All they have to do is pass all Power IRPs down the stack. Wdm1's Wdm1Power dispatch routine shown in Listing 10.1 does just this. Note the use of PoCallDriver rather than IoCallDriver. If you do not process Power IRPs, please include a default power handler like Wdm1Power so that these IRPs reach lower drivers.

Drivers for removable devices should check to see if its media is present; if not, it should call PoStartNextPowerIrp and complete the Power IRP straightaway with a status of STATUS_ DELETE_PENDING.

Listing 10.1 Passing on all Power IRPs

```
NTSTATUS Wdm1Power( IN PDEVICE_OBJECT fdo, IN PIRP Irp)
{
    DebugPrint("Power %I",Irp);
    PWDM1_DEVICE_EXTENSION dx = (PWDM1_DEVICE_EXTENSION)fdo->DeviceExtension;

    // Just pass to lower driver
    PoStartNextPowerIrp( Irp);
    IoSkipCurrentIrpStackLocation(Irp);
    return PoCallDriver( dx->NextStackDevice, Irp);
}
```

Noting Device Power State Changes

Suppose a function driver does not handle Power IRPs, but its bus driver does. In this case, the function driver might want to know when its device is powered down and take some appropriate action. Unfortunately, there is no simple call to discover the current device power state. Instead, it must remember the last device power state from a Set Power IRP.

The Power IRP handler can easily be enhanced to note any changes of power state if the minor function code is IRP_MN_SET_POWER. The IRP stack *Parameters.Power.Type* field is either SystemPowerState or DevicePowerState. The *Parameters.Power.State* field gives the system or device power state.

The following code snippet shows how to check for a Set Device Power IRP and store the new device power state in a DEVICE_POWER_STATE *PowerState* field in the device extension.

```
PIO_STACK_LOCATION IrpStack = IoGetCurrentIrpStackLocation(Irp);
POWER_STATE_TYPE PowerType = IrpStack->Parameters.Power.Type;
POWER_STATE PowerState = IrpStack->Parameters.Power.State;
if( IrpStack->MinorFunction==IRP_MN_SET_POWER &&
    PowerType==DevicePowerState)
    dx->PowerState = PowerState.DeviceState;
```

Device Power Policy Owner

In most device stacks, there is one driver that knows if the device is being used and its power state. This driver is deemed the device power policy owner. There is nothing to stop more than one driver trying to set power policy. In most cases, the first function driver for a device is the power policy owner. However, the bus driver may well play an important part in power management, or even be the power policy owner.

The device power policy owner should call PoSetPowerState to set the current device power state. This device power state is stored for a device by the Power Manager. However, there is no call to retrieve the current device power state, so you must keep a copy of this value.

The Wdm2 driver defines the following two extra fields in its device extension to support Power Management.

```
DEVICE_POWER_STATE PowerState;   // Our device power state
PULONG PowerIdleCounter;          // Device idle counter
```

The AddDevice routine initializes these variables as follows. PoSetPowerState can only be used by drivers to set the device power state. Do not try to set the system power state.

```
dx->PowerState = PowerDeviceD3;
dx->PowerIdleCounter = NULL;
POWER_STATE NewState;
NewState.DeviceState = dx->PowerState;
PoSetPowerState( fdo, DevicePowerState, NewState);
```

An AddDevice routine can also set some extra bits in the FDO *Flags*. If DO_POWER_INRUSH is set, this device needs an inrush of power when it is powered up. Devices with this bit set are powered up one at a time. If the DO_POWER_PAGABLE flag bit is set, the Power handling routines are called at PASSIVE_LEVEL IRQL. Otherwise, they are called at DISPATCH_LEVEL IRQL. Inrush Power IRPs are sent at DISPATCH_LEVEL, so the power routines for these devices must not be pageable.

The device power policy owner must process Query and Set Power IRPs using the technique described previously. The following Wdm2 example shows how to do this in detail.

It is recommended that a device is fully powered up by the time the Plug and Play *Start Device* IRP has completed. As you will get only the full list of resources with this *Start Device* IRP, you will have to power up the device once the resources have been found and allocated. There might well be no need to power down a device during resource reassignment.

You may decide that the device can be powered off when it has been idle for a while. The easiest way to do this is to call the PoRegisterDeviceForIdleDetection function. After a time-out, the Power Manager sends a Set device power IRP to power down the device. Alternatively, you can generate a similar IRP yourself whenever you deem it appropriate. To avoid an idle time-out, you must keep resetting an idle counter using PoSetDeviceBusy every time an I/O operation takes place.

The following line at the end of AddDevice shows how to register for idle notification. The two ULONG parameters are the time-outs in seconds. The first applies when the system is trying to conserve power. The second applies when best performance is the goal. The final parameter to PoRegisterDeviceForIdleDetection is the device state that must be set when a device powers down. Wdm2 powers down to device state D3. Call PoRegisterDeviceForIdleDetection with zeroes for the time-outs to disable idle notification.

```
dx->PowerIdleCounter = PoRegisterDeviceForIdleDetection( pdo, 30, 60,
    PowerDeviceD3);
```

The following code snippet shows how PoSetDeviceBusy is used to reset the idle counter at the beginning of each I/O operation.

```
if( dx->PowerIdleCounter)
    PoSetDeviceBusy(dx->PowerIdleCounter);
```

At the beginning of every I/O operation, you must check to see if the device is powered up. If it is not, most drivers will simply power up the device before starting the request. An alternative strategy is to queue the IRP until the device powers up at some other time.

Make sure that you do not accidentally power down a device during a lengthy I/O operation.

Handling Set Power IRPs

The parameters on the stack of a Set Power IRP indicate which system or device power state is being set.

If *Parameters.Power.Type* is SystemPowerState, *Parameters.Power.State.SystemState* is the desired system state. If *Parameters.Power.Type* is DevicePowerState, *Parameters.Power. State.DeviceState* is the desired device state. The *Parameters.Power.ShutdownType* field gives more information about system Shutdown messages. A final parameter, *Parameters. Power.SystemContext,* is not currently used.

Listing 10.2 shows the main Wdm2 handler for all Power IRPs, Wdm2Power. If this is a Set Power IRP, PowerSetPower is called. Other Power IRPs are passed to DefaultPowerHandler, which simply hands them down to the lower drivers in the same way as Wdm1. All Power IRPs are completed with an error status if the device is not in the started PnP state.

Listing 10.2 Basic Wdm2 **power handling**

```
NTSTATUS Wdm2Power( IN PDEVICE_OBJECT fdo, IN PIRP Irp)
{
    PWDM2_DEVICE_EXTENSION dx = (PWDM2_DEVICE_EXTENSION)fdo->DeviceExtension;
    if( dx->IODisabled)
        return CompleteIrp( Irp, STATUS_DEVICE_NOT_CONNECTED, 0);
    if (!LockDevice(dx))
        return CompleteIrp( Irp, STATUS_DELETE_PENDING, 0);

    NTSTATUS status = STATUS_SUCCESS;
    DebugPrint("Power %I",Irp);

    PIO_STACK_LOCATION IrpStack = IoGetCurrentIrpStackLocation(Irp);
    ULONG MinorFunction = IrpStack->MinorFunction;

    if( MinorFunction==IRP_MN_SET_POWER)
        status = PowerSetPower(dx,Irp);
    else
        status = DefaultPowerHandler(dx,Irp);

    UnlockDevice(dx);
    return status;
}

NTSTATUS DefaultPowerHandler( IN PWDM2_DEVICE_EXTENSION dx, IN PIRP Irp)
```

Listing 10.2 Basic Wdm2 **power handling (continued)**

```
{
    DebugPrintMsg("DefaultPowerHandler");
    // Just pass to lower driver
    PoStartNextPowerIrp( Irp);
    IoSkipCurrentIrpStackLocation(Irp);
    return PoCallDriver( dx->NextStackDevice, Irp);
}
```

Listing 10.3 shows the PowerSetPower routine. As can be seen, there are two main sections to this code. The first handles setting system power states and the second sets a device power state.

Listing 10.3 PowerSetPower **routine**

```
NTSTATUS PowerSetPower( IN PWDM2_DEVICE_EXTENSION dx, IN PIRP Irp)
{
    NTSTATUS status = STATUS_SUCCESS;
    PIO_STACK_LOCATION IrpStack = IoGetCurrentIrpStackLocation(Irp);
    POWER_STATE_TYPE PowerType = IrpStack->Parameters.Power.Type;
    POWER_STATE PowerState = IrpStack->Parameters.Power.State;


    ////////////////////////////////////////////////////////////////////////
    //    Set System Power

    if( PowerType==SystemPowerState)
    {
        DEVICE_POWER_STATE DesiredDevicePowerState =
            (PowerState.SystemState<=PowerSystemWorking ? PowerDeviceD0 : PowerDeviceD3);

        if( DesiredDevicePowerState<dx->PowerState)
        {
            // This system state means we have to increase device power
            DebugPrint("System state %d.  Increase device power to %d",
                        PowerState.SystemState, DesiredDevicePowerState);
            // Process on way up stack...
            PoStartNextPowerIrp(Irp);
            IoCopyCurrentIrpStackLocationToNext(Irp);
            IoSetCompletionRoutine( Irp, OnCompleteIncreaseSystemPower, NULL,
                        TRUE, TRUE, TRUE);
            return PoCallDriver( dx->NextStackDevice, Irp);
        }
        else if( DesiredDevicePowerState>dx->PowerState)
        {
```

Listing 10.3 PowerSetPower **routine (continued)**

```
        // This system state means we have to decrease device power
        DebugPrint("System state %d.  Decrease device power to %d",
                    PowerState.SystemState, DesiredDevicePowerState);
        // Send power down request to device
        status = SendDeviceSetPower( dx, DesiredDevicePowerState);
        if( !NT_SUCCESS(status))
        {
            PoStartNextPowerIrp(Irp);
            return CompleteIrp( Irp, status, 0);
        }
    }
}


////////////////////////////////////////////////////////////////////////
//    Set Device Power
else if( PowerType==DevicePowerState)
{
    DEVICE_POWER_STATE DesiredDevicePowerState = PowerState.DeviceState;

    if( DesiredDevicePowerState<dx->PowerState)
    {
        // Increase device power state
        DebugPrint("Increase device power to %d", DesiredDevicePowerState);
        // Process on way up stack...
        PoStartNextPowerIrp(Irp);
        IoCopyCurrentIrpStackLocationToNext(Irp);
        IoSetCompletionRoutine( Irp, OnCompleteIncreaseDevicePower, NULL,
            TRUE, TRUE, TRUE);
        return PoCallDriver( dx->NextStackDevice, Irp);
    }
    else if( DesiredDevicePowerState>dx->PowerState)
    {
        // Decrease device power state
        DebugPrint("Decrease device power to %d", DesiredDevicePowerState);
        // Set power state
        SetPowerState(dx,PowerState.DeviceState);
    }
}


////////////////////////////////////////////////////////////////////////
//    Unrecognised Set Power
```

Listing 10.3 `PowerSetPower` **routine (continued)**

```
#if DBG
    else
        DebugPrint("Power: unrecognised power type %d",PowerType);
#endif


    // Finally pass to lower drivers
    return DefaultPowerHandler(dx,Irp);
}
```

Setting System Power States

When a new system state is being set, the first task is determining to which device power state this corresponds. As mentioned previously, in `Wdm2` the fully on D0 device state is only used for the fully on S0 system power state. For all other system power states, the fully off D3 device state is used.

`PowerSetPower` works out the corresponding device power state. If this matches the current device power state, saved in the device extension, no more need be done, apart from falling through to call `DefaultPowerHandler`.

If the device needs to be powered up, the situation depicted in Figure 10.3 applies. The Set System Power IRP handler can start the device only after the lower drivers have processed the request. `PowerSetPower` installs `OnCompleteIncreaseSystemPower` as the completion routine.

Listing 10.4 shows how `OnCompleteIncreaseSystemPower` eventually calls `SendDeviceSetPower` to send a Set Device Power IRP to itself. `OnCompleteIncreaseSystemPower` works out the desired device power state again. It is possible that a Set Device Power IRP has arrived in the meantime to power up the device.

If the device needs to be powered down, as per Figure 10.2, `PowerSetPower` calls `SendDeviceSetPower` to send a Set device Power request first. It can then simply let `DefaultPowerHandler` pass the IRP down to the lower drivers.

Listing 10.4 `OnCompleteIncreaseSystemPower` **routine**

```
NTSTATUS OnCompleteIncreaseSystemPower( IN PDEVICE_OBJECT fdo, IN PIRP Irp,
    IN PVOID context)
{
    PWDM2_DEVICE_EXTENSION dx = (PWDM2_DEVICE_EXTENSION)fdo->DeviceExtension;
    if (Irp->PendingReturned)
        IoMarkIrpPending(Irp);
    NTSTATUS status = Irp->IoStatus.Status;
    DebugPrint("OnCompleteIncreaseSystemPower %x",status);
    if( !NT_SUCCESS(status))
        return status;

    PIO_STACK_LOCATION IrpStack = IoGetCurrentIrpStackLocation(Irp);
    POWER_STATE PowerState = IrpStack->Parameters.Power.State;
```

Listing 10.4 `OnCompleteIncreaseSystemPower` **routine (continued)**

```
    DEVICE_POWER_STATE DesiredDevicePowerState =
        (PowerState.SystemState<=PowerSystemWorking ? PowerDeviceD0 :
        PowerDeviceD3);
    if( DesiredDevicePowerState<dx->PowerState)
        status = SendDeviceSetPower( dx, DesiredDevicePowerState);

    PoStartNextPowerIrp(Irp);
    return status;
}
```

Sending a Set Device Power IRP

The Set System Power IRP has to send a Set Device Power IRP to itself to change its own device power state.

SendDeviceSetPower in Listing 10.5 does this job. The PoRequestPowerIrp call does most of the useful work. It allocates the required type of Power IRP and initializes the IRP stack parameters. You must pass the name of a completion routine and context pointer. Finally, you can receive a pointer to the allocated IRP using the last parameter.

The completion routine should not free the IRP memory. The Power Manager frees the IRP after the completion routine exits. You cannot use the allocated IRP pointer to inspect the results of the IRP.

The completion routine OnCompleteDeviceSetPower has a different prototype to normal but does the same job. SendDeviceSetPower uses an SDSP structure on the kernel stack as the context pointer for the completion routine. This contains an initialized event that is set into the signalled state by the completion routine and a field to return the IRP completion status.

When you send a Power IRP using PoRequestPowerIrp you must not start the next Power IRP using PoStartNextPowerIrp in the completion routine[2].

As usual, SendDeviceSetPower waits for the completion event to become signalled using KeWaitForSingleObject. It retrieves the Set device Power completion status from the SDSP structure.

I found a twist in this tale when running this routine in Windows 98. On my computer, the Set device Power state was sent and seemed to complete successfully. However, it was not actually received by the Wdm2 driver. The last section of code calls SetPowerState if the device power state is not what it should be. SetPowerState actually changes the device's power state. This looks like a bug in Windows 98.

Finally, on this topic, the DDK documentation says that you cannot wait using events for your own Power IRP to complete. SendDeviceSetPower makes a new IRP and so it is all right for it to wait using an event. However, when PowerSetPower sets OnCompleteIncreaseSystemPower as

---

2. I think that this is because you are already processing a Power IRP and only one such IRP can be active at a time.

the completion routine of the IRP it cannot wait using an event. Luckily "forward and wait" processing is not needed in this case.

Listing 10.5 SendDeviceSetPower **routine**

```
typedef struct _SDSP
{
    KEVENT event;
    NTSTATUS Status;
} SDSP, *PSDSP;

NTSTATUS SendDeviceSetPower( IN PWDM2_DEVICE_EXTENSION dx,
                             IN DEVICE_POWER_STATE NewDevicePowerState)
{
    DebugPrint("SendDeviceSetPower to %d", NewDevicePowerState);
    POWER_STATE NewState;
    NewState.DeviceState = NewDevicePowerState;
    SDSP sdsp;
    KeInitializeEvent( &sdsp.event, NotificationEvent, FALSE);
    sdsp.Status = STATUS_SUCCESS;
    NTSTATUS status = PoRequestPowerIrp( dx->pdo, IRP_MN_SET_POWER,
            NewState, OnCompleteDeviceSetPower, &sdsp, NULL);
    if( status==STATUS_PENDING)
    {
        KeWaitForSingleObject( &sdsp.event, Executive, KernelMode, FALSE, NULL);
        status = sdsp.Status;
    }

    // Cope with W98 not passing power irp to us
    if( NT_SUCCESS(status) && dx->PowerState!=NewDevicePowerState)
    {
        DebugPrintMsg("SendDeviceSetPower: Device state not set properly by us.  Setting
again");
        SetPowerState(dx,NewDevicePowerState);
    }

    return status;
}

VOID OnCompleteDeviceSetPower( IN PDEVICE_OBJECT fdo, IN UCHAR MinorFunction,
            IN POWER_STATE PowerState, IN PVOID Context, IN PIO_STATUS_BLOCK IoStatus)
{
    DebugPrintMsg("OnCompleteDeviceSetPower");
```

Listing 10.5 `SendDeviceSetPower` **routine (continued)**

```
    PSDSP psdsp = (PSDSP)Context;
    psdsp->Status = IoStatus->Status;
    KeSetEvent( &psdsp->event, 0, FALSE);
}
```

Setting Device Power States

PowerSetPower in Listing 10.3 handles Set device Power IRPs.

If the device must be powered up, the power state must be changed after all the lower drivers have processed the Set device Power IRP. As before, a completion routine, OnCompleteIncreaseDevicePower, is set and the lower drivers are called. OnCompleteIncreaseDevicePower shown in Listing 10.6 eventually calls SetPowerState to change the device power state.

If the device must be powered down, PowerSetPower can simply call SetPowerState and pass the IRP to the lower drivers.

Listing 10.6 `OnCompleteIncreaseDevicePower` **routine**

```
NTSTATUS OnCompleteIncreaseDevicePower( IN PDEVICE_OBJECT fdo, IN PIRP Irp,
    IN PVOID context)
{
    PWDM2_DEVICE_EXTENSION dx = (PWDM2_DEVICE_EXTENSION)fdo->DeviceExtension;
    if (Irp->PendingReturned)
        IoMarkIrpPending(Irp);
    NTSTATUS status = Irp->IoStatus.Status;
    DebugPrint("OnCompleteIncreaseDevicePower %x",status);
    if( !NT_SUCCESS(status))
        return status;

    PIO_STACK_LOCATION IrpStack = IoGetCurrentIrpStackLocation(Irp);
    POWER_STATE PowerState = IrpStack->Parameters.Power.State;
    SetPowerState(dx,PowerState.DeviceState);

    PoStartNextPowerIrp(Irp);
    return status;
}
```

SetPowerState

Listing 10.7 shows the SetPowerState routine in DeviceIo.cpp that actually changes the device power state. SetPowerState simply stores the new device state in the Wdm2 device extension and then calls PoSetPowerState to inform the Power Manager.

A SetPowerState routine for a real device will need to interact with the device to change the power state. This will need to synchronize itself with any other hardware activities. A Critical Section routine called via KeSynchronizeExecution might be used for this purpose.

If powering a device down, you may need to store some context information that the device normally holds. For example, you could store "the current volume settings" for a set of speakers. When power is restored, pass the stored volume settings to the device.

As stated before, you may want to queue up I/O request IRPs while a device is powered down.

Listing 10.7 SetPowerState **routine**

```
void SetPowerState( IN PWDM2_DEVICE_EXTENSION dx,
                    IN DEVICE_POWER_STATE NewDevicePowerState)

{
    DebugPrint("SetPowerState %d", NewDevicePowerState);
    // Use KeSynchronizeExecution if necessary
    // to actually change power in device
    // Remember new state
    dx->PowerState = NewDevicePowerState;
    POWER_STATE NewState;
    NewState.DeviceState = NewDevicePowerState;
    PoSetPowerState( dx->fdo, DevicePowerState, NewState);
}
```

Dispatch Routine Power Handling

Each Wdm2 I/O request dispatch routine must check that the Wdm2 device is powered up. Listing 10.8 shows how Wdm2Write calls PowerUpDevice to power the device up if necessary by calling SendDeviceSetPower. Finally, PowerUpDevice resets the idle counter using PoSetDeviceBusy.

Listing 10.8 Dispatch routine power handling

```
NTSTATUS Wdm2Write( IN PDEVICE_OBJECT fdo, IN PIRP Irp)
{
    // ...
    NTSTATUS status = PowerUpDevice(fdo);
    if( !NT_SUCCESS(status))
        return CompleteIrp(Irp, status, 0);
    // ...
}


NTSTATUS PowerUpDevice( IN PDEVICE_OBJECT fdo)
{
    PWDM2_DEVICE_EXTENSION dx = (PWDM2_DEVICE_EXTENSION)fdo->DeviceExtension;

    // If need be, increase power
    if( dx->PowerState>PowerDeviceD0)
```

Listing 10.8 Dispatch routine power handling (continued)

```
    {
        NTSTATUS status = SendDeviceSetPower( dx, PowerDeviceD0);
        if (!NT_SUCCESS(status))
            return status;
    }


    // Zero our idle counter
    if( dx->PowerIdleCounter)
        PoSetDeviceBusy(dx->PowerIdleCounter);


    return STATUS_SUCCESS;
}
```

Testing Wdm2 Power Capabilities

You can investigate some of Wdm2's power capabilities by inspecting the checked build Debug-Print output using the DebugPrint Monitor. Make sure that you have installed the DebugPrint driver and the checked build of the Wdm2 driver.

Under W2000, the idle detection process can be seen working. After 60 seconds on my AC powered desktop system, the Power Manager sends a Set device Power IRP to power down the device to D3. If you then run the **Wdm2Test** Win32 application, the first read or write powers the Wdm2 device back up to D0.

On the same computer running Windows 98, no idle Set device Power IRPs were seen. This must be another manifestation of the bug described earlier. I changed the code so that Wdm2 started in a powered off state. When **Wdm2Test** was run, the Set device Power IRP (to power the device up) completed successfully, but the driver was not sent the Power IRP. This is why SendDeviceSetPower shown in Listing 10.5 checks that the device has really changed power state, and forces the change if necessary.

I cannot explain why Set Power IRPs are not received by the Wdm2 driver in Windows 98. Perhaps the idle detection process is working, but the Set device IRP is simply not seen by the driver.

If your system can sleep or hibernate then use the *Wdm2Power* application to test these features.

Device Capabilities

Another piece of the Power Management puzzle is the IRP_MN_QUERY_CAPABILITIES Plug and Play IRP minor function code. The bus driver usually fills in a DEVICE_CAPABILITIES structure. Several of these fields relate to Power Management. Function or filter drivers can inspect or modify the values set by the bus driver.

Listing 10.9 shows how PnpQueryCapabilitiesHandler handles the Query Capabilities PnP IRP for Wdm2. It first forwards the IRP down the stack and waits for it to complete. It then alters the DEVICE_CAPABILITIES structure, if necessary.

The *DeviceState* field is an array that indicates the corresponding device power state for each system power state. Or more precisely, it specifies the most powered state that a device can be in at a system power level. For example, when the system is fully on in S0, the device can be fully on in D0 or it may have idled into D3.

PnpQueryCapabilitiesHandler ensures that the correct "most powered device state" is specified by the bus driver. The SetMostPoweredState macro checks to see if a device state has been set and ups the *DeviceState* entry if appropriate.

Listing 10.9 PnpQueryCapabilitiesHandler **routine**

```
#define SetMostPoweredState( SystemState, OurDeviceState)      \
    dps = deviceCapabilities->DeviceState[SystemState];        \
    if( dps==PowerDeviceUnspecified || dps>OurDeviceState)     \
        deviceCapabilities->DeviceState[SystemState] = OurDeviceState


NTSTATUS PnpQueryCapabilitiesHandler( IN PDEVICE_OBJECT fdo, IN PIRP Irp)
{
    NTSTATUS status = ForwardIrpAndWait( fdo, Irp);
    if( NT_SUCCESS(status))
    {
        PIO_STACK_LOCATION IrpStack = IoGetCurrentIrpStackLocation(Irp);
        PDEVICE_CAPABILITIES deviceCapabilities;
        deviceCapabilities =
            IrpStack->Parameters.DeviceCapabilities.Capabilities;
#if DBG
        for(int ds=PowerSystemWorking;ds<PowerSystemMaximum;ds++)
            DebugPrint("Capabilities from bus: DeviceState[%d]=%d",
                        ds, deviceCapabilities->DeviceState[ds]);
#endif
        DEVICE_POWER_STATE dps;

        SetMostPoweredState( PowerSystemWorking, PowerDeviceD0);   // S0
        SetMostPoweredState( PowerSystemSleeping1, PowerDeviceD3); // S1
        SetMostPoweredState( PowerSystemSleeping2, PowerDeviceD3); // S2
        SetMostPoweredState( PowerSystemSleeping3, PowerDeviceD3); // S3
        SetMostPoweredState( PowerSystemHibernate, PowerDeviceD3); // S4
        SetMostPoweredState( PowerSystemShutdown, PowerDeviceD3);  // S5
    }
    return CompleteIrp( Irp, status, Irp->IoStatus.Information);
}
```

The bus driver for the Wdm2 device sets different values in DeviceState in Windows 98 and Windows 2000 Beta 2, as shown in Table 10.4. Windows 98 sets the most powered device state for S0–S4 to D0 and for S5 to D3. The most powered values for Wdm2 do not alter these values. In contrast, Windows 2000 Beta 2 sets the most powered device state for S1–S3 to D3 but does not set values for the other system states. PnpQueryCapabilitiesHandler will, therefore, modify *DeviceState* for S0, S4, and S5.

Table 10.4   DeviceState **values**

System state	S0	S1	S2	S3	S4	S5
W98	D0	D0	D0	D0	D0	D3
W2000	?	D3	D3	D3	?	?
Wdm2	D0	D3	D3	D3	D3	D3

The *SystemWake* and *DeviceWake* fields in the DEVICE_CAPABILITIES structure indicate the lowest powered state from which a device can wake the system. These fields are set to PowerSystemUndefined (0) if the device cannot wake the system. If necessary, alter the values that the bus driver has set.

The *D1Latency*, *D2Latency*, and *D3Latency* fields indicate the approximate time in 100-microsecond units that the device takes to return to the fully on D0 state for each sleeping state. Increase the values set by the bus driver, if necessary.

Finally, note that another PnP IRP, IRP_MN_QUERY_RELATIONS, has a PowerRelations subtype that asks for the PDOs of the devices that are related to the current device. These other devices are powered down when this device powers down.

Advanced Power Management

Wake

The basic idea of IRP_MN_WAIT_WAKE Power IRP is quite simple. It lets devices wake the system up from a sleeping state. A classic example is letting an incoming call from a modem wake up the system. Alternatively, a soft power on/off switch could power the system up when it is pressed.

Send an IRP_MN_WAIT_WAKE IRP to yourself using PoRequestPowerIrp before you power down. The bus driver marks the IRP as pending. When the wake event occurs, the bus driver completes the IRP and wakes the system up or powers the device up.

The situation becomes complicated when your bus driver cannot wake up the system. The bus driver FDO must generate another IRP_MN_WAIT_WAKE IRP and send it to its PDO. This continues until the root ACPI driver is reached, which can wake the system up. Each IRP_MN_WAIT_WAKE IRP is completed in turn by the bus drivers when a wake event is detected.

You must be prepared to cancel an IRP_MN_WAIT_WAKE IRP when a device is removed. When a device is stopped, the IRP_MN_WAIT_WAKE IRP must be cancelled and sent again when the resources have been reassigned.

Power Sequence

A driver can send an IRP_MN_POWER_SEQUENCE IRP to its bus driver (device power policy manager) to see how many times the device has entered the device states D1, D2, and D3. This information is used to see if a device did power down and so needs to be powered up.

A bus driver zeroes these counts when the device is created. It then increments the relevant count every time it moves into that device state. It never zeroes the counts again.

Stopping System Power Down Events

The PoRegisterSystemState kernel call provides a function similar to the SetThreadExecutionState Win32 call. It lets a driver tell the Power Manager that the system is busy, that the display is required, or that a user is present. If successful, PoRegisterSystemState returns a non-NULL pointer. When you are finished, call PoUnregisterSystemState to cancel your busy notification.

The similar function, PoSetSystemState just resets the Power Manager idle counters for these same attributes. Calling PoSetSystemState does not make the settings persist.

These functions are only available in W2000.

Power Notification

A driver can register to receive power notification events on a device tree using the PoRegisterDeviceNotify function. A notification handle is returned that must be passed to PoCancelDeviceNotify to stop receiving power notification events. These functions are only available in W2000.

Detecting System Power State Changes

In Windows 2000, a driver can use a kernel callback object to detect when the system power policy changes or when a sleep or shutdown is imminent. Register a callback routine to receive notification of \Callback\PowerState events.

First, initialise an attribute block using InitializeObjectAttributes. Then obtain a callback object pointer using ExCreateCallback. Finally, register your callback routine using ExRegisterCallback. The arguments passed to your callback routine indicate what power event has happened — see the DDK for details. Call ExUnregisterCallback when you do not wish to receive any more notifications.

You can also register to receive \Callback\SetSystemTime notifications when the system time is changed. Finally, you can create your own callback object; calling ExNotifyCallback runs the callback routines in all the other drivers that have registered for your callback.

WMI Support

Chapter 12 shows that Windows Management Instrumentation (WMI) is a facility that lets users interact with drivers. In fact, it will be just system operators, developers, or network professionals who wish to use WMI to see what is happening in a system, possibly to diagnose a fault condition.

A driver supports WMI by handling the IRP_MJ_SYSTEM_CONTROL IRP. WMI lets drivers make information available to users or receive data from them. The data objects are in structures identified by GUIDs. There are various standard WMI data blocks, but drivers can define their own data blocks, if need be.

The Power Management portion of a driver can use two standard WMI data blocks to let a user enable or disable two features. The MSPower_DeviceEnable data block, identified by GUID_POWER_DEVICE_ENABLE, has an *Enable* Boolean that a driver should use to determine if it should dynamically power itself down when it is idle. The MSPower_DeviceWakeEnable data block, identified by GUID_POWER_DEVICE_WAKE_ENABLE, has a similar *Enable* Boolean whose value should be used to enable its Wake feature.

A driver that supports these WMI features may also let their settings be changed using IOCTLs.

GUID_POWER_DEVICE_TIMEOUT is also defined to let the user change the idle timeouts. This feature is not working in W2000 Beta 3.

If you implement WMI for the MSPower_DeviceEnable or MSPower_DeviceWakeEnable data blocks, the driver properties box gains an extra Power Management tab. Figure 10.4 shows how this looks for a Wdm3 device. The Wdm3 driver supports the MSPower_DeviceEnable data block, but not MSPower_DeviceWakeEnable, so only the power down option is available in the Power Management tab.

Figure 10.4   Wdm3 **Power Management device properties**

@ 1999 PHD Computer Consultants Ltd

Conclusion

This chapter has shown how to handle the various aspects of WDM Power Management. The Wdm2 driver provides a practical illustration of these techniques. The *Wdm2Power* user mode application can be used to suspend or hibernate a computer.

The next chapter shows how INF files are used to install WDM device drivers.

Chapter 11

Installation

This chapter looks at how to install WDM drivers using installation INF files and describes the process of finding INF files using Hardware and Compatible IDs. The chapter then looks at how to install NT style (non-WDM) drivers. Example Win32 installation code is listed.

WDM Driver Installation Process

A WDM driver is installed if a new device is detected by a bus driver, or if you install a device using the Add New Hardware wizard in the Control Panel. A later section describes how the right driver is found. In both cases, the Plug and Play (PnP) Manager adds entries for the device and its driver to its configuration tables in the registry.

A driver is installed by following the instructions in an INF file. The driver executable is copied to the right location, usually the Windows System32\Drivers directory. Then various registry entries have to be made.

Some devices may now need to have some system resources assigned to them, such as I/O Port addresses and interrupt numbers. The PnP Manager may need to juggle the resources that have been assigned to existing devices to make the desired resources available to the new device. If necessary, existing devices are stopped (if they agree) and their resources reassigned.

If the driver is not already present, it is loaded and its DriverEntry routine is called. The driver AddDevice routine is called for the new device. Various Plug and Play IRPs are then sent to the device as detailed in Chapter 9. This process culminates when the *Start Device* PnP IRP is sent to tell a driver which resources it has been assigned. The driver uses these resource assignments to start its device. Normal I/O requests can then proceed.

The INF file that was used is copied to the Windows INF subdirectory. In Windows 98, they are put in the INF\OTHER subdirectory, with the INF filename changed to include the Manufacturer name. In Windows 2000, the INF file is copied to the next available OEM*.INF

227

filename in the INF directory (e.g., OEM4.INF). If you reinstall a driver with the Have Disk option, the new INF file is copied to a new OEM*.INF file. It is probably best if you delete any out-of-date OEM*.INF files.

Please note a couple of important points about installation.

- The driver executable must have an 8.3 filename in Windows 98.
- Windows will not use the INF file if it contains invalid section details.

If you want to install a driver that is required in the text-mode setup phase of W2000 installation, you must provide a txtsetup.oem file on a floppy disk. See the DDK for details.

INF Files

An INF file contains all the necessary information to install a WDM device driver, including a list of the files to copy, the registry entries to make, etc.

Windows provides a standard installer for most classes of device and a default installer. The default class installer processes INF files in the manner described in the rest of this chapter. Installers for other classes of device may perform some additional installation steps.

An INF file is a text file that looks like an old INI file. It is made up of sections, each starting with a line with the section name in square brackets (e.g., [Version]). The section contents follow. Each line either has a simple entry (e.g., entry) or sets a value (e.g., entry=value). Sections can be in any order. The order of lines in a section is sometimes important. Comments can be added after a semicolon character. Section and entry names are not case sensitive. A backslash at the end of a line in a section indicates that the line continues onto the next. Double-quotes are used to ensure the correct interpretation of characters (e.g., a backslash in a filename or leading or trailing spaces in strings). INF files can be in Unicode or plain ANSI characters. In some sections, you can opt to include other INF files needed for the installation.

If you need to customize the installation process, consider writing a class installer or — for Windows 2000 — a class coinstaller or a device coinstaller. A class installer takes over the installation of a whole class of devices; it can choose which installation jobs it handles and which it leaves to the default class installer. A coinstaller helps an existing class installer to do its job. A class coinstaller helps install all new devices of a particular class, while a device coinstaller helps just the installation of the device currently being installed. Coinstallers should not interact with the user. Please see the W2000 DDK for details of how to write these DLLs.

Standard Sections

Table 11.1 shows the typical sections of an INF file. Sections that you must name are given in italics. The DDK has full details of these sections and the other ones that you can include. Most section names may also exist in a decorated form, as described in the following text. For example, Strings.LangIdSubLangId indicates a locale and SectionName.NT indicates a platform.

The Version section is usually placed at the start of the INF file. Any of the given values for Signature work for both Windows 98 and Windows 2000. See later for details of how to write cross-platform INF files for both Windows 98 and Windows 2000.

Once your driver has passed the Microsoft Hardware Compatibility Tests, you should include a reference to your digital signature file in the CatalogFile entry.

Table 11.1 Typical sections for an INF file

Section	Entry	Value description
Version	Signature	$Windows NT$, $Windows 95$ or $Chicago$
	Provider	INF file creator
	Class	One of the system-defined class names (listed in the DDK) or your new class name.
	ClassGuid	The matching class GUID.
	DriverVer	mm/dd/yyyy [,a.b.c.d].
	CatalogFile[.NTetc]	Digitally-signed catalog file.
Strings	%string%="value"	Specifies a string.
SourceDisksNames	For each distribution floppy disk or CD-ROM, specifies its description and possibly the cabinet file and directory.	
SourceDisksFiles	Specifies the filename, the source disk ID and optionally the subdirectory and file size. This section can be empty if all the files are in the root directory.	
DestinationDirs	DefaultDestDir=dirid[,subdir] filelist=dirid[,subdir]	Specifies the directory ID, and optionally the subdirectory, for default file copies and file copies in the filelist section. The dirid is a number indicating in which standard location to put the files (see the following).
Manufacturer	%manufacturer_name%=models	Specifies the manufacturer name and the name of the corresponding models section.
models	Specifies a product name, the name of the corresponding install section, a Hardware ID, and zero or more Compatible IDs.	
install.Interfaces		List of device interfaces to add. Further AddInterface sections can specify more details, such as registry entries to add to the device interface key.

Table 11.1	Typical sections for an INF file (continued)	
Section	**Entry**	**Value description**
install	CopyFiles=@filename | filelist	Specifies a file to copy, or the name of the filelist section where the files are listed.
	AddReg=addreg	Specifies the name of the addreg section.
	LogConfig=logconfig	For legacy devices, specifies the name of the logconfig section, in which I/O addresses, IRQ configurations, etc., are detailed.
	DriverVer	mm/dd/yyyy [,a.b.c.d]
	ProfileItems	List of ProfileItem sections specifying items to add to the Start Menu.
filelist	A list of the files to be installed	
addreg		Add new keys and values.
logconfig		Legacy device configurations.
install.AddService	ServiceType=1 StartType=start-code ErrorControl=error-control-level ServiceBinary=path-to-driver etc.	For W2000 drivers, specifies the driver service details.

Strings

The Strings section defines strings that are substituted wherever else they are used. For example, if the Strings section looks like this:

```
[Strings]
Msft="Microsoft"
```

Any instance elsewhere of %Msft% is replaced with Microsoft. Strings are particularly useful for representing GUIDs, but can be used for any type of string, even bit values.

You can internationalize your INF file using strings in two ways. One way is to create a base INF file without any strings in an .inx file, and have a separate .txt file for each locale with the appropriate Strings section. Alternatively, have a common INF file but different Strings sections for each locale, appending .LangIDSubLangID to detail which language and sublanguage applies to this section. LangID and SubLangID are both two digits, as defined in winnt.h. SubLangID 00 is a neutral sublanguage. The following example shows how a different string is defined for UK English.

```
[Strings]
PrinterName="Abc Color printer"
[Strings.0902]
PrinterName="Abc Colour printer"
```

INF File Section Hierarchy

The sections in an INF file are arranged in a hierarchy. Table 11.2 shows that the Version, Strings, SourceDisksNames, SourceDisksFiles, and DestinationDirs sections are at the top level.

The Manufacturer section has a list of manufacturers (i.e., a list of manufacturer names and their section names). Table 11.2 has two manufacturers, Abc Inc and Xyz Ltd. Each manufacturer has a section that lists the product models that this INF file describes. Each product model defines various IDs and an installation section base name.

The Models section for Abc, Inc. is called Abc.Inc. This section has entries for each product that it sells. The entry for Product 1 specifies that the Product1.Install section contains the main installation instructions. In this case, an optional Product1.Install.Services section has also been included. More optional install sections can be included.

The Product1.Install section has entries that point to yet more sections. The Product1.CopyFiles section lists the files to copy. The Product1.AddReg section lists the entries that must be made in the registry. Further sections at this level can be added for other installation options.

The Product1.Install.Services section is only used in Windows 2000 to specify a service entry.

Table 11.2   INF file section hierarchy

```
[Version]
[Strings]
[SourceDisksNames]
[SourceDisksFiles]
[DestinationDirs]

[Manufacturer]
Abc, Inc.=Abc.Inc
Xyz Ltd=Xyz.Ltd
    [Abc.Inc]
    Product 1=Product1.Install,HardwareId
    Product 2=Product2.Install,HardwareId
        [Product1.Install]
        CopyFiles=Product1.CopyFiles
        AddReg=Product1.AddReg
            [Product1.CopyFiles]
```

Table 11.2   INF file section hierarchy (continued)

```
        . . .
     [Product1.AddReg]
        . . .
   [Product1.Install.Services]
   AddService=Product1,0x00000002,Product1.Service
     [Product1.Service]
   . . .


     [Product2.Install]
        . . .

     . . .
[Xyz.Ltd]
   . . .
```

Wdm1Free.INF

The quickest way to explain how an INF file works is to use a real example. Listing 11.1 shows the INF file for the Wdm1 driver, Wdm1free.Inf.

It kicks off with a Version section, which says that all the drivers and devices installed by this INF file belong to the Unknown device class. The Provider entry is set to %WDMBook%. The Strings section at the end replaces %WDMBook% with its full name, WDM Book.

The SourceDisksNames section lists the installation disks. The first and only disk is labelled "Wdm1 build directory." The SourceDisksFiles section covers W98 and specifies that the only driver file, Wdm1.sys, is found on installation disk 1 and is found in subdirectory obj\i386\free. These options make it easy to install the Wdm1 driver directly from its development location. For a commercial release, it is simpler to put the driver files in the root directory. The SourceDisksFiles.x86 section covers W2000 and specifies that Wdm1.sys is found in the objfre\i386 subdirectory.

The Manufacturer section lists just one manufacturer, again called WDM Book. There is just one product model defined here in the WDM.Book section. The Wdm1 model name %Wdm1% is the one that is shown to the user, "WDM Book: Wdm1 Example, free build". The %Wdm1% model has a Hardware ID of *wdmBook\Wdm1. Hardware IDs are covered later.

The Wdm1.Install section has the instructions for installing the Wdm1 driver in Windows 98. The files to copy are listed in the Wdm1.Files.Driver section and the registry entries are listed in the Wdm1.AddReg section.

Legacy non-Plug and Play devices may also have LogConfig sections to specify the resources that a device needs. See Chapter 15 for an example of this.

The Wdm1.Files.Driver section simply lists the files that must be installed. The DestinationDirs section specifies where the files listed in the Wdm1.Files.Driver section should go.

The Wdm1.AddReg section specifies that two registry entries should be created for the driver, DevLoader with *ntkern and NTMPDriver with Wdm1.sys. These entries are described in detail later.

That wraps it up for a Windows 98 installation. I shall look at the remaining Windows 2000 sections later.

Listing 11.1   Wdm1free.inf

```
; Wdm1free.Inf - install information file
; Copyright © 1998,1999 Chris Cant, PHD Computer Consultants Ltd

[Version]
Signature="$Chicago$"
Class=Unknown
Provider=%WDMBook%
DriverVer=04/26/1999,1.0.6.0

[Manufacturer]
%WDMBook% = WDM.Book

[WDM.Book]
%Wdm1%=Wdm1.Install, *wdmBook\Wdm1

[DestinationDirs]
Wdm1.Files.Driver=10,System32\Drivers
Wdm1.Files.Driver.NTx86=10,System32\Drivers

[SourceDisksNames]
1="Wdm1 build directory",,,

[SourceDisksFiles]
Wdm1.sys=1,obj\i386\free

[SourceDisksFiles.x86]
Wdm1.sys=1,objfre\i386


;;;;;;;;;;;;;;;;;;;;;;;;;;;;;;;;;;;;;;;;
; Windows 98

[Wdm1.Install]
CopyFiles=Wdm1.Files.Driver
AddReg=Wdm1.AddReg

[Wdm1.AddReg]
HKR,,DevLoader,,*ntkern
HKR,,NTMPDriver,,Wdm1.sys
```

```
[Wdm1.Files.Driver]
Wdm1.sys


;;;;;;;;;;;;;;;;;;;;;;;;;;;;;;;;;;;;;;;;;
; Windows 2000

[Wdm1.Install.NTx86]
CopyFiles=Wdm1.Files.Driver.NTx86

[Wdm1.Files.Driver.NTx86]
Wdm1.sys,,,%COPYFLG_NOSKIP%

[Wdm1.Install.NTx86.Services]
AddService = Wdm1, %SPSVCINST_ASSOCSERVICE%, Wdm1.Service


[Wdm1.Service]
DisplayName     = %Wdm1.ServiceName%
ServiceType     = %SERVICE_KERNEL_DRIVER%
StartType       = %SERVICE_DEMAND_START%
ErrorControl    = %SERVICE_ERROR_NORMAL%
ServiceBinary   = %10%\System32\Drivers\Wdm1.sys


;;;;;;;;;;;;;;;;;;;;;;;;;;;;;;;;;;;;;;;;;
; Strings


[Strings]
WDMBook="WDM Book"
Wdm1="WDM Book: Wdm1 Example, free build"
Wdm1.ServiceName="WDM Book Wdm1 Driver"


SPSVCINST_ASSOCSERVICE=0x00000002    ; Driver service is associated with device being
installed
COPYFLG_NOSKIP=2    ; Do not allow user to skip file
SERVICE_KERNEL_DRIVER=1
SERVICE_AUTO_START=2
SERVICE_DEMAND_START=3
SERVICE_ERROR_NORMAL=1
```

DestinationDirs **Section**

Each line in the DestinationDirs section specifies the base directory for files listed in a Copy-Files section.

The W2000 DDK lists the directory codes. For Windows 2000, code 12 means System32\
drivers in the Windows directory, but in Windows 98, it means System32\ioSubsys. However, WDM drivers must be installed in the Windows System32\drivers directory in both
versions of Windows. Therefore, code 10 (for the Windows directory) must be used, with
System32\Drivers specified as a subdirectory.

The Wdm1.Files.Driver entry is therefore set to 10,System32\Drivers, where 10 is code
for the Windows directory (e.g., C:\Windows or C:\WINNT). The second field is the subdirectory to add to this base directory. Therefore, the Wdm1 driver files are stored in the Windows
System32\Drivers directory.

AddReg **Section**

An AddReg section specifies what entries to add to the registry. Each line specifies an entry of
the form:

```
reg_root,[subkey],[value_entry_name],[flags],[value]
```

reg_root specifies the root registry key, from the list in Table 11.3. The optional subkey
field specifies a subkey off the root key. value_entry_name is the entry name to add, flags
indicates the entry type, and value is its value. flags can be omitted for REG_SZ values.

Table 11.3   **Possible** AddReg reg_root **values**

HKCR	HKEY_CLASSES_ROOT
HKCU	HKEY_CURRENT_USER
HKLM	HKEY_LOCAL_MACHINE
HKU	HKEY_USERS
HKR	The most relevant relative registry key

HKR is the most useful root registry key. It specifies the "relevant" registry entry for the sec-
tion in which it appears. In the case of the Wdm1.AddReg section, it is the driver key. In
Windows 98 for the Wdm1 driver, this key is HKLM\System\CurrentControlSet\Services\
Class\Unknown\0000. The 0000 will be replaced with the appropriate Unknown device number.

For the Wdm1.AddReg section, the default installer consequently adds DevLoader and NTMP-
Driver values to this key. The DDK does not state why these particular entries are necessary;
just do it. This driver key ends up with other values, as you can see if you inspect it using
RegEdit.

In this INF file, Windows 2000 does not use the Wdm1.AddReg section. Instead, as I shall
show, the Wdm1.Service section makes some different registry entries for the driver (in HKLM\
System\CurrentControlSet\Services\Wdm1).

Other Registry Entries

The installation process also creates various entries for the Wdm1 device that was just installed.
The INF file does not have to alter or add to these entries.

In Windows 98, the device has a registry key at HKLM\Enum\Root\Unknown\0000, with 0000
changed to a different Unknown device number, if appropriate. In Windows 2000, the device

registry key is HKLM\System\CurrentControlSet\Control\Class\{4D36E97E...}\0000, where {4D36E97E...} is the GUID for Unknown type devices.

When the Wdm1 driver runs, it also registers its device interface with GUID {C0CF0640-5F6E-11d2-B677-00C0DFE4C1F3}, WDM1_GUID. In both Windows 98 and Windows 2000, the registry key for this is HKLM\System\CurrentControlSet\Control\ DeviceClasses\{C0CF0640...}. There are various subkeys for each device and their associated symbolic links.

Windows 98 remembers that the INF file has been installed in the registry. The HKLM\ Software\Microsoft\Windows\CurrentVersion\Setup\SetupX\INF\OEM Name key has a value C:\W98\INF\OTHER\WDM BookWDM1.INF set to "WDM BookWDM1.INF". This entry is not deleted if you remove the Wdm1 device.

InfEdit

The *InfEdit* program in the Windows 98 DDK can be used to edit simple INF files. *InfEdit* does not display its main menu in Windows 2000 and is, therefore, unusable there.

As Figure 11.1 shows, *InfEdit* displays the section names as folders in a tree in its left-hand pane. The entries in each section are shown in the right-hand pane. *InfEdit* lists all the possible section names and entries. Or, at least, all the ones that were possible when the program was written.

The *InfEdit* program has some distinct drawbacks, so I would advise against its use. I asked it to read the Wdm1Free.INF file. When it wrote the file out again, it had lost the Source-DisksNames subdirectory information. All the strings were replaced with new unhelpful names. It displayed all the Windows 2000 specific sections in a Miscellaneous Sections category. Worse, it did not recognize the strings within these sections and deleted them from the Strings section.

Figure 11.1 *InfEdit* program

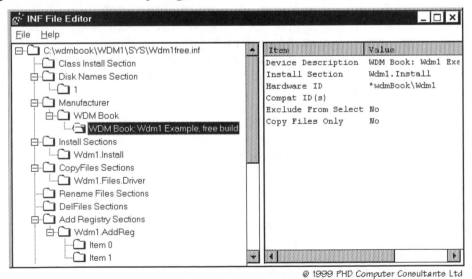

@ 1999 PHD Computer Consultants Ltd

Cross-Platform and WDM INF Files

You can create a single INF file that handles all the necessary details for the different Windows 2000 platforms and Windows 98.

The Version section Signature entry value is not used to determine the supported platforms. Instead, each platform has a different install section indicating what to do for that platform. The sections have a common base name, with additional "decoration" characters at the end of the section name to indicate the platform. If there are no section platform variants, the basic section covers all W2000 platforms and W98.

If there are sections with different platform variants, the following rules apply. If there is no decoration, the section covers W98. .NTx86 covers the W2000 x86 platform. .NTalpha covers the W2000 alpha platform. .NT covers all W2000 platforms. You need to supply only the relevant sections.

This install section covers all platforms:

```
[Abc_Install]
...
```

These two sections handle Windows 98 and Windows 2000 x86 cases.

```
[Def_Install]
; Windows 98
...
[Def_Install.NTx86]
; Windows 2000 x86 platform
...
```

Note that the Windows 98 setup code needs commas for all the optional values, while Windows 2000 can survive without them. For example, in the SourceDisksFiles section, each entry is documented as

```
filename = diskid[,[subdir][,size]]
```

The filename entry has two optional values. If you are going to leave these out in an INF file, make sure that you put two commas in, so that it works in Windows 98.

```
filename=diskid,,
```

Installing Wdm1 in Windows 2000

The Wdm1free.INF installation file in Listing 11.1 has separate installation instructions for Windows 2000.

The base section name for the Wdm1 device installation is Wdm1.Install. As there is a section named Wdm1.Install.NTx86 present, this section's instructions are used for the Windows 2000 x86 platform.

Wdm1.Install.NTx86 has just a CopyFiles directive. The Wdm1.Files.Driver.NTx86 section lists just Wdm1.sys, as before. This time, %COPYFLG_NOSKIP% is specified in the fourth field.

This indicates that the user cannot skip this file. There is no point in skipping the driver file in a Wdm1 installation. Note how the COPYFLG_NOSKIP string is used to hide the value 2.

```
[Wdm1.Files.Driver.NTx86]
Wdm1.sys,,,%COPYFLG_NOSKIP%


[Strings]
COPYFLG_NOSKIP=2    ; Do not allow user to skip file
```

Windows 2000 Service Registry Entries

A Windows 2000 device driver needs to be installed as a service. This means creating a registry key with the same name as the driver executable in HKLM\System\CurrentControlSet\ Services\.

The default installer finds the Wdm1.Install.NTx86.Services section. It first adds the relevant decoration .NTx86 and then adds .Services. This section lists the services that must be added. The service is called Wdm1. The %SPSVCINST_ASSOCSERVICE% value means that the service is associated with a device. The final field says that the Wdm1.Service section has the service registry details.

```
[Wdm1.Install.NTx86.Services]
AddService = Wdm1, %SPSVCINST_ASSOCSERVICE%, Wdm1.Service


[Wdm1.Service]
DisplayName     = %Wdm1.ServiceName%
ServiceType     = %SERVICE_KERNEL_DRIVER%
StartType       = %SERVICE_DEMAND_START%
ErrorControl    = %SERVICE_ERROR_NORMAL%
ServiceBinary   = %10%\System32\Drivers\Wdm1.sys
```

A Windows 2000 service key must have ServiceType, StartType, ErrorControl, and ServiceBinary entries. These eventually appear as Type, Start, ErrorControl, and ImagePath registry entries, respectively. A DisplayName value and registry entry are usually added, as well.

The ServiceType entry must be %SERVICE_KERNEL_DRIVER% (1) for a kernel mode driver. Various other bits can be set such as %SERVICE_FILE_SYSTEM_DRIVER% (2).

The StartType entry specifies when the driver must be started, as shown in the list in Table 11.4. WDM device drivers should specify %SERVICE_DEMAND_START% (3).

Table 11.4   Service StartType **values**

Value	Constant	Description
0	SERVICE_BOOT_START	Start when W2000 is loaded.
1	SERVICE_SYSTEM_START	Start when W2000 is initializing itself.
2	SERVICE_AUTO_START	Start when W2000 is up and running.
3	SERVICE_DEMAND_START	Start manually or when an associated device is added.
4	SERVICE_DISABLED	Never start.

The ErrorControl entry specifies how the system should respond if the driver cannot load, as shown in the list in Table 11.5. The Wdm1 installation opts for normal error logging.

Table 11.5   Service ErrorControl **values**

Value	Constant	Description
0	SERVICE_ERROR_IGNORE	Log error but do not display a message to the user.
1	SERVICE_ERROR_NORMAL	Log error and display a warning message.
2	SERVICE_ERROR_SEVERE	Log error and restart with last known good configuration.
3	SERVICE_ERROR_CRITICAL	Log error. Try last known good configuration. If this fails, force a bugcheck.

The AddService directive may also contain a further field, which specifies a section that is used to install NT event logging registries. I found that this did not work in the Beta 2 version of Windows 2000.

Several other entries may appear in a Services section that let you insert further registry entries. Some of these are used to determine the load order of drivers that are auto-started, as described in the following text.

Locating Drivers

The Windows New Device Wizard is called into action whenever a new device is found: on power up, when a hot-pluggable device is installed, or when a device is installed from the Control Panel. It scans through all the available INF files trying to find an appropriate driver.

Windows uses the device *Hardware ID* or device *Compatible ID* to select which driver to load. The bus driver that found the device provides the Hardware ID and optionally one or more Compatible IDs. Some bus drivers provide fixed IDs. Otherwise, the bus driver must interrogate the device to find its IDs.

The Hardware and Compatible IDs that an INF file supports are listed for each product in a manufacturer's Models section. Each product is defined in a line in this section, as follows.

```
device_description=install,hardware_id[,compatible_id...]
```

The device_description is displayed to the user. install is the name of the section with the installation instructions. One hardware_id and zero or more compatible_ids can then be given.

In the Wdm1 INF file, only one Hardware ID is given, *wdmBook\Wdm1. For Unknown devices installed from the Control Panel, only a Hardware ID is needed because the device is always installed.

```
%Wdm1%=Wdm1.Install, *wdmBook\Wdm1
```

The rules for selecting a device are quite complicated. By preference, Windows selects a driver for a device whose Hardware IDs match. Otherwise, it selects the driver whose Compatible ID best matches the device's Compatible ID, or prompts the user for an alternative driver INF file.

In Windows 2000, if more than one INF file has identical Hardware IDs or Compatible IDs, the DriverVer directive is used to work out the most recent driver. The INF file with the most recent date (in USAn format) is used. The version number is only used for display purposes. A DriverVer directive in the Version section provides the default for the whole file. A DriverVer directive in an Install section overrides this default. The wustamp tool can be used to insert DriverVer directives into INF files.

Hardware IDs

Hardware IDs come in almost any form. They can be *DeviceID, where DeviceID is a three-letter company ID followed by a 4 hex digit device ID. Many Windows Hardware IDs use pnp as the company ID (e.g., *pnp0700 is a standard floppy disk controller).

Other Hardware IDs are in a form specific to the enumerator and often start with an enumerator code and a backslash. Here are typical Hardware IDs.

```
PCI\VEN_1011&DEV_0021       ; DEC 21052 PCI to PCI bridge
USB\VID_045E&PID_000B       ; Microsoft USB Keyboard
USB\VID_046A&PID_0001       ; Cherry USB Keyboard
HID\VID_046A&PID_0001       ; Cherry USB Keyboard
Gameport\SideWinder3dPro    ; Microsoft SideWinder 3D Pro
PCMCIA\ACCTON-EN2212-C817   ; Accton EN2212 Ethernet PCMCIA Adapter
ISAPNP\ICI1995              ; CoreLogic NetViper-100 ISDN Adapter
```

These Hardware IDs often include the vendor code and a device or product ID. These Hardware IDs can come in versions that are more specific. For example, USB devices can have REV_ revision numbers, MC_ multi-configuration IDs or MI_ multi-interface numbers added to the basic Hardware ID. Complicated rules apply if a USB device has more than one configuration or interface[1]. In a simple case, if you produce a new driver for your version two

---

1. See the article on the Microsoft web site entitled "USB Plug and Play IDs and Selecting Device Drivers to Load", http://www.microsoft.com/hwdev/busbios/usbpnp.htm.

USB device, use REV_ in the Hardware ID to specify the latest driver. Make sure that the most desirable Hardware ID appears first in the Models list.

```
[models]
%USBDevice_V2%=V2Install,USB\VID_ABCD&PID_EF01&REV_0002
%USBDevice_V1%=V1Install,USB\VID_ABCD&PID_EF01
```

Compatible IDs

Compatible IDs are used if a matching Hardware ID is not found. If a suitable driver is not found for a device's Hardware ID, its Compatible ID is used in the lookup process. A standard driver might be able to control this device. If the Compatible IDs match, the standard driver is used.

Compatible IDs are primarily used by USB devices. A USB device has a vendor ID and product ID that are used to form its Hardware ID. However, a USB device should also have class, subclass, and protocol information for each interface it supports. For example, a USB HID device has an interface class code of 03 (constant USB_DEVICE_CLASS_HUMAN_INTERFACE). This forms a Compatible ID of USB\Class_03.

In Windows 2000, the INPUT.INF file has a line that matches this HID Compatible ID.

```
%HID.DeviceDesc% = HID_Inst,GENERIC_HID_DEVICE,
    USB\Class_03&SubClass_01,USB\Class_03
```

In Windows 98, this case is caught in HIDDEV.INF.

```
%USB\Class_03%=USBHIDDevice,USB\Class_03
```

When a new USB HID keyboard (with standard capabilities) is attached to the computer, the default installer might not match the keyboard's Hardware ID, but it should match the Compatible ID. The standard HID keyboard driver is used, as described in the two previous cases. The HID USB keyboard should, therefore, install satisfactorily.

When locating a driver for a HID device where the Hardware ID is not recognized, Compatible IDs are not used. Instead, KEYBOARD.INF uses the HID_DEVICE_SYSTEM_KEYBOARD Hardware ID to locate the default driver.

Repeated Enumeration

A new device may require several driver-loading steps before it becomes fully operational.

Consider what happens when a Cherry HID USB keyboard is inserted for the first time. The USB class drivers should already be up and running. They will detect the new device and interrogate it to obtain its vendor ID and product ID, along with various class values. The following basic device Hardware ID is formed.

```
USB\VID_046A&PID_0001
```

In Windows 2000, this Hardware ID is found in INPUT.INF. The installation section prompts the loading of the HID class drivers and the HID USB minidriver.

The HID USB minidriver is loaded for the new device. It generates a new Hardware ID to represent the HID device.

```
HID\VID_046A&PID_0001
```

In Windows 2000, this Hardware ID is found in KEYBOARD.INF. The installation section loads the system keyboard driver and the driver that provides the keyboard interface to the HID class drivers.

In Windows 98, the Cherry HID USB keyboard is installed in one step, because the KEYBOARD.INF detects the USB\VID_046A&PID_0001 Hardware ID.

NT Style Driver Installation

NT style (non-WDM) kernel mode drivers for NT 3.51, NT 4, and Windows 2000[2] can be installed by hand. This means copying the driver executable to the Windows System32\Drivers directory, making the right registry settings and rebooting the system. However, it is far safer to write an installation program to do the job[3].

You need to use the appropriate Win32 function calls to copy files and alter the registry. In addition, you must use the Windows 2000 Service Control Manager functions.

The code in install.cpp (on the book's CD-ROM) shows how to install an NT style driver. This is not a complete Win32 program. You must embed it in your own installation application.

The InstallDriver routine controls the whole installation process. It calls CreateDriver and StartDriver, when appropriate. FindInMultiSz is used to see if the driver name is in a REG_MULTI_SZ value. The code follows the logic laid out in the following sections to install a driver called "AbcDriver".

install.cpp installs two sets of registry values that I have not mentioned before.

- Event log registry entries are added. See Chapter 13 for details.
- The driver has a Parameters subkey. The values in this subkey are used to control some global features of the driver. You might like to write a Control Panel applet or other control application to let users modify these values.

Install Process

1. Get the Windows System32 directory using GetSystemDirectory.

   Use CopyFile to copy your driver to the System32\Drivers directory.

2. Create the driver service (or stop the existing driver) — see the following.

3. Make the appropriate driver registry key, e.g., HKLM\SYSTEM\CurrentControlSet\Services\AbcDriver using RegCreateKeyEx.

   Set ErrorControl, Start, and Type registry values in the previous key using RegSetValueEx.

   If desired, set Group, DependOnGroup, and DependOnService registry values, etc. as described later.

---

2. NT style drivers may well work in Windows 98, as well. See the next section for details.

---

3. NT 4 driver writers can write an OEMSETUP.INF or TXTSETUP.OEM script to install their drivers. See the NT 4 DDK for details.

4. If desired, make a `Parameters` registry key (e.g., `HKLM\SYSTEM\CurrentControlSet\Services\AbcDriver\Parameters`) and set any appropriate values.

5. Open the event log registry key at `HKLM\SYSTEM\CurrentControlSet\Services\EventLog\System`.

   Read the `Sources` value from this key.

   See if your driver is in this list. If not, add your null-terminated driver name to `Sources` and write it back to the registry.

6. Create the driver event log registry key at `HKLM\SYSTEM\CurrentControlSet\Services\EventLog\System\AbcDriver`.

   Set the `TypesSupported` and `EventMessageFile` values.

7. Start the driver — see the following.

Creating or Stopping a Driver Service

1. Open the Service Control Manager using `OpenSCManager`.
2. Try to open your driver service using `OpenService`.

   If `OpenService` exists, see if it is currently running using `ControlService SERVICE_CONTROL_INTERROGATE`.

   If `OpenService` is running, stop it using `ControlService SERVICE_CONTROL_STOP`.

   Give the driver 10 seconds to stop, checking every second with `ControlService SERVICE_CONTROL_INTERROGATE`.

   Close the driver service handle using `CloseServiceHandle`.

   Return.

3. Create the driver service using `CreateService`.
4. Close the Service Control Manager using `CloseServiceHandle`.

Starting a Driver

1. Open the Service Control Manager using `OpenSCManager`.
2. Open your driver service using `OpenService`.
3. Get the driver run state using `ControlService SERVICE_CONTROL_INTERROGATE`.
4. If need be, call `StartService` to start your driver.

   Give it 10 seconds to start, checking every second with `ControlService SERVICE_CONTROL_INTERROGATE`.

5. Close the driver service and Service Control Manager using `CloseServiceHandle`.

Driver Load Order

In NT 3.51, NT 4, and Windows 2000 various registry entries affect the load order of drivers. While these entries can be used for WDM device drivers, the Plug and Plug enumeration process usually ensures that drivers are loaded in the right order.

Each driver can be put in a *group*. *Tags* determine the driver load order within a group. Each driver can insist that it is loaded after a particular group or driver has loaded. Groups are loaded in a ServiceGroupOrder.

The DependOnGroup entry in a driver's service registry key is a REG_MULTI_SZ stating which groups must be loaded before this driver. Similarly, DependOnService is a REG_MULTI_SZ listing the drivers and services that must be loaded before this driver.

The Group entry is a REG_SZ giving the driver's group name. Tag is a REG_DWORD giving the driver tag number. The HKLM\SYSTEM\CurrentControlSet\Control\GroupOrderList registry key has a REG_BINARY entry for each group. The first byte of this binary data is the tag count. The following DWORDs contain the tags of the drivers in the order that they should be loaded. Three NULL bytes pad out the binary data.

The HKLM\SYSTEM\CurrentControlSet\Control\ServiceOrderList registry key has a REG_MULTI_SZ entry called List. The strings in the list are the driver group names in the order that they should be loaded.

A driver's Start setting, shown in Table 11.4, will override all these driver loading rules.

NT 4 Control Panel *Devices* Applet

In NT 4 and NT 3.51, the Control Panel *Devices* applet can be used to start and stop drivers and set the Startup option. Figure 11.2 shows the *Devices* applet in action.

Figure 11.2 NT 4 Control Panel *Devices*

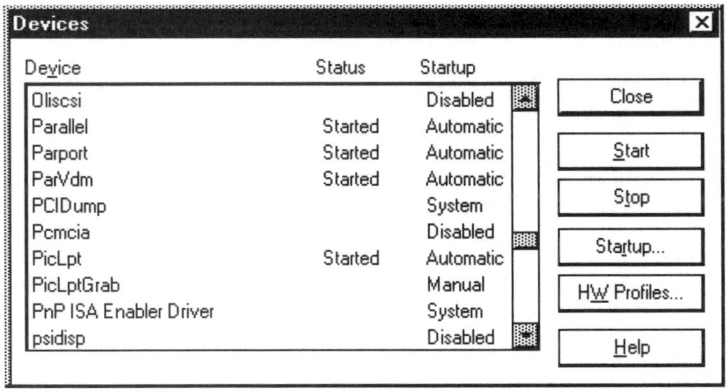

@ 1999 PHD Computer Consultants Ltd

Windows 2000 Device Management

In Windows 2000, the recommended tool for most device management tasks is the Computer Management console. This is found in the Start menu *Programs+Administrative tools+Computer Management* option. The Device Manager is also available from the Control Panel System applet.

Figure 11.3 shows the Computer Management console running on my computer with the Device Manager Devices selected. You can get this same view from the Control Panel System applet; select the Hardware tab and click on Device Manager.

You can see Wdm1 and Wdm2 devices in the "Other Devices" (Unknown) category, along with the DebugPrint driver. Right-click on the driver name to uninstall it or update the driver from its properties box.

You can see any non-WDM drivers by right-clicking on Devices. Select View in the pop up menu. Check the "Show hidden devices" option. The devices display now includes a "Non-Plug and Play Drivers" category. You can start and stop the NT style drivers and change their startup options.

Figure 11.3   W2000 Computer Management console

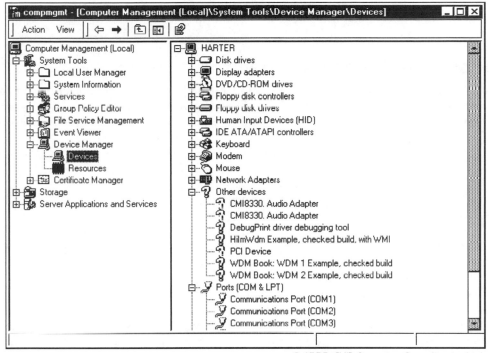

@ 1999 PHD Computer Consultants Ltd

The book software includes a small tool called *Servicer* in the Servicer subdirectory. Run this to see the display shown in Figure 11.4. "Parallel" has been typed into the Driver name box and the Lookup button has been pressed. The Parallel driver is found to be running. You can stop and start the Parallel driver using the appropriate buttons.

Servicer could be enhanced. With a bit of ingenuity, it could list all the available drivers in a list box. It could also be enhanced to change the Startup attributes in the same way as the NT 4 *Devices* Control Panel applet.

Do not try to stop a WDM driver that has a device attached. It will not work.

Figure 11.4 *Servicer* **program**

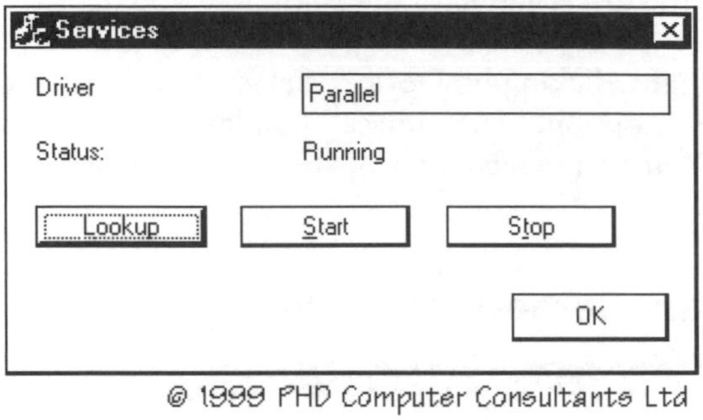

@ 1999 PHD Computer Consultants Ltd

Windows 98 Device Management

The familiar *Device Manager* display is used to manage WDM devices in Windows 98. Start the System applet in the Control Panel. The *Device Manager* tab shows a display very similar to the right-hand pane of the Computer Management Console Devices windows shown in Figure 11.3.

The Control Panel *Add New Hardware* wizard is used to add some types of new device. The Add New Hardware wizard is run automatically if a new device is detected by a bus driver.

REG Files

A quick and dirty way of installing a series of registry settings is to use .REG files. Use *RegEdit* to export a branch of the registry. Run *RegEdit* on another computer to import the .REG file to create the same registry structure, entries, and values.

Two other Windows 2000 command line tools also do the same job. *regdmp* writes registry information to a REG file. *regini* imports a REG file.

REG files help when you only have to do one or two driver installations and you do not want to write a complete installation program. Another possible use is in product support; ask a customer to use *RegEdit* to export a portion of their registry to send to you.

Installing NT Style Drivers in Windows 98

It is often possible to use an NT style driver in Windows 98. The NT style PHDIo driver described in Chapters 15–19 can be run in Windows 98 as well as the NT and Windows 2000 x86 platforms. PHDIo is a generic driver for handling simple devices and it is useful to be able to install it in Windows 98.

The DebugPrint driver described in Chapter 14 is currently installed as a WDM Plug and Play device. However, it would be perfectly sensible for it to be installed as an NT style driver. Quite a bit of its infrastructure would change. It would not have an AddDevice routine and it would not receive Plug and Play IRPs. This would actually make the driver quite a bit easier!

Contrary to my expectations, an NT style driver can use various kernel calls that are not supposed to be available to it (e.g., HalTranslateBusAddress for bus address translation, HalGetInterruptVector to get an interrupt vector, and IoReportResourceUsage to grab resource assignments). With these routines, an NT style driver such as PHDIo can use I/O ports and interrupts. You will have to experiment to see if DMA resources can also be used.

An NT style driver can use other techniques to talk to hardware. It could attach itself to another driver device and interact with hardware that way. Alternatively, I assume that it could use the available VxDs to interact with hardware.

You have to install an NT style driver "by hand" in Windows 98. In other words, the Add New Hardware wizard will not process an INF file nicely for you. Instead, a relatively straightforward program will be needed to copy the driver executable, set some registry values, and reboot the computer. For example, if you were trying to install a driver called *AbcXyz*, you would use the following steps.

- Copy the driver executable, AbcXyz.sys, to the Windows System32\Drivers directory.
- Make a new registry key HKLM\System\CurrentControlSet\Services\AbcXyz\, replacing AbcXyz with your driver name.
- In this registry key, set the values according to Table 11.6.
- Restart Windows 98.

Table 11.6   Windows 98 registry values

Value	Type	Contents	
Type	DWORD	1	
Start	DWORD	2	as per Table 11.4
DisplayName	REG_SZ	"AbcXyz"	driver name
ErrorControl	DWORD	1	as per Table 11.5

Conclusion

This chapter first looked at how to install WDM device drivers using INF files. It then described the installation process for NT style kernel mode drivers. The device driver management tools were discussed. The *Servicer* program can start and stop NT style drivers.

Our tour of the core device driver functionality continues in the next chapter by looking at how to interact with the Windows Management Instrumentation system.

Listing 11.2 NT\Install.cpp

```
// install.Inf - NT driver install program
// Copyright © 1998 Chris Cant, PHD Computer Consultants Ltd

// This is not a complete program

void InstallDriver( CString DriverName, CString DriverFromPath)
{
/////////////////////////////////////////////////////////////////////////
```

```
// Get System32 directory

_TCHAR System32Directory[_MAX_PATH];
if( 0==GetSystemDirectory(System32Directory,_MAX_PATH))
{
AfxMessageBox("Could not find Windows system directory");
return;
}

///////////////////////////////////////////////////////////////////////
// Copy driver .sys file across

CString DriverFullPath = System32Directory+"\\Drivers\\"+DriverName+".sys";
if( 0==CopyFile( DriverFromPath, DriverFullPath, FALSE)) // Overwrite OK
{
CString Msg;
Msg.Format("Could not copy %s to %s", DriverFullPath, Drivers);
AfxMessageBox(Msg);
return;
}

///////////////////////////////////////////////////////////////////////
// Create driver (or stop existing driver)

if( !CreateDriver( DriverName, DriverFullPath))
return;

///////////////////////////////////////////////////////////////////////
// Create/Open driver registry key and set its values
//Overwrite registry values written in driver creation

HKEY mru;
DWORD disposition;
if( RegCreateKeyEx( HKEY_LOCAL_MACHINE, "SYSTEM\\CurrentControlSet\\Services\\"+Driver-
Name,
0, "", 0, KEY_ALL_ACCESS, NULL, &mru, &disposition)
!= ERROR_SUCCESS)
{
AfxMessageBox("Could not create driver registry key");
return;
}
// Delete ImagePath
RegDeleteValue(mru,"ImagePath");
```

```
// Delete DisplayName
RegDeleteValue(mru,"DisplayName");
// ErrorControl
DWORD dwRegValue = SERVICE_ERROR_NORMAL;
if( RegSetValueEx(mru,"ErrorControl",0,REG_DWORD,(CONST BYTE*)&dwRegValue,sizeof(DWORD))
!= ERROR_SUCCESS)
{
AfxMessageBox("Could not create driver registry value ErrorControl");
return;
}
// Start
dwRegValue = SERVICE_AUTO_START;
if( RegSetValueEx(mru,"Start",0,REG_DWORD,(CONST BYTE*)&dwRegValue,sizeof(DWORD))
!= ERROR_SUCCESS)
{
AfxMessageBox("Could not create driver registry value Start");
return;
}
// Type
dwRegValue = SERVICE_KERNEL_DRIVER;
if( RegSetValueEx(mru,"Type",0,REG_DWORD,(CONST BYTE*)&dwRegValue,sizeof(DWORD))
!= ERROR_SUCCESS)
{
AfxMessageBox("Could not create driver registry value Type");
return;
}
// DependOnGroup
_TCHAR DependOnGroup[] = "Parallel arbitrator\0\0";
if( RegSetValueEx(mru,"DependOnGroup",0,REG_MULTI_SZ,(CONST BYTE*)&DependOn-
Group,strlen(DependOnGroup)+2)
!= ERROR_SUCCESS)
{
AfxMessageBox("Could not create driver registry value DependOnGroup");
return;
}
// DependOnService
_TCHAR DependOnService[] = "parport\0\0";
if( RegSetValueEx(mru,"DependOnService",0,REG_MULTI_SZ,(CONST BYTE*)&DependOnSer-
vice,strlen(DependOnService)+2)
!= ERROR_SUCCESS)
{
AfxMessageBox("Could not create driver registry value DependOnService");
return;
```

```
}
RegCloseKey(mru);


///////////////////////////////////////////////////////////////////////
// Create/Open driver\Parameters registry key and set its values

if( RegCreateKeyEx( HKEY_LOCAL_MACHINE, "SYSTEM\\CurrentControlSet\\Services\\"+Driver-
Name+"\\Parameters",
0, "", 0, KEY_ALL_ACCESS, NULL, &mru, &disposition)
!= ERROR_SUCCESS)
{
AfxMessageBox("Could not create driver\\Parameters registry key");
return;
}
// EventLogLevel
dwRegValue = 1;
if( RegSetValueEx(mru,"EventLogLevel",0,REG_DWORD,(CONST
BYTE*)&dwRegValue,sizeof(DWORD))
!= ERROR_SUCCESS)
{
AfxMessageBox("Could not create driver\\Parameters registry value EventLogLevel");
return;
}
// Default or No Name
CString DefaultName = DriverName;
int DeviceNameLen = DefaultName.GetLength()+1;
LPTSTR lpDefaultName = DefaultName.GetBuffer(DeviceNameLen);
if( RegSetValueEx(mru,"",0,REG_SZ,(CONST BYTE*)lpDefaultName,DeviceNameLen)
!= ERROR_SUCCESS)
{
AfxMessageBox("Could not create driver\\Parameters default registry value");
return;
}
DefaultName.ReleaseBuffer(0);
RegCloseKey(mru);


///////////////////////////////////////////////////////////////////////
// Open EventLog\System registry key and set its values

if( RegCreateKeyEx( HKEY_LOCAL_MACHINE, "SYSTEM\\CurrentControlSet\\Services\\EventLog\\
System",
0, "", 0, KEY_ALL_ACCESS, NULL, &mru, &disposition)
!= ERROR_SUCCESS)
```

```
{
AfxMessageBox("Could not open EventLog\\System registry key");
return;
}
// get Sources size
DWORD DataSize = 0;
DWORD Type;
if( RegQueryValueEx(mru,"Sources",NULL,&Type,NULL,&DataSize)
!= ERROR_SUCCESS)
{
AfxMessageBox("Could not read size of EventLog\\System registry value Sources");
return;
}
// read Sources
int DriverNameLen = strlen(DriverName);
DataSize += DriverNameLen+1;
LPTSTR Sources = new _TCHAR[DataSize];
if( RegQueryValueEx(mru,"Sources",NULL,&Type,(LPBYTE)Sources,&DataSize)
!= ERROR_SUCCESS)
{
AfxMessageBox("Could not read EventLog\\System registry value Sources");
return;
}
// If driver not there, add and write
if( FindInMultiSz(Sources,DataSize,DriverName)==-1)
{
strcpy(Sources+DataSize-1,DriverName);
DataSize += DriverNameLen;
*(Sources+DataSize) = '\0';

if( RegSetValueEx(mru,"Sources",0,REG_MULTI_SZ,(CONST BYTE*)Sources,DataSize)
!= ERROR_SUCCESS)
{
AfxMessageBox("Could not create driver registry value Sources");
return;
}
}


///////////////////////////////////////////////////////////////////////
// Create/Open EventLog\System\driver registry key and set its values

if( RegCreateKeyEx( HKEY_LOCAL_MACHINE, "SYSTEM\\CurrentControlSet\\Services\\EventLog\\
System\\"+DriverName,
```

```
0, "", 0, KEY_ALL_ACCESS, NULL, &mru, &disposition)
!= ERROR_SUCCESS)
{
AfxMessageBox("Could not create EventLog\\System\\driver registry key");
return;
}
// TypesSupported
dwRegValue = 7;
if( RegSetValueEx(mru,"TypesSupported",0,REG_DWORD,(CONST
BYTE*)&dwRegValue,sizeof(DWORD))
!= ERROR_SUCCESS)
{
AfxMessageBox("Could not create EventLog\\System\\driver registry value TypesSup-
ported");
return;
}
// EventMessageFile
LPTSTR EventMessageFile = "%SystemRoot%\\System32\\IoLogMsg.dll;%SystemRoot%\\System32\\
Drivers\\"+DriverName+".sys";
if( RegSetValueEx(mru,"EventMessageFile",0,REG_EXPAND_SZ,(CONST BYTE*)EventMessage-
File,strlen(EventMessageFile)+1)
!= ERROR_SUCCESS)
{
AfxMessageBox("Could not create EventLog\\System\\driver registry value EventMessage-
File");
return;
}
RegCloseKey(mru);


///////////////////////////////////////////////////////////////////////////
// Start driver service

if( !StartDriver(DriverName))
return;
}



///////////////////////////////////////////////////////////////////////////


BOOL CreateDriver( CString DriverName, CString FullDriver)
{
///////////////////////////////////////////////////////////////////////////
// Open service control manager
```

```
SC_HANDLE hSCManager = OpenSCManager(NULL,NULL,SC_MANAGER_ALL_ACCESS);
if( hSCManager==NULL)
{
AfxMessageBox("Could not open Service Control Manager");
return FALSE;
}


//////////////////////////////////////////////////////////////////////
// If driver is running, stop it

SC_HANDLE hDriver = OpenService(hSCManager,DriverName,SERVICE_ALL_ACCESS);
if( hDriver!=NULL)
{
SERVICE_STATUS ss;
if( ControlService(hDriver,SERVICE_CONTROL_INTERROGATE,&ss))
{
if( ss.dwCurrentState!=SERVICE_STOPPED)
{
if( !ControlService(hDriver,SERVICE_CONTROL_STOP,&ss))
{
AfxMessageBox("Could not stop driver");
CloseServiceHandle(hSCManager);
CloseServiceHandle(hDriver);
return FALSE;
}
// Give it 10 seconds to stop
BOOL Stopped = FALSE;
for(int seconds=0;seconds<10;seconds++)
{
Sleep(1000);
if( ControlService(hDriver,SERVICE_CONTROL_INTERROGATE,&ss) &&
ss.dwCurrentState==SERVICE_STOPPED)
{
Stopped = TRUE;
break;
}
}
if( !Stopped)
{
AfxMessageBox("Could not stop driver");
CloseServiceHandle(hSCManager);
CloseServiceHandle(hDriver);
```

```
return FALSE;
}
}
CloseServiceHandle(hDriver);
}
return TRUE;
}


///////////////////////////////////////////////////////////////////
// Create driver service

hDriver = CreateService(hSCManager,DriverName,DriverName,SERVICE_ALL_ACCESS,
SERVICE_KERNEL_DRIVER,SERVICE_AUTO_START,SERVICE_ERROR_NORMAL,
Drivers,NULL,NULL,"parport\0\0",NULL,NULL);
if( hDriver==NULL)
{
AfxMessageBox("Could not install driver with Service Control Manager");
CloseServiceHandle(hSCManager);
return FALSE;
}


///////////////////////////////////////////////////////////////////
CloseServiceHandle(hSCManager);
return TRUE;
}


///////////////////////////////////////////////////////////////////

BOOL StartDriver(CString DriverName)
{
///////////////////////////////////////////////////////////////////
// Open service control manager

SC_HANDLE hSCManager = OpenSCManager(NULL,NULL,SC_MANAGER_ALL_ACCESS);
if( hSCManager==NULL)
{
AfxMessageBox("Could not open Service Control Manager");
return FALSE;
}


///////////////////////////////////////////////////////////////////
// Driver isn't there
```

```
SC_HANDLE hDriver = OpenService(hSCManager,DriverName,SERVICE_ALL_ACCESS);
if( hDriver==NULL)
{
AfxMessageBox("Could not open driver service");
CloseServiceHandle(hSCManager);
return FALSE;
}


SERVICE_STATUS ss;
if( !ControlService(hDriver,SERVICE_CONTROL_INTERROGATE,&ss) ||
ss.dwCurrentState!=SERVICE_STOPPED)
{
AfxMessageBox("Could not interrogate driver service");
CloseServiceHandle(hSCManager);
CloseServiceHandle(hDriver);
return FALSE;
}
if( !StartService(hDriver,0,NULL))
{
AfxMessageBox("Could not start driver");
CloseServiceHandle(hSCManager);
CloseServiceHandle(hDriver);
return FALSE;
}
// Give it 10 seconds to start
BOOL Started = FALSE;
for(int seconds=0;seconds<10;seconds++)
{
Sleep(1000);
if( ControlService(hDriver,SERVICE_CONTROL_INTERROGATE,&ss) &&
ss.dwCurrentState==SERVICE_RUNNING)
{
Started = TRUE;
break;
}
}
if( !Started)
{
AfxMessageBox("Could not start driver");
CloseServiceHandle(hSCManager);
CloseServiceHandle(hDriver);
```

```
return FALSE;
}
CloseServiceHandle(hDriver);
CloseServiceHandle(hSCManager);
return TRUE;
}


///////////////////////////////////////////////////////////////////////////
//Try to find Match in MultiSz, including Match's terminating \0

int FindInMultiSz(LPTSTR MultiSz, int MultiSzLen, LPTSTR Match)
{
int MatchLen = strlen(Match);
_TCHAR FirstChar = *Match;
for(int i=0;i<MultiSzLen-MatchLen;i++)
{
if( *MultiSz++ == FirstChar)
{
BOOL Found = TRUE;
LPTSTR Try = MultiSz;
for(int j=1;j<=MatchLen;j++)
if( *Try++ != Match[j])
{
Found = FALSE;
break;
}
if( Found)
return i;
}
}
return -1;
}


///////////////////////////////////////////////////////////////////////////
```

Chapter 12

Windows Management Instrumentation

This chapter looks at Windows Management Instrumentation (WMI), the first of two methods of reporting management information to system administrators. The alternative — NT events for NT 3.51, NT 4, and Windows 2000 drivers — is covered in the next chapter. NT events are not recorded in Windows 98.

Windows Management Instrumentation is supposed to work in most Windows platforms. However, it did not work for me in Windows 98. As well as reporting information and firing events, WMI lets users set values that control the operation of a device or even invoke driver methods.

If possible, a driver should support the "WMI extensions for WDM". Although not mandatory, it can make it easier to debug or control your driver in the field. WMI support is required for some system functions (e.g., if you want to support user control of Power Management options in the device properties tab). If you do not support WMI, please provide a default WMI IRP handler such as in the Wdm1SystemControl routine.

The example Wdm3 driver builds on the Wdm2 driver, adding WMI and NT event support. It handles the standard MSPower_DeviceEnable WMI data block and defines its own custom WMI data block called Wdm3Information. The Wdm3Information data block returns three pieces of information to a user for each Wdm3 device: the length of the buffer, the first DWORD of the buffer contents, and the symbolic link name of the device.

I was not able to get these two further aspects of WMI working in the Beta version of W2000. A user is supposed to be able to invoke a function within the driver. The driver is supposed to be able to fire a WMI event that will be displayed in a user application.

I must first start with an overview of WMI. Although this overview appears quite complicated, it is actually fairly straightforward to support WMI in a driver.

Overview

A driver that implements the WMI extensions for WDM provides information and events to user applications, and these applications can invoke driver methods. The emphasis is on providing diagnostic and management information and tools, not the regular control of a driver.

WMI is part of the Web-Based Enterprise Management (WBEM) initiative that aims to reduce the total cost of computer ownership. WBEM is on-line at `www.microsoft.com/management/wbem/`. WBEM gives network professionals easy access to all the resources under their control. As the name suggests, the emphasis is on using browsers to access the information across a whole enterprise, either using COM ActiveX controls or a Java API on top of the HyperMedia Management Protocol (HMMP). Standard Win32 user programs can access WBEM repositories using COM APIs.

The WBEM core components are installed by default in Windows 2000 and are available in Windows 98[1]. To inspect the WBEM repository, you will need the WBEM SDK, available on-line and on the MSDN CDs.

WBEM embraces existing technologies, such as HMMP, Simple Network Management Protocol (SNMP), Desktop Management Interface (DMI), and Common Management Information Protocol (CMIP).

Microsoft has defined a Win32 implementation of WBEM and supplies various standard *providers* of information, giving access to the registry, the NT event log, Win32 information, and WDM drivers. In addition to the COM APIs, you can access the repository through an ODBC driver. Microsoft's WBEM SDK includes tools to browse and edit the WBEM repository.

The WBEM Query Language (WQL) is a subset of SQL with some extensions and can only be used to read information.

WBEM Model

WBEM is based on the Common Information Model (CIM), detailed at `www.dmtf.org/work/cim.html`. CIM is a structure for defining objects that need to be managed. A CIM Object Manager (CIMOM) stores these objects for the current computer in a repository. CIMOM is the heart of the WBEM system. It responds to requests from user mode management application clients and obtains information from providers.

Servers and Namespaces

Each WBEM enabled computer has its own CIM object database. Each computer is called a server and is named after the computer (e.g., \\MyComputer\). The current computer can be called \\.\.

---

1. In Windows 98, select the Control Panel "Add/Remove Programs" applet. Click the Windows Setup tab. Highlight the Internet Tools options. Click on Details. Check the "Web-based Enterprise Mgmt" box. Click OK to proceed with the installation. You will need the WBEM core kit and possibly other components for NT 4 and Windows 95.

The objects in the CIM database are grouped into namespaces, which can be arranged into a hierarchy, although a namespace does not inherit anything from others higher in the tree. There is always a `Root` namespace. The `Root\Default` and `Root\Security` namespaces have standard contents. Microsoft defines the Win32 namespaces `Root\CimV2` and `Root\WMI`. If need be, you can define new namespaces.

CIM Objects

The objects in a namespace are defined as being *classes*, arranged in a hierarchy so that classes do inherit properties from base classes, possibly in different namespaces. A class is defined in a text file in the Managed Object Format (**MOF**). The compiled MOF data is presented by a provider so that CIMOM knows what object types can be created. Use the *mofcomp* tool to compile a custom WMI class definition and include it as a resource in your driver.

CIM class definitions look vaguely similar to C++ classes. Classes have properties and methods. A class can override a base class definition, including standard classes in the `Default` namespace. You can have abstract classes that cannot be instantiated. Singleton classes support only a single instance. See the WBEM documentation or CIM specification for full class details.

One of the properties of a class is called the *key* property, used to differentiate separate instances of a class. If the key property is called `InstanceName`, the full object path of a class instance might be `\\Server\Namespace:Class.InstanceName="PCI\VEN.."`.

Each class can have zero or more instances. For example, the `Win32 LogicalDisk` class may have two instances to represent drives C and D, differentiated by the `DeviceId` key property. Locally, the full object path for drive C is `\\.\Root\CimV2:Win32_LogicalDisk.DeviceId="C"`. You can use relative paths once attached to a namespace (e.g., `Win32_LogicalDisk.DeviceId="C"`).

You can denote a class as being "expensive to collect" if it takes significant extra processing to collect the information. In this case, a user application must specifically request that the information be collected.

CIM object classes can be static or dynamic. Static classes can have static or dynamic instances. However, dynamic classes have only dynamic instances. Static instances are preserved across reboots. A static instance can have static or dynamic properties. Static information is provided from the CIMOM registry while dynamic information is supplied by a provider.

The WDM Provider

The WDM Provider is a Windows service that interacts with drivers and the CIM Object Manager. It is a WMI class, instance, method, and event provider. The WDM class provider retrieves class definitions in binary MOF format from drivers and sends them to CIMOM. It updates the definitions as driver changes occur.

The WDM instance provider is a dynamic provider, creating instances on demand. The WDM method provider lets applications invoke methods in a WMI driver. The WDM event provider receives events from WMI devices and translates them into an instance derived from `WMIEvent`, itself a subclass of the CIM `__ExtrinisicEvent` system class.

There are standard WMI classes defined that you can and should use. However, you can define your own if need be, each identified by a new GUID. In this case, you write a MOF file and compile it using *mofcomp*. You can include the binary MOF data either in your driver's resource or in a separate DLL. You must write a separate class for each *WMI data block* and *WMI event block*. An event block class must be derived from the WMIEvent class.

All WMI blocks appear in the Root\WMI CIMOM namespace.

Standard WMI Objects

The Windows WDM provider has several standard Win32 WMI blocks defined. You can find these using the WBEM Object Browser. The MOF definitions for most of the standard Win32 WMI blocks are in the W2000 DDK file src\storage\class\disk\WMICORE.MOF.

The W2000 DDK sources also have plenty of examples of how to implement WMI. For example, the standard Windows 2000 serial driver provides five WMI data blocks of information, MSSerial_CommInfo, MSSerial_CommProperties, MSSerial_HardwareConfiguration, MSSerial_PerformanceInformation and MSSerial_PortName. In each case the InstanceName property is the key, named after the PnP driver instance, not the serial port name. However, the MSSerial_PortName data object lets you retrieve the real port name in its *PortName* property. As another example, the MSSerial_CommInfo *BaudRate* property has the serial port's baud rate.

If you are writing a driver that is similar to a system driver, then consider reporting the standard system WMI blocks. The Wdm3 driver implements the standard MSPower_DeviceEnable WMI data block. As described in Chapter 10, this is used by the Device Manager to let users stop a device from powering down. This means that the Wdm3 must be able to accept a changed value for the Enable property, as well as reporting the current setting.

Listing 12.1 shows the source for the MSPower_DeviceEnable WMI data block, taken from WMICORE.MOF. A GUID is used to identify the block. This is {827c0a6f-feb0-11d0-bd26-00aa00b7b32a}, which is defined as GUID_POWER_DEVICE_ENABLE in the standard header WDMGUID.H.

All WMI classes have a key string property called *InstanceName* and Boolean property called *Active*. The MSPower_DeviceEnable class has only one "real" property, called *Enable*, that can be read and written and is identified as the first WmiDataId.

Listing 12.1 MSPower_DeviceEnable **WMI data block**

```
[Dynamic, Provider("WMIProv"),
 WMI,
 Description("The buffer for this control is a BOOLEAN and indicates if the device should
dynamically power on and off while the system is working. A driver would only support
such a setting if there is a significant user noticeable effect for powering off the
device. E.g., turning on the device may cause a user noticeable delay. Regardless of this
setting, the driver is still required to support system sleeping states irps (which
likely translates to powering off the device when a system sleep occurs)."),
 guid("827c0a6f-feb0-11d0-bd26-00aa00b7b32a"),
 locale("MS\\0x409")]
class MSPower_DeviceEnable
{
    [key, read]
    string InstanceName;

    [read]
```

Listing 12.1 `MSPower_DeviceEnable` **WMI data block (continued)**

```
    boolean Active;

    [WmiDataId(1),
     read,
     write]
    boolean Enable;
};
```

A WMI Driver

A WDM driver can use WMI mechanisms to publish information, permit configuration of its device, and supply notifications of events. It should continue to report standard Windows 2000 events, as many system administrators will still be looking for this event information. However, WMI is the way forward, as it allows information to be made available on non-W2000 systems, and it allows remote inspection and control of the device.

WDM drivers, NT style drivers, and miniport drivers/minidriver can support WMI. In the latter case, you must fit in with the WMI reporting mechanism supported by the class driver. For example, a SCSI miniport must set the *WmiDataProvider* BOOLEAN in the PORT_ CONFIGURATION_INFORMATION structure and handle SRB_FUNCTION_WMI requests.

The Wdm3 driver defines two custom WMI blocks. Listing 12.2 shows how the Wdm3Information data block and the Wdm3Event event block are defined in Wdm3.mof. The identifying GUID for each block is defined in the GUIDs.h header: WDM3_WMI_GUID and WDM3_ WMI_EVENT_GUID.

The Wdm3Information data block contains two 32-bit unsigned properties followed by a counted wide string property. The first real property, *BufferLen*, is the length of the shared memory buffer. *BufferFirstWord* is the first 32 bits of the shared memory buffer, or zero if the buffer is not long enough. Finally, *SymbolicLinkName* is the symbolic link name of the Wdm3 device interface.

The Wdm3Information data block has a definition for a PowerDown function. I found that this did not compile, so I commented it out.

The Wdm3Event WMI event block must be derived from the standard WMIEvent class. It simply defines a Message property.

Listing 12.2 `Wdm3Information` **and** `Wdm3Event` **block**

```
/*  WMI data block: Information about Wdm3 device
    Wdm3Information, identified by WDM3_WMI_GUID
*/

[WMI, Dynamic, Provider("WMIProv"),
 Description("Wdm3 information"),
 guid("{C0CF0643-5F6E-11d2-B677-00C0DFE4C1F3}"),
 locale("MS\\0x409")]

class Wdm3Information
```

Listing 12.2 `Wdm3Information` **and** `Wdm3Event` **block (continued)**

```
{
    [key, read]
        string InstanceName;
    [read]
        boolean Active;

    [WmiDataId(1),
        read,
        Description("Shared memory buffer length")
    ]
        uint32 BufferLen;

    [WmiDataId(2),
        read,
        Description("First ULONG of shared memory buffer")
    ]
        uint32 BufferFirstWord;

    [WmiDataId(3),
        read,
        Description("Symbolic link name")
    ]
        string SymbolicLinkName;

/*  Doesn't compile
    [Implemented]
        void PowerDown();
*/
};

/* WMI event block: Wdm3 device event
    Wdm3Event, identified by WDM3_WMI_EVENT_GUID
*/

[WMI, Dynamic, Provider("WMIProv"),
 guid("{C0CF0644-5F6E-11d2-B677-00C0DFE4C1F3}"),
 locale("MS\\0x409"),
 Description("Wdm3 event message")]

class Wdm3Event : WMIEvent
{
```

Listing 12.2 `Wdm3Information` **and** `Wdm3Event` **block (continued)**

```
    [key, read]
        string InstanceName;

    [read]
        boolean Active;

    [WmiDataId(1),
        read,
        Description("Message")
    ]
        string Message;
};
```

This example MOF file shows how a driver defines static instance names when it registers a WMI block. Alternatively, if necessary, the driver can define dynamic instance names if the instances change frequently at run time. Handling dynamic instance names is harder and is not a common requirement, so it is not discussed here.

WMI Build Environment

Quite a few changes must be made to a project to support WMI.

1. All the WMI functions, including support for the `Wdm3SystemControl` WMI IRP, are in `Wmi.cpp`. Include this file in the `SOURCES` list of files to compile in the `SOURCES` file.

2. I found that this line had to be included in `SOURCES` to ensure that the WMI library file was found.

```
TARGETLIBS=C:\NTDDK\LIBFRE\I386\WmiLib.Lib
```

3. The prebuild steps had to be altered to persuade *build* to run the *mofcomp* tool to compile `Wdm3.mof`. The `SOURCES` file `NTTARGETFILE0` macro is changed to ensure that `makefile.inc` runs the following command. The complete `makefile.inc` is given in the next chapter.

```
mofcomp -B:Wdm3.bmf -WMI Wdm3.mof
```

This command line compiles MOF source file `Wdm3.mof` into the binary MOF `Wdm3.bmf` file. The options to *mofcomp* ensure that the source code is checked for WMI compatibility and that the output goes in the correct file.

4. The main resource file `Wdm3.rc` must now include the binary MOF file `Wdm3.bmf`. The following line in the resource script identifies the data using the name `MofResource`.

```
MOFRESOURCE  MOFDATA MOVEABLE PURE  "Wdm3.bmf"
```

5. The main header in `Wdm3.h` had to use standard headers `wmilib.h` and `wmistr.h`. It also included `wdmguid.h` to get the definition of the standard GUID, `GUID_POWER_DEVICE_ENABLE`.

6. The device extension has these extra fields added.

```
WMILIB_CONTEXT WmiLibInfo;    // WMI Context
BOOLEAN IdlePowerDownEnable; // Enable power down option
BOOLEAN WMIEventEnabled;      // Enable WMI events
```

WmiLibInfo is used to pass the Wdm3 WMI GUIDs and various callbacks, as described later.

IdlePowerDownEnable implements the MSPower_DeviceEnable WMI data block. If *IdlePowerDownEnable* is TRUE, the driver can power down when it is idle.

Finally, *WMIEventEnabled* is used to enable WMI event reporting.

7. The main WMI code needs to refer to the driver registry path that was passed to Driver-Entry. Therefore, DriverEntry saves a copy in its Wdm3RegistryPath global variable.

```
// Save a copy of our RegistryPath for WMI
Wdm3RegistryPath.MaximumLength = RegistryPath->MaximumLength;
Wdm3RegistryPath.Length = 0;
Wdm3RegistryPath.Buffer =
    (PWSTR)ExAllocatePool( PagedPool, Wdm3RegistryPath.MaximumLength);
if( Wdm3RegistryPath.Buffer==NULL) return STATUS_INSUFFICIENT_RESOURCES;
RtlCopyUnicodeString( &Wdm3RegistryPath, RegistryPath);
```

The driver unload routine Wdm3Unload deletes the Wdm3RegistryPath buffer.

Registering as a WMI Data Provider

You must register as a WMI provider by calling IoWmiRegistrationControl with a WMIREG_ACTION_REGISTER command for each device when it is ready to handle WMI IRPs. The Wdm3 driver makes this call in its RegisterWmi routine. RegisterWmi is called at the end of Wdm3AddDevice. The corresponding DeregisterWmi routine deregisters the WMI support by calling IoWmiRegistrationControl with the WMIREG_ACTION_DEREGISTER command. DeregisterWmi is called when the device is removed.

Listing 12.3 shows the RegisterWmi and DeregisterWMI routines. RegisterWmi also sets up the *WmiLibInfo* structure in the device extension, ready for processing by the System Control WMI IRP. Table 12.1 shows the *WmiLibInfo* WMILIB_CONTEXT structure fields.

WmiLibInfo first contains the list of WMI block GUIDs that are handled by this driver. Wdm3GuidList is an array of these WMIGUIDREGINFO structures. Each specifies the GUID pointer, a count of instances, and optionally some flags. There is only one instance of each WMI block for this device. No flags are specified, as the appropriate flags are used later in the call to QueryWmiRegInfo.

The WMILIB_CONTEXT *WmiLibInfo* field also sets up several callback routines that we shall meet later. These are used to help the processing of the IRP_MJ_SYSTEM_CONTROL WMI IRP.

After calling `IoWmiRegistrationControl`, your driver will then be sent an `IRP_MN_REGINFO` System Control IRP to obtain the device's WMI MOF information, as described in the following.

Listing 12.3 `RegisterWmi` **and** `DeregisterWMI` **code**

```
const int GUID_COUNT = 3;

WMIGUIDREGINFO Wdm3GuidList[GUID_COUNT] =
{
    { &WDM3_WMI_GUID, 1, 0 },              // Wdm3Information
    { &GUID_POWER_DEVICE_ENABLE, 1, 0},    // MSPower_DeviceEnable
    { &WDM3_WMI_EVENT_GUID, 1, 0},         // Wdm3Event
};

const ULONG WDM3_WMI_GUID_INDEX = 0;
const ULONG GUID_POWER_DEVICE_ENABLE_INDEX = 1;
const ULONG WDM3_WMI_EVENT_GUID_INDEX = 2;

void RegisterWmi( IN PDEVICE_OBJECT fdo)
{
    PWDM3_DEVICE_EXTENSION dx=(PWDM3_DEVICE_EXTENSION)fdo->DeviceExtension;

    dx->WmiLibInfo.GuidCount = GUID_COUNT;
    dx->WmiLibInfo.GuidList = Wdm3GuidList;

    dx->WmiLibInfo.QueryWmiRegInfo = QueryWmiRegInfo;
    dx->WmiLibInfo.QueryWmiDataBlock = QueryWmiDataBlock;
    dx->WmiLibInfo.SetWmiDataBlock = SetWmiDataBlock;
    dx->WmiLibInfo.SetWmiDataItem = SetWmiDataItem;
    dx->WmiLibInfo.ExecuteWmiMethod = ExecuteWmiMethod;
    dx->WmiLibInfo.WmiFunctionControl = WmiFunctionControl;

    NTSTATUS status = IoWMIRegistrationControl( fdo, WMIREG_ACTION_REGISTER);
    DebugPrint("RegisterWmi %x",status);
}

void DeregisterWmi( IN PDEVICE_OBJECT fdo)
{
    IoWMIRegistrationControl( fdo, WMIREG_ACTION_DEREGISTER);
    DebugPrintMsg("DeregisterWmi");
}
```

Table 12.1 `WMILIB_CONTEXT` **structure**

`GuidCount`	ULONG	Required	Count of WMI blocks
`GuidList`		Required	Array with the GUIDs of the WMI blocks supported, etc.
`QueryWmiRegInfo`		Required	Provide further information about the WMI blocks you are registering
`QueryWmiDataBlock`	Callback	Required	Return a single instance or all instances of a data block
`SetWmiDataBlock`	Callback	Optional	Set all data items in a single instance of a data block
`SetWmiDataItem`	Callback	Optional	Set a single data item in a single instance of a data block
`ExecuteWmiMethod`	Callback	Optional	Execute a method associated with a data block
`WmiFunctionControl`	Callback	Optional	Enable and disable event notification and expensive data block collection

Handling System Control IRPs

Your driver must then handle System Control IRPs in your `SystemControl` routine. Table 12.2 shows the various minor codes associated with this IRP.

Table 12.2   **System control minor request codes**

`IRP_MN_REGINFO`	Query a driver's registration information
`IRP_MN_QUERY_ALL_DATA`	Get all instances in a given data block
`IRP_MN_QUERY_SINGLE_INSTANCE`	Get a single instance in a given data block
`IRP_MN_CHANGE_SINGLE_INSTANCE`	Change all data items in a given data block
`IRP_MN_CHANGE_SINGLE_ITEM`	Change a single data item in a given data block
`IRP_MN_ENABLE_EVENTS`	Enable event notification
`IRP_MN_DISABLE_EVENTS`	Disable event notification
`IRP_MN_ENABLE_COLLECTION`	Start collection of data that is expensive to collect
`IRP_MN_DISABLE_COLLECTION`	Stop collection of data that is expensive to collect
`IRP_MN_EXECUTE_METHOD`	Execute a method in a data block

If you have dynamic instance names, you must process the IRP and call `WmiCompleteRequest`, or pass it onto the next driver if you do not recognize the GUID.

For static instances, as used by Wdm3, you simply have to call `WmiSystemControl`, passing a pointer to your `WMILIB_CONTEXT` structure. `WmiSystemControl` processes the IRPs as much as possible. If the request is destined for your driver, it invokes the appropriate callback to let you process the IRP. On return, `WmiSystemControl` sets its `IrpDisposition` parameter to tell you how to continue processing the IRP, as shown in Table 12.3. For example, if the IRP was not destined for your driver, IrpDisposition is set to IrpForward and the IRP is sent down the stack.

Table 12.3 WmiSystemControl IrpDisposition **handling**

IrpProcessed	Either call WmiCompleteRequest here or in your call-back routine.
IrpNotComplete	IRP processed but an error detected, so just call IoCompleteRequest.
IrpNotWMI or IrpForward	Forward to next driver: call IoSkipCurrentIrpStackLocation and IoCallDriver

Listing 12.4 shows how the Wdm3SystemControl routine processes System Control IRPs. The main processing is carried out by the system function WmiSystemControl. Wdm3SystemControl acts on the IrpDispostion parameter in the recommended way.

Listing 12.4 Wdm3SystemControl **routine**

```
NTSTATUS Wdm3SystemControl( IN PDEVICE_OBJECT fdo, IN PIRP Irp)
{
    DebugPrintMsg("Wdm3SystemControl");
    PWDM3_DEVICE_EXTENSION dx = (PWDM3_DEVICE_EXTENSION)fdo->DeviceExtension;
    SYSCTL_IRP_DISPOSITION disposition;

    NTSTATUS status =
        WmiSystemControl( &dx->WmiLibInfo, fdo, Irp, &disposition);
    switch(disposition)
    {
    case IrpProcessed:
        // This irp has been processed and may be completed or pending.
        break;

    case IrpNotCompleted:
        // This irp has not been completed, but has been fully processed.
        // we will complete it now
        IoCompleteRequest( Irp, IO_NO_INCREMENT);
        break;

    case IrpForward:
    case IrpNotWmi:
        // This irp is either not a WMI irp or is a WMI irp targetted
        // at a device lower in the stack.
        IoSkipCurrentIrpStackLocation(Irp);
        status = IoCallDriver( dx->NextStackDevice, Irp);
        break;
```

Listing 12.4 `Wdm3SystemControl` **routine (continued)**

```
    default:
        DebugPrint("Wdm3SystemControl bad disposition %d",disposition);
//      ASSERT(FALSE);
    }


    return status;
}
```

At a minimum, you should write `QueryWmiRegInfo`, `QueryWmiDataBlock`, `SetWmiDat-aBlock`, and `SetWmiDataItem` callbacks. Even though the last two callbacks are optional, it is best to implement them to ensure that `WmiCompleteRequest` is called with an appropriate error code.

QueryWmiRegInfo **Handler**

The `Wdm3` `QueryWmiRegInfo` callback routine shown in Listing 12.5 provides further information about the WMI blocks you are registering, in addition to the *GuidList* provided in your `WMILIB_CONTEXT`.

Return your MOF resource name and your registry path.

If you are writing a WDM driver, you will usually set the `WMIREG_FLAG_INSTANCE_PDO` flag and pass the device's PDO. See the DDK documentation for details of other flags that you can set.

Listing 12.5 `QueryWmiRegInfo` **routine**

```
#define MofResourceNameText L"MofResource"

NTSTATUS QueryWmiRegInfo(
    IN PDEVICE_OBJECT fdo, OUT PULONG PRegFlags,
    OUT PUNICODE_STRING PInstanceName,
    OUT PUNICODE_STRING *PRegistryPath,
    OUT PUNICODE_STRING MofResourceName,
    OUT PDEVICE_OBJECT *Pdo)
{
    DebugPrintMsg("QueryWmiRegInfo");
    PWDM3_DEVICE_EXTENSION dx = (PWDM3_DEVICE_EXTENSION)fdo->DeviceExtension;

    *PRegFlags = WMIREG_FLAG_INSTANCE_PDO;
    *PRegistryPath = &Wdm3RegistryPath;
    RtlInitUnicodeString( MofResourceName, MofResourceNameText);
    *Pdo = dx->pdo;


    return STATUS_SUCCESS;
}
```

QueryWmiDataBlock **Handler**

A QueryWmiDataBlock callback routine returns one or more instances of a data block. When finished, it should call WmiCompleteRequest. A driver can return STATUS_PENDING if the IRP cannot be completed immediately.

QueryWmiDataBlock is passed GuidIndex, the relevant index into your GuidList array, the InstanceIndex, and the InstanceCount required. You have to fill the Buffer (that has a maximum size of BufferAvail). Fill in InstanceLengthArray, an array of ULONGs giving the length of each instance.

In the buffer, each instance's data should be aligned on an 8-byte boundary. The actual data should correspond to the class MOF definition. A MOF string should be returned as a counted Unicode, MOF Boolean returned as a BOOLEAN, MOF uint32 returned as a ULONG, MOF uint64 as a ULONGLONG, etc.

Listing 12.6 shows how the Wdm3 QueryWmiDataBlock routine handles requests for the Wdm3Information and MSPower_DeviceEnable WMI data blocks. For Wdm3Information, QueryWmiDataBlock first works out the size of buffer required and checks that the given output buffer is large enough. It then stores the *BufferLen*, *BufferFirstWord*, and *SymbolicLinkName* property values. The *SymbolicLinkName* string is stored in counted Unicode (i.e., a USHORT count followed by the wide char characters). Finally, the number of bytes set is stored. Only one instance of a Wdm3Information data block is ever processed, so this size is stored in the first ULONG pointed to by InstanceLengthArray.

QueryWmiDataBlock handles a MSPower_DeviceEnable WMI data block request in a similar way.

Listing 12.6 QueryWmiDataBlock **routine**

```
NTSTATUS QueryWmiDataBlock( IN PDEVICE_OBJECT fdo, IN PIRP Irp,
    IN ULONG GuidIndex, IN ULONG InstanceIndex,
    IN ULONG InstanceCount, IN OUT PULONG InstanceLengthArray,
    IN ULONG OutBufferSize, OUT PUCHAR PBuffer)
{

    DebugPrint("QueryWmiDataBlock: GuidIndex %d, InstanceIndex %d, "
        "InstanceCount %d, OutBufferSize %d",
        GuidIndex,InstanceIndex,InstanceCount,OutBufferSize);
    PWDM3_DEVICE_EXTENSION dx = (PWDM3_DEVICE_EXTENSION)fdo->DeviceExtension;
    NTSTATUS status;
    ULONG size = 0;


    switch( GuidIndex)
    {
    case WDM3_WMI_GUID_INDEX:      // Wdm3Information
    {
        ULONG SymLinkNameLen = dx->ifSymLinkName.Length;
        size = sizeof(ULONG)+sizeof(ULONG)+SymLinkNameLen+sizeof(USHORT);


        // Check output buffer size
```

Listing 12.6   `QueryWmiDataBlock` **routine (continued)**

```
        if( OutBufferSize<size)
        {
            status = STATUS_BUFFER_TOO_SMALL;
            break;
        }


        // Store uint32 BufferLen
        *(ULONG *)PBuffer = BufferSize;
        PBuffer += sizeof(ULONG);


        // Store uint32 BufferFirstWord
        ULONG FirstWord = 0;
        if( Buffer!=NULL && BufferSize>=4)
            FirstWord = *(ULONG*)Buffer;
        *(ULONG *)PBuffer = FirstWord;
        PBuffer += sizeof(ULONG);


        // Store string SymbolicLinkName as counted Unicode
        *(USHORT *)PBuffer = (USHORT)SymLinkNameLen;
        PBuffer += sizeof(USHORT);
        RtlCopyMemory( PBuffer, dx->ifSymLinkName.Buffer, SymLinkNameLen);


        // Store total size
        *InstanceLengthArray = size;
        status = STATUS_SUCCESS;


        break;
    }
    case GUID_POWER_DEVICE_ENABLE_INDEX:      // MSPower_DeviceEnable
    {
        size = sizeof(BOOLEAN);


        // Check output buffer size
        if( OutBufferSize<size)
        {
            status = STATUS_BUFFER_TOO_SMALL;
            break;
        }


        // Store boolean IdlePowerDownEnable in Enable property
        *(BOOLEAN *)PBuffer = dx->IdlePowerDownEnable;
```

Listing 12.6 QueryWmiDataBlock **routine (continued)**

```
        // Store total size
        *InstanceLengthArray = size;
        status = STATUS_SUCCESS;

        break;
    }
default:
    DebugPrintMsg("QueryWmiDataBlock: Bad GUID index");
    status = STATUS_WMI_GUID_NOT_FOUND;
    break;
    }

    return WmiCompleteRequest( fdo, Irp, status, size, IO_NO_INCREMENT);
}
```

SetWmiDataBlock **Handler**

The Wdm3 SetWmiDataBlock routine shown in Listing 12.7 lets a user change the device extension *IdlePowerDownEnable* settings. If the MSPower_DeviceEnable GUID index is given and the input buffer size is large enough, the BOOLEAN at the start of Pbuffer is stored in *IdlePowerDownEnable*.

SetWmiDataBlock then acts on the new power down enable setting. If powering down is now enabled, PoRegisterDeviceForIdleDetection is called, if necessary. If now disabled, PoRegisterDeviceForIdleDetection is called to turn off idle detection; if the Wdm3 device is powered down, SendDeviceSetPower is called to power the device up.

Listing 12.7 SetWmiDataBlock **routine**

```
NTSTATUS SetWmiDataBlock(
    IN PDEVICE_OBJECT fdo, IN PIRP Irp,
    IN ULONG GuidIndex, IN ULONG InstanceIndex,
    IN ULONG BufferSize,
    IN PUCHAR PBuffer)
{

    DebugPrint("SetWmiDataBlock: GuidIndex %d, InstanceIndex %d, BufferSize %d",
        GuidIndex,InstanceIndex,BufferSize);
    PWDM3_DEVICE_EXTENSION dx = (PWDM3_DEVICE_EXTENSION)fdo->DeviceExtension;

    if( GuidIndex==GUID_POWER_DEVICE_ENABLE_INDEX)  // MSPower_DeviceEnable
    {
        if( BufferSize<sizeof(BOOLEAN))
            return WmiCompleteRequest( fdo, Irp, STATUS_BUFFER_TOO_SMALL, 0,
                IO_NO_INCREMENT);
```

Listing 12.7 `SetWmiDataBlock` **routine (continued)**

```
    // Get Enable property into IdlePowerDownEnable
    dx->IdlePowerDownEnable = *(BOOLEAN*)PBuffer;

    // Action IdlePowerDownEnable
    if( dx->IdlePowerDownEnable)
    {
        DebugPrintMsg("SetWmiDataBlock: Enabling power down");
        // Enable power down idling
        if( dx->PowerIdleCounter==NULL)
            dx->PowerIdleCounter =
                PoRegisterDeviceForIdleDetection( dx->pdo, 30, 60,
                PowerDeviceD3);
    }
    else
    {
        DebugPrintMsg("SetWmiDataBlock: Disabling power down");
        // Disable power down idling
        if( dx->PowerIdleCounter!=NULL)
            dx->PowerIdleCounter =
                PoRegisterDeviceForIdleDetection( dx->pdo, 0, 0,
                PowerDeviceD3);
        if( dx->PowerState>PowerDeviceD0)
        {
            DebugPrintMsg("SetWmiDataBlock: Disabling power down: power up");
            SendDeviceSetPower( dx, PowerDeviceD0);
        }
    }

    return WmiCompleteRequest( fdo, Irp, STATUS_SUCCESS, 0,
        IO_NO_INCREMENT);
    }
    return FailWMIRequest( fdo, Irp, GuidIndex);
}
```

`SetWmiDataItem` **Handler**

A `SetWmiDataItem` callback handles the setting of one data item in an instance. This routine is not called in Wdm3. Listing 12.8 shows how `SetWmiDataItem` calls `FailWMIRequest` to reject the

WMI IRP. FailWMIRequest fails the IRP with STATUS_WMI_GUID_NOT_FOUND if an invalid GUID was given, or STATUS_INVALID_DEVICE_REQUEST otherwise.

Listing 12.8 SetWmiDataItem **routine**

```
NTSTATUS SetWmiDataItem(
    IN PDEVICE_OBJECT fdo, IN PIRP Irp,
    IN ULONG GuidIndex, IN ULONG InstanceIndex,
    IN ULONG DataItemId,
    IN ULONG BufferSize, IN PUCHAR PBuffer)
{
    DebugPrint("SetWmiDataItem: GuidIndex %d, InstanceIndex %d, DataItemId %d,
        BufferSize %d", GuidIndex,InstanceIndex,DataItemId,BufferSize);
    return FailWMIRequest( fdo, Irp, GuidIndex);
}


NTSTATUS FailWMIRequest(
    IN PDEVICE_OBJECT fdo,
    IN PIRP Irp,
    IN ULONG GuidIndex)
{
    DebugPrint("FailWMIRequest: GuidIndex %d",GuidIndex);
    NTSTATUS status;

    if( GuidIndex<0 || GuidIndex>=GUID_COUNT)
        status = STATUS_WMI_GUID_NOT_FOUND;
    else
        status = STATUS_INVALID_DEVICE_REQUEST;

    status = WmiCompleteRequest( fdo, Irp, status, 0, IO_NO_INCREMENT);

    return status;
}
```

ExecuteWmiMethod **Handler**

Listing 12.9 shows how I think the optional ExecuteWmiMethod routine should be implemented. As I stated earlier, I could not get *mofcomp* to compile the *PowerDown* method in the Wdm3Information WMI data block. Therefore, I have not so far been able to test this method.

Listing 12.9 ExecuteWmiMethod **routine**

```
NTSTATUS ExecuteWmiMethod(
    IN PDEVICE_OBJECT fdo, IN PIRP Irp, IN ULONG GuidIndex,
    IN ULONG InstanceIndex, IN ULONG MethodId, IN ULONG InBufferSize,
    IN ULONG OutBufferSize, IN OUT PUCHAR Buffer)
```

Listing 12.9 `ExecuteWmiMethod` **routine (continued)**

```
{
    DebugPrint("ExecuteWmiMethod: GuidIndex %d, InstanceIndex %d, "
        "MethodId %d, InBufferSize %d OutBufferSize %d",
        GuidIndex,InstanceIndex,MethodId,InBufferSize,OutBufferSize);
    PWDM3_DEVICE_EXTENSION dx = (PWDM3_DEVICE_EXTENSION)fdo->DeviceExtension;

    if( GuidIndex==WDM3_WMI_GUID_INDEX && MethodId==0)
    {
        DebugPrintMsg("ExecuteWmiMethod: PowerDown method");

        // Power Down
        if( dx->PowerState<PowerDeviceD3)
            SendDeviceSetPower( dx, PowerDeviceD3);
        return WmiCompleteRequest( fdo, Irp, STATUS_SUCCESS, 0,
            IO_NO_INCREMENT);
    }
    return FailWMIRequest( fdo, Irp, GuidIndex);
}
```

Firing WMI Events

To fire a WMI event, simply call `WmiFireEvent`, passing the relevant WMI event block GUID, the `InstanceIndex`, and any event data. `WmiFireEvent` makes the appropriate call to `IoWmiWriteEvent`.

The Wdm3 driver provides a helper function `Wdm3FireEvent`, shown in Listing 12.10. This takes a NULL-terminated wide string, formats it into a `Wdm3Event` message, and calls `WmiFireEvent`.

A user application must ask for events first by setting an `Enable` flag. This request arrives at the driver in its `WmiFunctionControl` optional callback routine. If the Function parameter is `WmiEventControl`, the new `Enable` setting is stored in the *WMIEventEnabled* flag in the device extension. `WmiFunctionControl` is also called to ask a driver to collect WMI data blocks that were marked as being expensive to collect.

The *WMIEventEnabled* flag is initially FALSE, which means that no events are generated. As I have found no way of registering for events, the `Wdm3FireEvent` and `WmiFunctionControl` functions have not been tested.

Listing 12.10 `Wdm3FireEvent` **and** `WmiFunctionControl`
                  routines

```
void Wdm3FireEvent( IN PDEVICE_OBJECT fdo, wchar_t* Msg)
{
    DebugPrint("Wdm3FireEvent: Msg %S", Msg);
    PWDM3_DEVICE_EXTENSION dx = (PWDM3_DEVICE_EXTENSION)fdo->DeviceExtension;
```

Listing 12.10 `Wdm3FireEvent` **and** `WmiFunctionControl` routines (continued)

```
    if( !dx->WMIEventEnabled) return;

    // Get MsgLen in bytes
    int MsgLen = 0;
    wchar_t* Msg2 = Msg;
    while( *Msg2++!=0)
        MsgLen += sizeof(wchar_t);

    // Allocate event memory
    PUSHORT pData = (PUSHORT)ExAllocatePool( NonPagedPool, MsgLen+2);
    if( pData==NULL) return;
    PUSHORT pData2 = pData;
    *pData2++ = MsgLen;
    RtlMoveMemory( pData2, Msg, MsgLen);
    WmiFireEvent( fdo, (LPGUID)&WDM3_WMI_EVENT_GUID, 0, MsgLen+2, pData);
}


NTSTATUS WmiFunctionControl( IN PDEVICE_OBJECT fdo, IN PIRP Irp,
        IN ULONG GuidIndex, IN WMIENABLEDISABLECONTROL Function,
        IN BOOLEAN Enable)
{
    DebugPrint("WmiFunctionControl: GuidIndex %d, Function %d, Enable %d",
        GuidIndex,Function,Enable);
    PWDM3_DEVICE_EXTENSION dx = (PWDM3_DEVICE_EXTENSION)fdo->DeviceExtension;

    if( GuidIndex==WDM3_WMI_EVENT_GUID_INDEX && Function==WmiEventControl)
    {
        DebugPrint("WmiFunctionControl: Event enable %d", Enable);
        dx->WMIEventEnabled = Enable;
        return WmiCompleteRequest( fdo, Irp, STATUS_SUCCESS, 0,
            IO_NO_INCREMENT);
    }
    return FailWMIRequest( fdo, Irp, GuidIndex);
}
```

WMI in Action

The WMI features of the Wdm3 driver can be tested in three ways. In all cases, the DebugPrint output of the checked build gives copious trace information about the requests as they happen.

The first thing I must say is that I could not get the driver to compile or run in Windows 98. I am sure that I could have got the Wdm3 driver to compile if I had copied the correct headers and libraries into the W98 DDK, but that did not seem appropriate. It is possible that the one test Windows 98 computer available had its WBEM/WMI runtime corrupted. In the Windows 2000 beta 2, I obtained an update to the WBEM/WMI SDK. Perhaps this update was not meant for Windows 98 and therefore messed up the runtime environment. Anyway, I was able to do the rest of my testing only in Windows 2000.

The MSPower_DeviceEnable WMI data block is used by the Device Manager to display the Power Management tab in the device properties box, as shown in Figure 10.3 in Chapter 10. It retrieves the *Enable* property, eventually in a call to QueryWmiDataBlock. If the user changes the setting, it is changed in SetWmiDataBlock when the device properties box is closed.

Alternatively, the standard *WBEM Object Browser* can be used to inspect the MSPower_ DeviceEnable or Wdm3Information WMI data blocks. Log on to the \Root\WMI namespace as the current user, select the Wdm3Information class, for instance, and select one of the Wdm3 device instances.

Figure 12.1 shows the Wdm3Information display for the Wdm3 device that has an instance name of Root\\Unknown\\0004_0 as its key property. The *Wdm3Test* application has been run, so the current buffer length is 4. The *BufferFirstWord* property has a decimal value of 2882400001, which is hex 0xabcdef01.

W2000 Beta 3 includes the *WBEMTest* tool (System32\WBEM\wbemtest.exe) that can also be used to inspect WBEM classes and instances.

By the way, I found that the WBEM Object Browser seemed a bit sluggish. This is possibly because it is using WQL to query the database. In contrast, the Device Manager's use of WMI when dealing with a device's Power Management tab seems to run quickly.

Figure 12.1 WBEM Object Browser inspecting a Wdm3Information instance

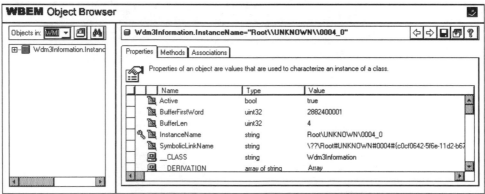

@ 1999 PHD Computer Consultants Ltd

You can use the *WBEM CIM Studio* to look at the class definitions. If you change a WMI data block, use WBEM CIM Studio to delete the old class definition. A reboot of the computer seems to be necessary to reregister the class correctly.

I could not work out how to use the *WBEM Event Registration* utility to enable events in the Wdm3 driver.

Conclusion

This chapter has shown that it is reasonably straightforward to add administrative control facilities to a driver using the Windows Management Instrumentation extensions for WDM. Despite claims to the contrary, I could not get WMI to run in Windows 98, reducing its usefulness. I also could not get WMI events and methods to work in the Windows 2000 beta 2.

The next chapter looks at NT events, another way of reporting important events to NT 3.51, NT 4, and Windows 2000 administrators. It also concludes my look at the features of the Wdm3 example driver.

Chapter 13

Event Reporting

This chapter looks at the second method of reporting information to system administrators, NT events. The last chapter looked at the first method, Windows Management Instrumentation (WMI), which ought to work in Windows 98, as well as the NT and Windows 2000 platforms.

Drivers can generate NT events in NT 3.51, NT 4, and Windows 2000. Events are stored in a system event log that can be viewed by a user mode Event Viewer.

The Wdm3 driver generates NT events in a few places. Although the Wdm3 example is a WDM device driver, NT events can and should be generated by NT style NT 3.51 and NT 4 drivers.

Overview

In NT 3.51, NT 4, and Windows 2000, drivers should report any problems to the system event log. Windows 98 WDM device drivers can make the relevant kernel calls, but they do nothing.

Once events are firmly in the event log, they are preserved even if a system crashes. Events can, therefore, be useful in some debugging circumstances (e.g., where DebugPrint information is lost as a driver crashes).

In NT 3.51 and NT 4, use the *EventVwr* tool to view events. In Windows 2000 use the *Event Viewer* System Log portion of the *Computer Management Console*. In both cases, you must double-click on a record to bring up the full details of the event, as shown in Figure 13.1.

The Event Detail tab shows most of the event information. Events are categorized as either Informational, Warning, or Error. The message text is taken from a resource in the driver's executable. A driver can specify some small extra strings that are inserted into the message text.

The Record Data tab shows (in hex) any additional data bytes that were passed by the driver. In Windows 2000, most drivers always seem to show at least 0x28 bytes of record data. Any data that your driver provides starts at offset 0x28.

Do not swamp the event log with superfluous information. Obviously, try to report errors in a meaningful way. Remember that the event log will only be useful when a problem arises. Some informational messages may be useful for displaying status information, such as network addresses.

If you are being clever, you could dynamically adjust the amount of information that you produce. You might start off by reporting transactions that need to be retried as warning messages. If these keep occurring, you could stop reporting these retry messages.

Other drivers inspect a registry value when they start up to determine the level of reporting. During debugging or diagnostic testing, the registry value could be set in such as way as to generate lots of useful reports. This may be the only way to obtain debugging information in the field.

Figure 13.1 Event Viewer in action

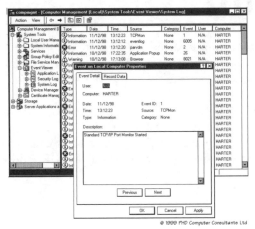

@ 1999 PHD Computer Consultants Ltd

Message Files

When you log an event, you pass an EventId, a number specifying the event that you are reporting. You must include a message resource in your driver if you want the event viewer to display the appropriate description.

The Wdm3Msg.mc message file for the Wdm3 driver is shown in Listing 13.1. The Message-IdTypedef, SeverityNames, and FacilityNames sections are fairly standard. A facility identifies the type of driver. Most driver writers use the spare facility number of 0x7 for the Wdm3 facility. Microsoft defined facility numbers are defined in NTSTATUS.H.

The following blocks of lines define one message at a time. The contents of each line are self-explanatory. The actual event message is on one or more lines, ending with a line that contains just a period. The following escape codes have special meaning in the message text:

%b is a space, %t is a tab, %r is a carriage return, and %n is a linefeed. In addition, %1 to %99 are where driver-supplied strings are inserted. Actually, %1 is always the driver name, so the driver strings start with %2.

Listing 13.1 Wdm3Msg.mc **message file**

```
MessageIdTypedef = NTSTATUS


SeverityNames = (
    Success       = 0x0:STATUS_SEVERITY_SUCCESS
    Informational = 0x1:STATUS_SEVERITY_INFORMATIONAL
    Warning       = 0x2:STATUS_SEVERITY_WARNING
    Error         = 0x3:STATUS_SEVERITY_ERROR
    )


FacilityNames = (
    System        = 0x0
    Wdm3          = 0x7:FACILITY_WDM3_ERROR_CODE
    )



MessageId=0x0001
Facility=Wdm3
Severity=Informational
SymbolicName=WDM3_MSG_LOGGING_STARTED
Language=English
Event logging enabled for Wdm3 Driver.
.


MessageId=+1
Facility=Wdm3
Severity=Informational
SymbolicName=WDM3_MESSAGE
Language=English
Message: %2.
.
```

The mc command is used to compile the message definition file. It produces three or more output files. In this case, these are the Wdm3Msg.rc resource script, the Wdm3Msg.h header file, and the MSG00001.BIN message data file. Further message files are produced if you support more than one language. The Wdm3Msg.rc resource script contains just a reference to the MSG00001.BIN message data file (or files), as follows.

```
LANGUAGE 0x9,0x1
1 11 MSG00001.bin
```

The `Wdm3Msg.h` header file contains the message symbolic names defined in a form that can be used by the driver code, as shown in Listing 13.2. The message ID, severity, and facility code have been combined, with the "customer" bit set to make a suitable NTSTATUS value. The main Wdm3 header, `Wdm3.h`, now also includes `Wdm3Msg.h`.

Listing 13.2 `Wdm3Msg.h` file

```
// MessageId: WDM3_MSG_LOGGING_STARTED
//
// MessageText:
//
//  Event logging enabled for Wdm3 Driver.
//
#define WDM3_MSG_LOGGING_STARTED        ((NTSTATUS)0x60070001L)


//
// MessageId: WDM3_MESSAGE
//
// MessageText:
//
//  Message: %2.
//
#define WDM3_MESSAGE                    ((NTSTATUS)0x60070002L)
```

The message file must be compiled before the main driver code is built. The `NTTARGETFILE0` macro in the `SOURCES` file is used to specify any prebuild steps.

```
NTTARGETFILE0=prebuild
```

As described in Chapter 4, this invokes *nmake* on the `makefile.inc` makefile before the main compile. The `prebuild` step compiles the WMI MOF file and the event message definition file, as shown in Listing 13.3. The `mc` command is run, if necessary, using the `-c` option to set the "customer" bit and the `-v` option for verbose output.

Listing 13.3 New `makefile.inc`

```
prebuild:    Wdm3Msg.h Wdm3.bmf

Wdm3.bmf:    Wdm3.mof
    mofcomp -B:Wdm3.bmf -WMI Wdm3.mof

Wdm3Msg.rc Wdm3Msg.h: Wdm3Msg.mc
    mc -v -c Wdm3Msg.mc

PostBuildSteps: $(TARGET)
```

Listing 13.3 New `makefile.inc` **(continued)**

```
!if "$(DDKBUILDENV)"=="free"
    rebase -B 0x10000 -X . $(TARGET)
!endif
    copy $(TARGET) $(WINDIR)\system32\drivers
```

The final change to the build process is to make the main resource file, `Wdm3.rc`, include the message resource script, `Wdm3Msg.rc`. In Visual C++, select the *View+Resource Includes...* menu and add the following line to the "Read-only symbol directives" box.

```
#include "Wdm3Msg.rc"
```

In Windows 2000, I found that building the driver from a changed message definition file initially reported an error but then went on to compile successfully.

Registering as an Event Source

The final hurdle to overcome is registering your driver as an event source so that the event viewer knows where to find your message text resource. Two registry changes must be made.

First, the `HKLM\System\CurrentControlSet\Services\EventLog\System` key has an existing `REG_MULTI_SZ` value called `Sources`. Add the name of your driver's executable (without the extension) as a line in `Sources`.

In this same registry key, make a new subkey with this same driver name. In this subkey, add a `REG_EXPAND_SZ` value called `EventMessageFile` and set a `REG_DWORD` called `TypesSup-ported` with a value of 0x7. For `Wdm3`, set `EventMessageFile` to the following value.

```
%SystemRoot%\System32\IoLogMsg.dll;%SystemRoot%\System32\Drivers\Wdm3.sys
```

The `Wdm3` installation INF file supposedly has the correct information to make these registry changes when a `Wdm3` device is installed. Listing 13.4 shows the amendments made to the standard installation file. The `AddService` directive's last field specifies the name of the section containing the error logging registry values. There are optional fields to specify the log type (`System`, `Security`, or `Application`) and a log name.

The `Wdm3 Service.EventLog` section specifies the values for the `EventMessageFile` and `TypesSupported` values. The `EventMessageFile` entry is on one long line.

However, I found that this did not work completely in Windows 2000 Beta 3. "Wdm3" was correctly added to the `Sources` value and the `HKLM\System\CurrentControlSet\Services\EventLog\System\Wdm3` key was correctly made, but no values were placed in the key.

It is simplest just to add these registry entries by hand. You will have to use *RegEdt32* to use the required registry types.

For NT 3.51 and NT 4 drivers, you cannot use an INF installation file. Instead, you will have to amend your installation program to set up the registry entries. The example installation code, install.cpp (on the book's CD-ROM) shows how to do this job.

Listing 13.4 Wdm3free.inf **installation file event logging sections**

```
[Wdm3.Install.NT.Services]
AddService =
    Wdm3, %SPSVCINST_ASSOCSERVICE%, Wdm3.Service, Wdm3.Service.EventLog


; ...


[Wdm3.Service.EventLog]
HKR,,EventMessageFile,%FLG_ADDREG_TYPE_EXPAND_SZ%,
    "%%SystemRoot%%\System32\IoLogMsg .dll;
        %%SystemRoot%%\System32\drivers\Wdm3.sys"
HKR,,TypesSupported,%FLG_ADDREG_TYPE_DWORD%,7
```

Generating Events

Listing 13.5 shows the routines that provide the event logging, InitializeEventLog, and LogEvent. Later, I shall describe the Wdm3EventMessage function that provides a simpler interface.

InitializeEventLog is simply used to store a pointer to the main DriverObject. It then calls LogEvent to send a WDM3_MSG_LOGGING_STARTED event. If all's well, this should be displayed in the Event Viewer with the corresponding description, "Event logging enabled for Wdm3 Driver".

LogEvent does the bulk of the work in logging an NT event. It may be called at DISPATCH_LEVEL or lower. Its first task is to decide the size of the error log packet. It then calls IoAllocateErrorLogEntry to obtain a suitably sized packet. It then fills the packet and sends it off using IoWriteErrorLogEntry.

LogEvent has parameters for the message ID (from the list in Wdm3Msg.h) and optionally an IRP pointer. If the IRP pointer is given, various fields in the packet are filled in with details of the IRP. LogEvent can also accept DumpData and Strings as parameters, for insertion into the event packet.

The basic IO_ERROR_LOG_PACKET structure contains one dump data ULONG, but no insertion strings. It is an extendible structure. Zero or more dump data ULONGs can be provided, followed immediately by any NULL terminated wide strings. This makes calculating and filling the packet size slightly involved. LogEvent saves each string length in a temporary array of integers called StringSizes. Note that the maximum packet size, ERROR_LOG_MAXIMUM_SIZE, is only 0x98 bytes, so do not try to pass large insertion strings.

Listing 13.5 InitializeEventLog **and** LogEvent **routines**

```
void InitializeEventLog( IN PDRIVER_OBJECT DriverObject)
{
    SavedDriverObject = DriverObject;
```

Listing 13.5 InitializeEventLog **and** LogEvent **routines (continued)**

```
    // Log a message saying that logging is started.
    LogEvent( WDM3_MSG_LOGGING_STARTED, NULL,  // IRP
              NULL, 0,                          // dump data
              NULL, 0);                         // strings
}


bool LogEvent(  IN NTSTATUS ErrorCode,
                IN PIRP     Irp,
                IN ULONG    DumpData[],
                IN int      DumpDataCount,
                IN PWSTR    Strings[],
                IN int      StringCount)
{
    if( SavedDriverObject==NULL ) return false;

    // Start working out size of complete event packet
    int size = sizeof(IO_ERROR_LOG_PACKET);

    // Add in dump data size.
    // Less one as DumpData already has 1 ULONG in IO_ERROR_LOG_PACKET
    if( DumpDataCount>0)
        size += sizeof(ULONG) * (DumpDataCount-1);

    // Add in space needed for insertion strings (inc terminating NULLs)
    int* StringSizes = NULL;
    if( StringCount>0)
    {
        StringSizes =
            (int*)ExAllocatePool(NonPagedPool,StringCount*sizeof(int));
        if( StringSizes==NULL) return false;

        // Remember each string size
        for( int i=0; i<StringCount; i++)
        {
            StringSizes[i] = (int)GetWideStringSize(Strings[i]);
            size += StringSizes[i];
        }
    }

    if( size>ERROR_LOG_MAXIMUM_SIZE)     // 0x98!
```

Listing 13.5 `InitializeEventLog` **and** `LogEvent` **routines (continued)**

```
{
    if( StringSizes!=NULL) ExFreePool(StringSizes);
    return false;
}


// Try to allocate the packet
PIO_ERROR_LOG_PACKET Packet = (PIO_ERROR_LOG_PACKET)
        IoAllocateErrorLogEntry( SavedDriverObject, size);
if( Packet==NULL)
{
    if( StringSizes!=NULL) ExFreePool(StringSizes);
    return false;
}


// Fill in standard parts of the packet
Packet->ErrorCode = ErrorCode;
Packet->UniqueErrorValue = 0;


// Fill in IRP related fields
Packet->MajorFunctionCode = 0;
Packet->RetryCount = 0;
Packet->FinalStatus = 0;
Packet->SequenceNumber = 0;
Packet->IoControlCode = 0;
if( Irp!=NULL)
{
    PIO_STACK_LOCATION IrpStack = IoGetCurrentIrpStackLocation(Irp);

    Packet->MajorFunctionCode = IrpStack->MajorFunction;
    Packet->FinalStatus = Irp->IoStatus.Status;

    if( IrpStack->MajorFunction==IRP_MJ_DEVICE_CONTROL ||
        IrpStack->MajorFunction==IRP_MJ_INTERNAL_DEVICE_CONTROL)
        Packet->IoControlCode =
            IrpStack->Parameters.DeviceIoControl.IoControlCode;
}

// Fill in dump data
if( DumpDataCount>0)
{
    Packet->DumpDataSize = (USHORT)(sizeof(ULONG)*DumpDataCount);
```

Listing 13.5 `InitializeEventLog` **and** `LogEvent` **routines (continued)**

```
        for( int i=0; i<DumpDataCount; i++)
            Packet->DumpData[i] = DumpData[i];
    }
    else Packet->DumpDataSize = 0;


    // Fill in insertion strings after DumpData
    Packet->NumberOfStrings = (USHORT)StringCount;

    if( StringCount>0)
    {
        Packet->StringOffset = sizeof(IO_ERROR_LOG_PACKET) +
                              (DumpDataCount-1) * sizeof(ULONG);

        PUCHAR pInsertionString = (PUCHAR)Packet + Packet->StringOffset;

        // Add each new string to the end
        for( int i=0; i<StringCount; i++)
        {
            RtlMoveMemory( pInsertionString, Strings[i], StringSizes[i]);
            pInsertionString += StringSizes[i];
        }
    }


    // Log the message
    IoWriteErrorLogEntry(Packet);


    if( StringSizes!=NULL) ExFreePool(StringSizes);
    return true;
}
```

To make `LogEvent` easier to use, `Wdm3` provides a function called `Wdm3EventMessage` that simply takes an ANSI string as an argument. This is passed as a wide string to `LogEvent` to be inserted in the `WDM3_MESSAGE` message. `Wdm3EventMessage` is shown in Listing 13.6.

Listing 13.6 `Wdm3EventMessage` **routines**

```
void Wdm3EventMessage( const char* Msg)
{
    int MsgLen = GetAnsiStringSize(Msg);
    int wMsgLen = MsgLen*2;
    PWSTR wMsg = (PWSTR)ExAllocatePool(NonPagedPool,wMsgLen);
    if( wMsg==NULL) return;
```

Listing 13.6 Wdm3EventMessage **routines (continued)**

```
    // Brutally make into a wide string
    for( int i=0;i<MsgLen;i++)
        wMsg[i] = (WCHAR)(unsigned char)Msg[i];


    PWSTR Strings[1] = { wMsg };
    LogEvent( WDM3_MESSAGE, NULL,    // IRP
              NULL, 0,               // dump data
              Strings, 1);           // strings


    ExFreePool(wMsg);
}
```

Testing Wdm3 Events

The Wdm3EventMessage function is called in a few places in the Wdm3 driver as a test. Wdm3Unload sends a message "Unload", Wdm3AddDevice sends "AddDevice", and Wdm3Pnp sends "PnP xx" where "xx" represents the Plug and Play minor function code, in hex.

Installing and reinstalling a Wdm3 device should generate several events in the System event log. The events are generated in both the free and checked build versions. Remember to refresh the Event Viewer display to see any new events.

Conclusion

You should use NT events to report any problem events in NT 3.51, NT 4, and Windows 2000 drivers. The two relevant kernel function calls are just stubs in Windows 98, so you can safely generate events in WDM device drivers.

Chapter 14

DebugPrint

This chapter looks at a fully-fledged driver, DebugPrint. The *DebugPrint* software lets you use formatted print trace statements in your code and view the output in the *DebugPrint Monitor* Win32 program.

Along the way, I cover the following important device driver topics.

- System threads
- Dispatcher objects: events, Mutexes, and semaphores
- Linked lists
- File I/O in drivers
- Queuing IRPs
- Basic cancel routines
- Win32 overlapped I/O requests

Design Specification

The main design requirement for the *DebugPrint* software is that test drivers can do formatted prints that appear in a user mode application running on the same PC as the driver. The software must work under Windows 98 and Windows 2000. It should be easy for developers to include trace statements in their code. They should be able to use trace statements in most types of driver code.

This specification does not include source level debugging or breakpoints.

Design Implementation

Figure 14.1 shows the design used in the *DebugPrint* software. The figure shows that more than one test driver can run at the same time. In each test driver, the trace output is first stored internally in an EventList doubly-linked list. A DebugPrint system thread in the driver code reads the EventList and writes the events to the DebugPrint driver.

The DebugPrint driver write routine stores the trace events in its own EventList. The *DebugPrint Monitor* application issues read requests to the DebugPrint driver to read the trace events. These are then displayed to the user in the *DebugPrint Monitor* window.

An alternative design might have removed the DebugPrint driver. The test driver Debug-Print system thread could then have written the trace events directly to a disk file. However, it is unclear whether it is possible to lock such a file so that "simultaneous" accesses by the test drivers and *DebugPrint Monitor* would be handled properly. Writing to a disk file would also be slower.

Figure 14.1 *DebugPrint* design

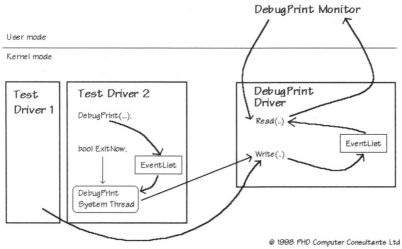

@ 1998 PHD Computer Consultants Ltd

The *DebugPrint* software, therefore, consists of the following three different pieces of code.

- Code added to the test driver to produce events
- The DebugPrint driver
- The *DebugPrint Monitor* application

Test Driver Code

As explained in Chapter 6, a driver writer has to add DebugPrint.c[1] and DebugPrint.h source files to their driver project to support DebugPrint calls. These routines ensure that the

---

1.   Since writing this chapter, the test driver DebugPrint code has changed from C++ to C. The code on the book software disk is slightly different from the code printed in this chapter, though they are functionally identical.

trace statement output is sent to the DebugPrint driver. It should eventually be possible to put all the code in DebugPrint.c into a static library or DLL that is linked with test drivers.

The main job performed by the DebugPrint code in each test driver is to write events to the DebugPrint driver. The kernel provides several routines to call other drivers, including ZwCreateFile, ZwWriteFile, and ZwClose. The documentation for these functions says that these functions can only be called at PASSIVE_LEVEL IRQL. This means that they cannot be called directly from some sorts of driver code.

In my initial design, I ignored this issue. However, I soon ran into another problem that was more difficult to track down. Eventually, I worked out that calls to ZwWriteFile have to be made in the same thread context as ZwCreateFile. The DDK documentation did not make this point prominently. In my initial design, I used ZwCreateFile in the test driver DriverEntry routine. However, the print routines using ZwWriteFile could naturally be called in dispatch routines, which are usually not running in the same thread context.

Thread contexts are not normally a problem for most device drivers. When you process a normal user request, the IRP has all the information you need. The fact that you may be running in an arbitrary thread context does not effect the job you have to do. The Zw... file access functions are one case when the thread context is significant. If you use "neither" Buffered I/O nor Direct I/O or use METHOD NEITHER IOCTLs, the thread context is also important.

System Threads

The solution to both these problems is to use a system thread. A driver can create a system thread (or threads) that runs in kernel mode. This thread runs at PASSIVE_LEVEL IRQL. If the DebugPrint system thread makes all the Zw... calls, the "same thread" problem is fixed. A doubly-linked list is used to store the events for processing by the system thread. Inserting events into this list can be done safely at IRQL levels up to and including DISPATCH_LEVEL. This means that most types of driver code can generate DebugPrint trace output.

The DebugPrintInit routine calls PsCreateSystemThread (at PASSIVE_LEVEL IRQL) to create its system thread as shown in the partial code in Listing 14.1. PsCreateSystemThread is passed the name of the function to run and a context to pass to it. The DebugPrint system thread function is defined as follows.

```
void DebugPrintSystemThread( IN PVOID Context)
```

DebugPrintInit passes just a NULL context to the thread function. Some drivers may wish to create one thread per device, and so will usually pass a pointer to the device extension as the context. A DebugPrint test driver has just one system thread for all its devices.

The DebugPrint system thread does not need a high priority. Therefore, DebugPrintSystemThread calls KeSetPriorityThread straightaway to set its priority to the lowest real-time priority. It uses KeGetCurrentThread to get a pointer to its own thread object.

If PsCreateSystemThread succeeds, it returns a handle to the thread. Later, I shall show that a driver can wait for certain objects to be set. It can wait for the completion of a system thread if it has a pointer to the system thread object, not a handle. Use ObReferenceObjectByHandle to retrieve the thread object pointer. If this succeeds, call ZwClose to close the thread handle. Technically speaking, the call to ZwClose reduces the reference count to the thread handle. When the thread completes, its handle reference count will be decremented to zero and the handle can be discarded. Eventually the thread object pointer must be dereferenced similarly using ObDereferenceObject.

A system thread must terminate itself using PsTerminateSystemThread. I am not sure what happens if a system thread function simply returns without calling PsTerminateSystemThread. The main driver code cannot force a system thread to terminate. For this reason, a global Boolean variable called ExitNow is used. This is set true when the driver wants the system thread to exit.

DebugPrintClose waits for the ThreadExiting event to be set into the signalled state by the thread as it exits. This ensures that the thread has completed by the time DebugPrintClose exits. DebugPrintClose is commonly called from a driver unload routine. All driver code must have stopped running when the unload routine exits, as the driver address space may soon be deleted. Make sure that any other call-backs are disabled, e.g., interrupt handlers, Deferred Procedure Calls, and timers.

Listing 14.1 DebugPrint **test driver thread and event handling**

```
KEVENT ThreadEvent;
KEVENT ThreadExiting;
PVOID ThreadObjectPointer=NULL;

void DebugPrintInit(char* _DriverName)
{
    // ...
    ExitNow = false;
    KeInitializeEvent(&ThreadEvent, SynchronizationEvent, FALSE);
    KeInitializeEvent(&ThreadExiting, SynchronizationEvent, FALSE);
    HANDLE threadHandle;

    status = PsCreateSystemThread( &threadHandle, THREAD_ALL_ACCESS, NULL, NULL,
        NULL, DebugPrintSystemThread, NULL);
    if( !NT_SUCCESS(status))
        return;

    status = ObReferenceObjectByHandle( threadHandle, THREAD_ALL_ACCESS, NULL,
        KernelMode, &ThreadObjectPointer, NULL);
    if( NT_SUCCESS(status))
        ZwClose(threadHandle);
    // ...
}

void DebugPrintClose()
{
    // ...
    ExitNow = true;
    KeSetEvent(&ThreadEvent, 0, FALSE);
    KeWaitForSingleObject( &ThreadExiting, Executive, KernelMode, FALSE, NULL);
```

Listing 14.1 DebugPrint **test driver thread and event handling (continued)**

```
    // ...
}


void DebugPrintSystemThread( IN PVOID Context)
{
    // Lower thread priority
    KeSetPriorityThread( KeGetCurrentThread(), LOW_REALTIME_PRIORITY);

    // Make One second relative timeout
    LARGE_INTEGER OneSecondTimeout;
    OneSecondTimeout.QuadPart = -1i64 * 1000000i64 * 10i64;

    // Loop waiting for events or ExitNow
    while(true)
    {
        KeWaitForSingleObject( &ThreadEvent, Executive,
                KernelMode, FALSE, &OneSecondTimeout);

        // Process events
        // ...

        if( ExitNow)
            break;
    }

    // Tidy up
    if(ThreadObjectPointer!=NULL)
    {
        ObDereferenceObject(&ThreadObjectPointer);
        ThreadObjectPointer = NULL;
    }
    DebugPrintStarted = FALSE;
    ClearEvents();
    KeSetEvent( &ThreadExiting, 0, FALSE);
    PsTerminateSystemThread(STATUS_SUCCESS);
}
```

System Worker Threads

An alternative technique is available if you want to perform occasional short tasks at PASSIVE_LEVEL IRQL. However, this system worker thread method was not suitable for the *DebugPrint* software because the thread context may change between calls.

To use the system worker thread method, first allocate a WORK_QUEUE_ITEM structure from nonpaged memory. Call ExInitializeWorkItem passing this pointer, a callback routine, and a context for it. When you want your function to be run, call ExQueueWorkItem. In due course, your callback routine is called at PASSIVE_LEVEL in the context of a system thread. Do not forget to free the WORK_QUEUE_ITEM structure memory when finished with it. System worker threads have a lower priority than system threads running at the lowest real time priority, but higher than most user mode threads.

The WdmIo and PHDIo drivers, described in Chapters 15-18, show how to use a system worker thread.

In W2000, it is recommended that you use the IoAllocateWorkItem, IoQueueWorkItem, IoFreeWorkItem functions instead.

Events

The main DebugPrint test driver code sets the ExitNow Boolean to true when it wants its system thread to terminate. However, it is not a good idea for the system thread to spin continuously waiting for this value to become true.

Instead, a kernel event called ThreadEvent signals when to check ExitNow. A KEVENT must be defined for the event in nonpaged memory. Kernel events are very similar to their user mode Win32 cousins.

Listing 14.1 shows how ThreadEvent is initialized using KeInitializeEvent, at PASSIVE_LEVEL IRQL. Two types of events are supported, SynchronizationEvent and NotificationEvent. The last parameter sets the initial state of the event, which is nonsignalled in this case.

When set, a Synchronization event only releases one waiting thread before reverting to the nonsignalled state. A Notification event stays signalled until explicitly reset.

DebugPrintClose uses KeSetEvent to set an event into the signalled state, after setting ExitNow to true. The third parameter to KeSetEvent specifies whether you are going to call one of the KeWait... routines straightaway. If not, you can call KeSetEvent at any IRQL up to and including DISPATCH_LEVEL. If waiting, you must be running at PASSIVE_LEVEL.

If you need to put an event into the nonsignalled state, call KeClearEvent or call KeResetEvent to determine the previous event state. You can use KeReadStateEvent to read the event state. All these routines can be called at DISPATCH_LEVEL or lower.

For NT and W2000 drivers you can use IoCreateNotificationEvent and IoCreateSynchronizationEvent to share an event between two or more drivers.

Synchronization

A thread running at PASSIVE_LEVEL can synchronize with other activities by waiting for *dispatcher* objects such as *events*, *Mutex* objects, and *semaphores*. You can wait for *timer* and *thread* objects. Finally, you can also wait on *file objects* if they have been opened in ZwCreateFile for overlapped I/O.

Although driver dispatch routines run at PASSIVE_LEVEL, they should not wait on kernel dispatcher objects, other than with a zero time-out. You can wait for inherently synchronous operations to complete using nonzero time-outs. Plug and Play handlers can wait on dispatcher objects. For example, the ForwardIrpAndWait routine described in Chapter 9 uses an event to signal when lower drivers have finished processing an IRP.

A thread waits for dispatcher objects to become signalled using KeWaitForSingleObject or KeWaitForMultipleObjects, which are similar to the Win32 equivalents. As Table 14.1 shows, KeWaitForSingleObject waits on just one dispatcher object, or until a time-out has expired. A negative timeout value is used for relative periods, as a LARGE_INTEGER in units of 100 nanoseconds. The DebugPrint system thread calls KeWaitForSingleObject with a relative time-out of one second. Positive time-out values represent an absolute system time, in 100-nanosecond units since January 1, 1601[2] in the GMT time zone.

The KeWaitForMultipleObjects routine works in a similar way, except that you can pass an array of dispatcher objects. You can opt to wait for just one of the objects to become signalled, or all of them.

Table 14.1   KeWaitForSingleObject **function**

NTSTATUS KeWaitForSingleObject	(IRQL—PASSIVE_LEVEL) **or at** DISPATCH_LEVEL **if a zero time-out is given**
Parameter	Description
IN PVOID Object	Pointer to dispatcher object
IN KWAIT_REASON WaitReason	Usually Executive for drivers, but can be UserRequest if running for user in a user thread.
IN KPROCESSOR_MODE WaitMode	KernelMode for drivers
IN BOOLEAN Alertable	FALSE for drivers
IN PLARGE_INTEGER Timeout	NULL for an infinite time-out. Negative time-outs are relative. Positive time-outs are absolute.
Returns	STATUS_SUCCESS STATUS_TIMEOUT

Mutex Objects

A Mutex is a mutual exclusion dispatcher object that can only be owned by one thread at a time. Mutexes are sometimes called "mutants." Initialize a KMUTEX object in nonpaged memory using KeInitializeMutex; the Level parameter is used to ensure that multiprocessor Windows 2000 systems can acquire multiple Mutexes safely.

A Mutex object is in the signalled state when it is available. A thread requests ownership using one of the KeWaitFor... routines. If two or more threads are waiting for a Mutex, only one thread will wake up and become its owner. Call KeReleaseMutex to release ownership.

If you already own a Mutex and ask for it again, the KeWaitFor... routine will return immediately. An internal counter is incremented, so call KeReleaseMutex once for each time you requested ownership of the Mutex.

---

2. I.e., soon after the Gregorian calendar was introduced in 1582.

The kernel causes a bugcheck if you do not release a Mutex before returning control to the I/O Manager. KeInitializeMutex and KeReleaseMutex must be called at PASSIVE_LEVEL. You can also inspect the Mutex state using KeReadStateMutex at an IRQL up to and including DISPATCH_LEVEL.

A *Fast Mutex* is a variation on an ordinary Mutex that is faster because it does not permit multiple ownership requests. An *Executive Resource* is another similar synchronization object, available in W2000 only. See the DDK documentation for more details of these objects.

Semaphores

A semaphore is a dispatcher object that maintains a count. Call KeInitializeSemaphore at PASSIVE_LEVEL IRQL to initialize a KSEMAPHORE object in nonpaged memory. You must specify maximum and initial counts.

A semaphore is nonsignalled when zero and signalled with any count greater than zero. A thread that calls one of the KeWaitFor... routines and finds a signalled semaphore will decrement its count and the thread will proceed. If a semaphore's count is 2 and three threads simultaneously attempt to wait for the semaphore, only two will proceed. The semaphore count ends up as 0 with one thread still waiting.

Call KeReleaseSemaphore, at DISPATCH_LEVEL or lower, to add a value to a semaphore count. You can read the semaphore count at any IRQL using KeReadStateSemaphore.

Timer, Thread, and File Objects

A timer is a dispatcher object that becomes signalled when its timer expires. A file object becomes signalled when an overlapped I/O operation has completed. The file must have been opened in ZwCreateFile with the DesiredAccess SYNCHRONIZE flag set. You can also wait for thread completion.

DebugPrint **System Thread Function**

Listing 14.1 shows that the DebugPrint system thread for drivers under test primarily consists of a loop that waits for the ExitNow flag to become true or for trace events to arrive.

At the top of this main loop, the thread function calls KeWaitForSingleObject to wait for the ThreadEvent to become signalled. As stated previously, DebugPrintClose sets the ExitNow flag to true and sets the ThreadEvent into the signalled state. The thread function is released; if it finds ExitNow true, it exits its main loop, tidies up, and terminates.

The call to KeWaitForSingleObject includes a one-second time-out. This is used to let the thread function look for and process trace events in the EventList buffer, as described in the next sections.

Generating Trace Events

The two formatted print functions, DebugPrint and DebugPrint2, eventually call DebugPrint-Msg. I will not go into the details of how the formatted prints work. You can work it out for yourself by looking at the code in the Print... and DebugSprintf routines in DebugPrint.c. The only point to note is that the DebugPrint routines can accept a variable number of arguments. I have assumed — successfully so far — that the va_list macros defined in stdarg.h work satisfactorily in kernel mode drivers.

Listing 14.2 shows how DebugPrintMsg builds a trace event and puts it in a DEBUGPRINT_ EVENT structure allocated from nonpaged memory. The DEBUGPRINT_EVENT structure is added into the EventList doubly-linked list. DebugPrintMsg is passed a NULL-terminated ANSI message string.

The event data consists of the following three items:

- the current system time in GMT,
- the driver name (specified in DebugPrintInit), and
- the message.

DebugPrintMsg first gets the current system time in GMT using KeQuerySystemTime. It converts the number of 100-nanosecond intervals since January 1, 1601 into a more recognizable TIME_FIELDS structure using RtlTimeToTimeFields. The ExSystemTimeToLocalTime function (which converts from GMT to local time) is only available in W2000, so it is not used here. The time is converted to the local timezone in the *DebugPrint Monitor* application.

It would reduce the event structure size if the LARGE_INTEGER output from KeQuerySystem-Time were stored directly. However, there is no equivalent of the RtlTimeToTimeFields routine in Win32, so the event structure holds the time as time fields.

DebugPrintMsg now works out the size of the event data (i.e., the three previous data items, including the strings' terminating NULLs). It then determines the size of the DEBUGPRINT_EVENT structure that envelops the event data. It allocates some nonpaged memory for this structure and fills it in. It then uses ExInterlockedInsertTailList to insert the DEBUGPRINT_EVENT structure at the end of EventList.

Listing 14.2 DebugPrint **test driver** DebugPrintMsg **function**

```
void DebugPrintMsg(char* Msg)
{
    if( !DebugPrintStarted) return;

    // Get current time
    LARGE_INTEGER Now, NowLocal;
    KeQuerySystemTime(&Now);
    TIME_FIELDS NowTF;
    RtlTimeToTimeFields( &Now, &NowTF);

    // Get size of Msg and complete event
    USHORT MsgLen = ANSIstrlen(Msg)+1;
    ULONG EventDataLen = sizeof(TIME_FIELDS) + DriverNameLen + MsgLen;
    ULONG len = sizeof(LIST_ENTRY)+sizeof(ULONG)+EventDataLen;

    // Allocate event buffer
    PDEBUGPRINT_EVENT pEvent =
        (PDEBUGPRINT_EVENT)ExAllocatePool(NonPagedPool,len);
    if( pEvent!=NULL)
    {
        PUCHAR buffer = (PUCHAR)pEvent->EventData;
```

Listing 14.2 DebugPrint **test driver** DebugPrintMsg

```
        // Copy event info to buffer
        RtlCopyMemory( buffer, &NowTF, sizeof(TIME_FIELDS));
        buffer += sizeof(TIME_FIELDS);
        RtlCopyMemory( buffer, DriverName, DriverNameLen);
        buffer += DriverNameLen;
        RtlCopyMemory( buffer, Msg, MsgLen);

        // Insert event into event list for processing by system thread
        pEvent->Len = EventDataLen;
        ExInterlockedInsertTailList(&EventList,&pEvent->
            ListEntry,&EventListLock);
    }
}
```

Linked Lists

Doubly-Linked Lists

A doubly-linked list is a slightly complicated beast to use safely. First, you need to declare a LIST_ENTRY structure in nonpaged memory for the list head. Drivers that need one list per device declare the list head in the device extension. However, the DebugPrint test driver code declares just its EventList variable as a global, as it is available to all devices.

```
LIST_ENTRY EventList;
```

Next, define a structure that you want to put in your doubly-linked list. Include a LIST_ ENTRY field in this structure to provide the links in both directions of the list. The DebugPrint structure is called DEBUGPRINT_EVENT. *EventData* is a variable length field, as it is not always 1-byte long. The *Len* field gives its length.

```
typedef struct _DEBUGPRINT_EVENT
{
    LIST_ENTRY ListEntry;
    ULONG Len;
    UCHAR EventData[1];
} DEBUGPRINT_EVENT, *PDEBUGPRINT_EVENT;
```

Initialize a doubly-linked list using InitializeListHead, passing a pointer to the list head variable. You can now insert DEBUGPRINT_EVENT structures at the head or tail of the list using the InsertHeadList and InsertTailList routines. The corresponding RemoveHeadList and RemoveTailList routines remove entries from the list[3]. Find out if the list is empty first using IsListEmpty.

---

3.   The NT 4 DDK wrongly states that RemoveHeadList and RemoveTailList return NULL if the list is empty.

All well and good. However, it is important that attempts to access the list are carried out safely so that the links are not corrupted in a multiprocessor environment. The kernel provides *interlocked* versions of the add and remove routines that use a spin lock to guard access to the link structure. The DebugPrint test driver code uses a spin lock called EventListLock and initializes it as normal.

```
KeInitializeSpinLock(&EventListLock);
InitializeListHead(&EventList);
```

Listing 14.2 shows how to use one of the interlocked linked list routines, ExInterlocked-InsertTailList. It is passed pointers to the list head, the LIST_ENTRY field in your structure, and the spin lock.

Listing 14.3 shows an extract from the DebugPrint test driver system thread function. This is the code that is run every second to see if any events have been produced by calls to DebugPrintMsg.

Listing 14.3 DebugPrint **test driver system thread event processing**

```
// Loop until all available events have been removed
while(true)
{
    PLIST_ENTRY pListEntry =
        ExInterlockedRemoveHeadList( &EventList, &EventListLock);
    if( pListEntry==NULL)
        break;

    // Get event as DEBUGPRINT_EVENT
    PDEBUGPRINT_EVENT pEvent =
        CONTAINING_RECORD( pListEntry, DEBUGPRINT_EVENT, ListEntry);

    // Get length of event data
    ULONG EventDataLen = pEvent->Len;

    // Send event to DebugPrint
    NTSTATUS status = ZwWriteFile(
        DebugPrintDeviceHandle, NULL, NULL, NULL,
        &IoStatus, pEvent->EventData, EventDataLen, &ByteOffset, NULL);
    // Ignore error returns

    // Free our event buffer
    ExFreePool(pEvent);
}
```

The code loops until all the events in EventList have been removed and sent to the Debug-Print driver. It removes the first entry from the doubly-linked list using ExInterlocke-dRemoveHeadList, passing pointers to the list head and the guarding spin lock. The return value is NULL if there is nothing left in the list.

ExInterlockedRemoveHeadList returns a pointer to the *ListEntry* field in the DEBUGPRINT_ EVENT structure. What is really needed is not this, but a pointer to the DEBUGPRINT_EVENT structure itself. For this particular structure, a simple cast would suffice. However, there is a way to deal correctly with the general case in which the LIST_ENTRY variable is not at the beginning of the structure. The system header files provide the appropriate macro, CONTAINING_RECORD. Pass the LIST_ENTRY pointer, the data type of your structure and the name of its LIST_ENTRY field. The returned value is the pointer to the DEBUGPRINT_EVENT structure.

Having got the correct event pointer in pEvent, the system thread extracts the length of the event data and writes the event data itself to the DebugPrint driver using ZwWriteFile. Finally, it frees the memory that was allocated for the DEBUGPRINT_EVENT structure, before checking to see if any more events are available.

The ClearEvents routine is called to clear any remaining events when the system thread finishes or DebugPrintClose is called. ClearEvents removes any events using ExInterlockedRemoveHeadList and frees the event memory.

Singly-Linked Lists

There are kernel functions for singly-linked lists, which are really stacks. Declare a list head as a variable of type SINGLE_LIST_ENTRY. Initialize it by setting its *Next* field to NULL. PushEntryList puts an entry onto the front of the list, while PopEntryList removes an entry from the front of the list. You must use the same technique as before to get the correct pointer from a popped entry: include a SINGLE_LIST_ENTRY field in the structure you put on the list and use CONTAINING_RECORD.

There are also interlocked versions of these routines, using a spin lock as before to ensure accesses are carried out safely.

Device Queues

A device queue is an enhanced form of a doubly-linked list, usually used for storing IRPs. These are covered in glorious detail in Chapter 16.

This is how we make our faster version.

Final Pieces

Let's draw together the final pieces of the DebugPrint code for test drivers.

DebugPrintInit makes a copy of the driver name passed to it. Why does it bother to do this? Well, initially it did not. However, the call to DebugPrintInit usually comes from a

DriverEntry routine. DriverEntry code is often marked as being discardable once the driver has been initialized. This meant that the code with the driver name string might be discarded. I found that this did indeed happen and so the original pointer referred to invalid memory. Watch out for this problem in your drivers.

Listing 14.4 shows how the system thread opens a connection to the DebugPrint driver using ZwCreateFile. It opens a connection to a DebugPrint driver device called \Device\PHD-DebugPrint. The DebugPrint driver also uses a driver interface so that user mode programs, such as *DebugPrint Monitor*, can find this device.

While drivers can use Plug and Play Notification to find other drivers that support a device interface, this is a complicated approach. Instead, the one and only DebugPrint device is given a kernel name of PHDDebugPrint. Calls to ZwCreateFile do not use Win32 symbolic link names for devices. Instead, as in this example, they must use the full kernel device name.

The filename must be specified as a UNICODE_STRING. RtlInitUnicodeString is used to initialize the DebugPrintName variable. The filename is set into an OBJECT_ATTRIBUTES structure using InitializeObjectAttributes. A pointer to this structure is finally passed to ZwCreateFile. Like its Win32 equivalent, you must specify access and share parameters to ZwCreateFile. There are a host of other options, so consult the documentation for full details. Finally, you get a file HANDLE to use in further I/O requests.

As mentioned earlier, subsequent calls to ZwReadFile, ZwWriteFile, ZwQueryInformationFile, and ZwClose must take place in the same thread context as the call to ZwCreateFile. All these file I/O routines must be called at PASSIVE_LEVEL.

A typical call to ZwWriteFile is illustrated in Listing 14.3. As well as the file handle, simply specify the data pointer and a transfer count. The completion details can be found in an IO_STATUS_BLOCK structure. Specify a file pointer byte offset.

The DebugPrint test driver code eventually closes the file handle using ZwClose just before it terminates.

The system thread code tried to open its connection to the DebugPrint driver for 5 minutes. This ensures that the DebugPrint driver has started during system startup.

Listing 14.4 DebugPrint **test driver system thread file opening**

```
// Make appropriate ObjectAttributes for ZwCreateFile
UNICODE_STRING DebugPrintName;
RtlInitUnicodeString( &DebugPrintName, L"\\Device\\PHDDebugPrint");
OBJECT_ATTRIBUTES ObjectAttributes;
InitializeObjectAttributes( &ObjectAttributes, &DebugPrintName,
        OBJ_CASE_INSENSITIVE, NULL, NULL);

// Open handle to DebugPrint device

IO_STATUS_BLOCK IoStatus;
HANDLE DebugPrintDeviceHandle = NULL;
NTSTATUS status = ZwCreateFile( &DebugPrintDeviceHandle,
    GENERIC_READ | GENERIC_WRITE,
    &ObjectAttributes,
    &IoStatus,
```

Listing 14.4 DebugPrint **test driver system thread file**

```
    0, // alloc size = none
    FILE_ATTRIBUTE_NORMAL,
    FILE_SHARE_READ|FILE_SHARE_WRITE,
    FILE_OPEN,
    0,
    NULL,  // eabuffer
    0 );   // ealength

if( !NT_SUCCESS(status) || DebugPrintDeviceHandle==NULL)
    goto exit1;
```

DebugPrint **Driver**

Referring to Figure 14.1, you can see that the job of the DebugPrint driver is to store trace events from all the drivers under test and make them available to the *DebugPrint Monitor* user mode application. As the figure illustrates, the test drivers only write to the DebugPrint driver, while the Monitor only reads.

This section will not look at all the DebugPrint driver code. The Plug and Play and Initialization code is largely the same as earlier WDM drivers. You should install just one Debug-Print device in the Other Devices category.

The interesting code is in the dispatch routines in Dispatch.cpp, along with a main header file DebugPrint.h. These files and the other source files can be found on the book's software disk.

Design

The DebugPrint driver uses a similar technique of storing all written trace events in a doubly-linked list called EventList. When the *DebugPrint Monitor* program starts, it reads all the available events. It then leaves one read request outstanding. When a new trace event is written by a test driver, the Monitor read request is satisfied straightaway. This design ensures that trace events get to the Monitor application as soon as possible.

This means that the DebugPrint driver has to be able to queue up incoming read requests. In fact, it only allows one read request to be queued. Any further read requests are rejected. As DebugPrint has an IRP queue, it must be prepared to cancel IRPs. The DbpCancelIrp routine described later does this job.

The DbpCreate and DbpClose routines simply complete their IRPs successfully. Note that these IRPs are issued from both test drivers and the Monitor application.

The DebugPrint driver uses Buffered I/O.

DebugPrint **Devices**

As mentioned previously, the DebugPrint driver uses both a device interface and a Windows 2000 device name. The *DebugPrint Monitor* application identifies a DebugPrint device using the device interface. The test drivers identify the DebugPrint device using the kernel name.

Listing 14.5 shows how a named Functional Device Object is created in DbpAddDevice in Pnp.cpp. The DebugPrintName variable is a UNICODE_STRING that is initialized with the desired kernel device name, \Device\PHDDebugPrint. This is passed to the IoCreateDevice call.

Later in DbpAddDevice, the DebugPrint device interface is registered in the same way as before. The device interface GUID is defined in DbgPrtGUID.h as {ED6026A2-6813-11d2-AE43-00C0DFE4C1F3}. This header file is also included by the *Debug-Print Monitor* project.

Listing 14.5 DebugPrint **driver device creation**

```
UNICODE_STRING DebugPrintName;
RtlInitUnicodeString( &DebugPrintName, L"\\Device\\PHDDebugPrint");

// Create our Functional Device Object in fdo
status = IoCreateDevice (DriverObject,
    sizeof(DEBUGPRINT_DEVICE_EXTENSION),
    &DebugPrintName,
    FILE_DEVICE_UNKNOWN,
    0,
    FALSE,    // Not exclusive
    &fdo);
```

If you try to create a second DebugPrint device, the same kernel device name is passed to IoCreateDevice. As this device name is already being used, the call will fail. This problem does not occur if you pass NULL for the device name, as in earlier examples. If you definitely need more than one named kernel device, you must keep a (zero-based) count of devices and append it as a device number to the kernel device name.

No special processing is required when you remove a device with a kernel device name.

Read Queue

The Read IRP queue is implemented using variables in the device extension, shown in Listing 14.6. If *ReadIrp* is NULL, no IRP is queued. Otherwise, it contains a pointer to the queued IRP. A spin lock called *ReadIrpLock* is used to protect access to the read queue. These fields are initialized as follows in the DbpAddDevice routine in Pnp.cpp.

```
// Initialise "read queue"
KeInitializeSpinLock(&dx->ReadIrpLock);
dx->ReadIrp = NULL;
```

Listing 14.6 DebugPrint **driver device extension**

```
typedef struct _DEBUGPRINT_DEVICE_EXTENSION
{
    PDEVICE_OBJECT    fdo;
    PDEVICE_OBJECT    NextStackDevice;
    UNICODE_STRING    ifSymLinkName;
```

Listing 14.6 DebugPrint **driver device extension (continued)**

```
    bool GotResources;         // Not stopped
    bool Paused;               // Stop or remove pending
    bool IODisabled;           // Paused or stopped


    LONG OpenHandleCount;      // Count of open handles


    LONG UsageCount;           // Pending I/O Count
    bool Stopping;             // In process of stopping
    KEVENT StoppingEvent;      // Set when all pending I/O complete


    PIRP ReadIrp;              // "Read queue" of 1 IRP
    KSPIN_LOCK ReadIrpLock;    // Spin lock to guard access to ReadIrp


    LIST_ENTRY EventList;      // Doubly-linked list of written Events
    KSPIN_LOCK EventListLock;  // Spin lock to guard access to EventList


} DEBUGPRINT_DEVICE_EXTENSION, *PDEBUGPRINT_DEVICE_EXTENSION;
```

Listing 14.7 shows the complete Read IRP handler. Its first job is to acquire the *ReadIrpLock* spin lock. If the *ReadIrp* field is not NULL, it means another Read IRP has been queued. The IRP is failed, not forgetting to release the spin lock first. Even though the Monitor program will never issue more than one read IRP, it is best to be on the safe side.

The ReadEvent routine is then called. If there are any trace events available straightaway, ReadEvent returns the event data in the IRP, completes the IRP, and returns true. If this happens, the read routine can just return straightaway without queuing the IRP.

If there are no trace events available, the IRP is queued. This simply means storing the IRP pointer in *ReadIrp*. The final job is to mark the IRP as pending using IoMarkIrpPending and set its cancel routine using IoSetCancelRoutine. The read routine must return STATUS_PENDING because it has queued its IRP.

IRPs that are not cancelled must remove their cancel routine using IoSetCancelRoutine before completing the IRP.

Listing 14.7 DebugPrint **driver read dispatch routine**

```
NTSTATUS DbpRead( IN PDEVICE_OBJECT fdo, IN PIRP Irp)
{
    PDEBUGPRINT_DEVICE_EXTENSION dx =
        (PDEBUGPRINT_DEVICE_EXTENSION)fdo->DeviceExtension;
    if( !dx->IODisabled)
        return CompleteIrp( Irp, STATUS_DEVICE_NOT_CONNECTED, 0);
    if( !LockDevice(dx))
        return CompleteIrp( Irp, STATUS_DELETE_PENDING, 0);


    // Get access to our Read IRP queue
```

Listing 14.7 DebugPrint **driver read dispatch routine (continued)**

```
    KIRQL irql;
    KeAcquireSpinLock(&dx->ReadIrpLock,&irql);

    // Only one listening read allowed at a time.
    if( dx->ReadIrp!=NULL)
    {
        KeReleaseSpinLock(&dx->ReadIrpLock,irql);
        UnlockDevice(dx);
        return CompleteIrp(Irp,STATUS_UNSUCCESSFUL,0);
    }


    // See if there's data available
    if( ReadEvent( dx, Irp))
    {
        KeReleaseSpinLock(&dx->ReadIrpLock,irql);
        UnlockDevice(dx);
        return STATUS_SUCCESS;
    }


    // No event is available, queue this read Irp
    dx->ReadIrp = Irp;
    KeReleaseSpinLock(&dx->ReadIrpLock,irql);

    // Mark Irp as pending and set Cancel routine
    Irp->IoStatus.Information = 0;
    IoMarkIrpPending(Irp);
    IoSetCancelRoutine(Irp,DbpCancelIrp);


    return STATUS_PENDING;
}
```

Cancelling IRPs

Any IRPs that are queued must have a cancel routine. A driver ought to also handle the IRP_
MJ_CLEANUP Cleanup IRP.

Cancel and Cleanup Circumstances

First, let's be clear when IRP cancel routines are called and when the Cleanup IRP is sent.

Case 1   is a situation in which a user application calls the CancelIo Win32 function on a file
handle. All IRPs with cancel routines have their cancel routine called. Only IRPs that have
been issued by the current thread are effected.

Case 2 covers these three situations:

- a user mode program crashes with IRPs pending,
- it exits with overlapped I/O requests pending and without closing its file handle, and
- if Ctrl+C is pressed in console applications.

In this case, all IRPs with cancel routines have their cancel routines called first. If there are outstanding IRPs without cancel routines, the I/O Manager simply sets the IRPs' *Cancel* flag and waits until the IRPs complete. Finally, the Cleanup IRP is sent.

If an uncancellable IRP does not complete within five minutes, the IRP is forcibly detached from the user process so that it can terminate. However, the IRP is still left uncompleted. You will not be able to reinstall the driver, so a reboot will be necessary to try the fixed version of your driver.

Issuing the Cleanup IRP seems to perform no useful function in this case.

Case 3 is a situation in which a user mode programs closes its file handle with overlapped I/O requests pending

In this case, IRPs with cancel routines do not have their cancel routines called. Instead, the Cleanup IRP is issued to cancel all pending IRPs.

The Implications

If you give each queued IRP a cancel routine, most normal cases are covered (i.e., the afore-mentioned Cases 1 and 2). To be thorough, however, you ought to provide a Cleanup handler to cover Case 3. As you well know, programmers are bound to forget to close a file handle sometime or other.

If you do not provide a cancel routine for IRPs but do provide a Cleanup handler, only Case 3 is handled correctly. This is an unsatisfactory solution. In addition, user mode applications will not be able to cancel IRPs with CancelIo.

Some drivers work by providing a cancel routine only for an IRP while it is queued. Once the IRP actually begins processing, the cancel routine is removed. If you use this technique, be sure to provide a reasonable timeout for real IRP processing. Otherwise, crashed programs will not be able to exit.

DebugPrint **IRP Cancelling**

Full Cancel and Cleanup support can be quite complicated. This is particularly the case for IRPs put in the device queue for processing in a StartIo routine. A full example for this case is given in Chapter 16.

The DebugPrint driver uses just a cancel routine for its one queued IRP. It does not handle the Cleanup IRP.

The DbpCancelIrp routine shown in Listing 14.8 is called if an IRP is cancelled. An I/O request is cancelled when a user mode application calls the Win32 CancelIo function. This technique is used by the *DebugPrint Monitor* application. The kernel will also cancel any outstanding IRPs if a process terminates unexpectedly or when the file handle is closed.

The I/O Manager uses its *Cancel spin lock* to ensure that cancel operations happen safely. A cancel routine is always called at DISPATCH_LEVEL IRQL holding this Cancel spin lock. The DbpCancelIrp routine can simply release this straightaway.

DbpCancelIrp then goes on to acquire the DebugPrint device extension *ReadIrpLock* spin lock. It checks to see if the given IRP pointer matches the one in the queue. If it does, it clears the queue. Regardless of whether the given IRP matches the one in the list, DbpCancelIrp just cancels the IRP by calling CompleteIrp, passing a status of STATUS_CANCELLED.

Listing 14.8 DebugPrint **driver cancel routine**

```
VOID DbpCancelIrp( IN PDEVICE_OBJECT fdo, IN PIRP Irp)
{
    PDEBUGPRINT_DEVICE_EXTENSION dx =
        (PDEBUGPRINT_DEVICE_EXTENSION)fdo->DeviceExtension;
    IoReleaseCancelSpinLock(Irp->CancelIrql);

    // If this is our queued read, then unqueue it
    KIRQL irql;
    KeAcquireSpinLock(&dx->ReadIrpLock,&irql);
    if( Irp==dx->ReadIrp)
    {
        UnlockDevice(dx);
        dx->ReadIrp = NULL;
    }
    KeReleaseSpinLock(&dx->ReadIrpLock,irql);

    // Whatever Irp it is, just cancel it
    CompleteIrp(Irp,STATUS_CANCELLED,0);
}
```

Write Algorithm

The DebugPrint driver write routine DbpWrite, shown in Listing 14.9, at first sight looks pretty similar to the DebugPrintMsg routine described earlier. Its job is to insert the trace event data into an interlocked doubly-linked list. If there is a Read IRP queued up, DbpWrite goes on to satisfy this read request.

DbpWrite first gets the write parameters, and completes Write IRPs with a zero transfer length straightaway. The device file pointer is ignored.

As shown previously, the event list is stored in the device extension in field *EventList*, protected by spin lock *EventListLock*. Each event is stored in a DEBUGPRINT_EVENT structure.

DbpWrite determines the correct size for the DEBUGPRINT_EVENT structure and tries to allocate some memory for it from the nonpaged pool. It fails the IRP with STATUS_INSUFFICIENT_ RESOURCES if no memory is available. Next, it copies the event data into the event and stores the data length, before calling ExInterlockedInsertTailList in the same way as before to insert the event safely into *EventList*.

DbpWrite now checks to see if there is a queued Read IRP. It must first grab the *ReadIrpLock* spin lock. If the *ReadIrp* field is not NULL, ReadEvent is called to complete the Read IRP and the *ReadIrp* field is reset to NULL.

Finally, DbpWrite completes its own Write IRP, returning STATUS_SUCCESS.

Listing 14.9 DebugPrint **driver write routine**

```
NTSTATUS DbpWrite( IN PDEVICE_OBJECT fdo, IN PIRP Irp)
{
    PDEBUGPRINT_DEVICE_EXTENSION dx =
        (PDEBUGPRINT_DEVICE_EXTENSION)fdo->DeviceExtension;
    if( !dx->IODisabled)
        return CompleteIrp( Irp, STATUS_DEVICE_NOT_CONNECTED, 0);
    if( !LockDevice(dx))
        return CompleteIrp( Irp, STATUS_DELETE_PENDING, 0);

    PIO_STACK_LOCATION IrpStack = IoGetCurrentIrpStackLocation(Irp);
    ULONG BytesTxd = 0;


    // Check write len
    ULONG WriteLen = IrpStack->Parameters.Write.Length;
    if( WriteLen==0)
    {
        UnlockDevice(dx);
        return CompleteIrp(Irp,STATUS_SUCCESS,0);
    }


    // Copy write data into an event
    ULONG Len = sizeof(LIST_ENTRY)+sizeof(ULONG)+WriteLen;
    PDEBUGPRINT_EVENT pEvent =
        (PDEBUGPRINT_EVENT)ExAllocatePool(NonPagedPool,Len);
    if( pEvent==NULL)
    {
        UnlockDevice(dx);
        return CompleteIrp(Irp,STATUS_INSUFFICIENT_RESOURCES,0);
    }


    pEvent->Len = WriteLen;
    RtlCopyMemory( pEvent->EventData,
        Irp->AssociatedIrp.SystemBuffer, WriteLen);


    // Insert event into event list
    ExInterlockedInsertTailList(&dx->EventList,&pEvent->
        ListEntry,&dx->EventListLock);


    // If read pending, then read it
    KIRQL irql;
    KeAcquireSpinLock(&dx->ReadIrpLock,&irql);
```

Listing 14.9 DebugPrint **driver write routine (continued)**

```
    if( dx->ReadIrp!=NULL)
        if( ReadEvent( dx, dx->ReadIrp))
        {
            UnlockDevice(dx);
            dx->ReadIrp = NULL;
        }
    KeReleaseSpinLock(&dx->ReadIrpLock,irql);


    // Complete IRP
    UnlockDevice(dx);
    return CompleteIrp(Irp,STATUS_SUCCESS,WriteLen);
}
```

Read Algorithm

The ReadEvent routine shown in Listing 14.10 is called by the read and write dispatch routines. It is called while holding the *ReadIrpLock* spin lock. ReadEvent returns true if a trace event was found.

ReadEvent tries to remove an entry from the event list using ExInterlockedRemoveHead-List. If it finds an entry, it obtains a pointer to the DEBUGPRINT_EVENT structure using CONTAINING_RECORD. It now checks the event data length against the size of the Read IRP buffer and shortens the transfer count, if necessary. The event data is copied to the Read I/O buffer and the Read IRP is completed. Finally, the event buffer memory is freed.

Listing 14.10 DebugPrint **driver** ReadEvent **routine**

```
bool ReadEvent( PDEBUGPRINT_DEVICE_EXTENSION dx, PIRP Irp)
{
    // Try to remove Event from EventList
    PLIST_ENTRY pListEntry =
        ExInterlockedRemoveHeadList( &dx->EventList, &dx->EventListLock);
    if( pListEntry==NULL)
        return false;


    // Get event as DEBUGPRINT_EVENT
    PDEBUGPRINT_EVENT pEvent =
        CONTAINING_RECORD( pListEntry, DEBUGPRINT_EVENT, ListEntry);


    // Get length of event data
    ULONG EventDataLen = pEvent->Len;


    // Get max read length acceptible
    PIO_STACK_LOCATION IrpStack = IoGetCurrentIrpStackLocation(Irp);
    ULONG ReadLen = IrpStack->Parameters.Read.Length;
```

Listing 14.10 DebugPrint **driver** ReadEvent **routine (continued)**

```
    // Shorten event length if necessary
    if( EventDataLen>ReadLen)
        EventDataLen = ReadLen;

    // Copy data to Irp and complete it
    RtlCopyMemory( Irp->AssociatedIrp.SystemBuffer,
        pEvent->EventData, EventDataLen);
    IoSetCancelRoutine(Irp,NUUL);
    CompleteIrp(Irp,STATUS_SUCCESS,EventDataLen);

    // Free event memory
    ExFreePool(pEvent);
    return true;
}
```

DebugPrint Monitor

The *DebugPrint Monitor* is a user mode application that keeps reading events from the DebugPrint driver and displays them to the user. The Monitor is a standard MFC application, so this chapter will not discuss all the code in detail. The source files can be found on the book software disk.

The Monitor program saves its current screen position and column widths in the registry. Consult the code to find out how this is done.

Design

The Monitor uses a single document interface. The view class CDebugPrintMonitorView is derived from CListView, which encapsulates a list view control. It is this list view control that stores the event information once it has been received by the Monitor. The corresponding document class CDebugPrintMonitorDoc does not store the event information. It simply implements the OnSaveDocument document method to save event information from the view to a .dpm file, and OnOpenDocument to load data.

The main work is carried out in a worker thread, which runs ListenThreadFunction in Listener.cpp. This thread chugs away in the background, reading any available events from the DebugPrint driver. It posts a message to the view class for each event, passing a pointer to the event data. The view class handles these messages in its OnDebugPrintEvent routine by inserting an appropriate event item at the end of the list control.

Win32 Worker Threads

The *DebugPrint Monitor* application class calls StartListener in its InitInstance method to start the listen worker thread. The ExitInstance method calls StopListener to stop it.

The code uses a LISTENER_INFO structure to pass information to the worker thread. There is only one instance of this object called ListenerInfo.

```
typedef struct _LISTENER_INFO
{
    HANDLE DebugPrintDriver;
    bool KeepGoing;
} LISTENER_INFO, *PLISTENER_INFO;


LISTENER_INFO ListenerInfo;
```

StartListener first stores a safe HWND handle to the view window, to be used later.

It then finds the first DebugPrint driver device and opens a handle to it using GetDevice-ViaInterface01. This is a variant on GetDeviceViaInterface, shown in Chapter 5. GetDeviceViaInterface01 opens the device for overlapped I/O by specifying the FILE_FLAG_OVERLAPPED bit in the dwFlagsAndAttributes parameter to CreateFile. The DebugPrint device handle is stored in ListenerInfo.DebugPrintDriver.

The code uses the *KeepGoing* field in ListenerInfo to signal when the worker thread should stop. *KeepGoing* is therefore set to true before the thread is started. The thread is started using AfxBeginThread, passing a pointer to the function to run, ListenThreadFunction, and a context to pass to it.

ListenThreadFunction loops, waiting either for event data or for *KeepGoing* to return false. ListenThreadFunction is in the file Dispatch.cpp on the CD-ROM.

DebugPrint_Event Class

The DebugPrint_Event class shown in Listing 14.11 is used to communicate events to the view class. Each class instance has *Driver* and *Message* strings and an MFC CTime *Timestamp*. A static method called SendEvent is used to post a DebugPrint_Event message to the view class.

```
DebugPrint_Event::SendEvent( "Monitor", "Starting to listen",
    CTime::GetCurrentTime(), false);
```

Listing 14.11 DebugPrint **driver** DebugPrint_Event **class**

```
const UINT WM_DEBUGPRINTEVENT = (WM_USER+1);


class DebugPrint_Event
{
public:
    CString Driver;
    CTime Timestamp;
    CString Message;
    bool SetModified;    // false to reset document SetModifiedFlag.


    static HWND ViewHwnd;    // View Hwnd
```

Listing 14.11 DebugPrint **driver** DebugPrint_Event **class (continued)**

```
    // Generate and send an event
    static void SendEvent( CString d, CString m, CTime t = 0, bool sm=true)
    {
        if( ViewHwnd==NULL) return;
        DebugPrint_Event* pEvent = new DebugPrint_Event;
        pEvent->Driver = d;
        if( t==0) t = CTime::GetCurrentTime();
        pEvent->Timestamp = t;
        pEvent->Message = m;
        pEvent->SetModified = sm;
        ::PostMessage( ViewHwnd, WM_DEBUGPRINTEVENT, 0, (LPARAM)pEvent);
    }
};
```

Win32 Overlapped I/O

The code in ListenThreadFunction mainly deals with overlapped I/O to the DebugPrint device. It must check the state of the *KeepGoing* flag. Overlapped I/O lets us issue a read request and get on with other tasks.

To do overlapped I/O, a Win32 event and an OVERLAPPED structure are needed. CreateEvent is used to initialize the FileIOWaiter manual event into the nonsignalled state. The OVERLAPPED structure (ol) stores the file pointer offset and FileIOWaiter event handle.

A standard ReadFile call is used to initiate a read request. The read buffer is 1024 bytes, which should be large enough for any DebugPrint event. ReadFile is passed a pointer to the OVERLAPPED structure. Overlapped I/O does work with device files in Windows 98, but does not work on ordinary file I/O.

ReadFile returns true if the read request completes straightaway. The number of bytes transferred is stored in TxdBytes. If the read request is held in the DebugPrint read queue, ReadFile returns false and GetLastError returns ERROR_IO_PENDING. The code checks for a real error return from ReadFile. If the *DebugPrint Monitor* application is run twice, the second incarnation will get an error here when the DebugPrint driver read routine fails an attempt to queue a second Read IRP.

If the read request is pending, ListenThreadFunction loops calling WaitForSingleObject with a timeout of 100ms. If WaitForSingleObject times-out, ListenThreadFunction checks to see if *KeepGoing* has returned false. If so, it calls CancelIo to cancel the pending Read IRP and exits. Incidentally, calling CancelIo in another thread, such as the StopListener routine, does not work.

WaitForSingleObject detects when the read request has finished when the FileIOWaiter event becomes signalled. In Windows 2000, ListenThreadFunction could wait on the file object instead of an event. However, this does not work in Windows 98, so I use events that work in both operating systems. ListenThreadFunction calls GetOverlappedResult to retrieve the number of bytes that were received.

The remaining code extracts the event information from the read buffer and builds an event object to post to the Monitor view class. The event timestamp that was generated in the

DebugPrintMsg routine is a time in GMT. The GMTtoLocalTime function does all the necessary grovelling around to convert this into a local time.

ListenThreadFunction finally closes the event and file handles using CloseHandle.

Listing 14.12 DebugPrint **driver** ListenThreadFunction function

```
UINT ListenThreadFunction( LPVOID pParam)
{
    PLISTENER_INFO pListenerInfo = (PLISTENER_INFO)pParam;
    if (pListenerInfo==NULL) return -1;

    CString StartMsg = "Starting to listen";
    // ...
    DebugPrint_Event::SendEvent( "Monitor", StartMsg,
                                CTime::GetCurrentTime(), false);

    // Buffer for events
    const int MAX_EVENT_LEN = 1024;
    char Event[MAX_EVENT_LEN+1];

    // Create Overlapped read structure and event
    HANDLE FileIOWaiter = CreateEvent( NULL, TRUE, FALSE, NULL);
    if( FileIOWaiter==NULL)
        goto Exit2;
    OVERLAPPED ol;
    ol.Offset = 0;
    ol.OffsetHigh = 0;
    ol.hEvent = FileIOWaiter;

    // Keep looping, waiting for events, until KeepGoing goes false
    for(;;)
    {
        // Initiate overlapped read
        DWORD TxdBytes;
        ResetEvent(FileIOWaiter);
        memset(Event,0,MAX_EVENT_LEN+1);
        if( !ReadFile( ListenerInfo.DebugPrintDriver, Event, MAX_EVENT_LEN,
            &TxdBytes, &ol))
        {
            // Check for read errors
            if( GetLastError()!=ERROR_IO_PENDING)
            {
                CString Msg;
```

Listing 14.12 DebugPrint **driver** ListenThreadFunction
function (continued)

```
            Msg.Format("Read didn't return pending %d", GetLastError());
            DebugPrint_Event::SendEvent( "Monitor", Msg);
            goto Exit;
        }


        // Wait for read to complete (check for KeepGoing
        // going false every 100ms)
        while( WaitForSingleObject( FileIOWaiter, 100)==WAIT_TIMEOUT)
        {
            if( !ListenerInfo.KeepGoing)
            {
                // Cancel the pending read
                CancelIo(ListenerInfo.DebugPrintDriver);
                goto Exit;
            }
        }


        // Get read result, ie bytes transferred
        if( !GetOverlappedResult( ListenerInfo.DebugPrintDriver, &ol,
            &TxdBytes, FALSE))
        {
            DebugPrint_Event::SendEvent( "Monitor",
                "GetOverlappedResult failed" );
            continue;
        }
    }

    // Check there's something there
    if( TxdBytes < sizeof(TIME_FIELDS)+2)
    {
        DebugPrint_Event::SendEvent( "Monitor", "Short read msg" );
        continue;
    }
    // Extract Timestamp, Driver and Msg, and post to View
    Event[MAX_EVENT_LEN] = '\0';
    PTIME_FIELDS pTF = (PTIME_FIELDS)Event;
    CTime gmtEventTime( pTF->Year,pTF->Month,pTF->Day,pTF->Hour,
                        pTF->Minute,pTF->Second);
    CTime EventTime = GMTtoLocalTime(gmtEventTime);
    char* DriverName = Event+sizeof(TIME_FIELDS);
    CString CSDriverName = DriverName;
```

Listing 14.12 DebugPrint **driver** ListenThreadFunction
              function (continued)

```
        CString CSDriverMsg = Event+sizeof(TIME_FIELDS)+strlen(DriverName)+1;
        DebugPrint_Event::SendEvent( CSDriverName, CSDriverMsg, EventTime);
    }


Exit:
    CloseHandle(FileIOWaiter);
Exit2:
    CloseHandle(ListenerInfo.DebugPrintDriver);
    ListenerInfo.DebugPrintDriver = NULL;
    DebugPrint_Event::SendEvent( "Monitor", "Stopped listening");
    return 0;
}
```

Conclusion

This chapter has built a full working driver, which can be used to generate debug trace prints that can be seen in a user application. It has covered system threads, dispatcher objects, linked lists, file I/O, a simple IRP queue, and IRP cancel routines.

Chapter 15

WdmIo and PHDIo Drivers

Let's finally talk to some real hardware. The next chapters discuss the WdmIo and PHDIo drivers. These general-purpose drivers give Win32 programmers access to simple hardware devices. You can do basic reads and writes of hardware ports. The drivers also let you perform interrupt driven I/O. There is just enough functionality to let a user mode application send output to a standard LPT printer port. Although this is not the recommended way to talk to a printer port[1], it does show that these drivers can be used in a real application.

The generic nature of the WdmIo and PHDIo drivers does make them slightly more complicated to understand. However, a complete and genuinely useful driver is worth having. Using a generic driver as an example also shows all the steps that a you need to use it as the basis of a more advanced driver.

To make these drivers easier to understand, this chapter looks at the interface they make available to Win32 programmers. I will use the test programs *WdmIoTest* and *PHDIoTest*, which send a test message out a parallel port to a printer, as an example. This chapter also looks at the installation and resource allocation issues for these drivers.

The following chapters explain how the WdmIo and PHDIo drivers work. They explain the following topics in detail.

- Queuing IRPs for serial hardware I/O
- StartIo routines
- Cancel routines for queued IRPs
- Handling the Cleanup IRP
- Interrupt driven I/O

---

1. See Chapter 19 for the best way to talk to a parallel port in NT and W2000.

- Deferred Procedure Calls
- Timers for time-outs
- Custom timers
- NT hardware detection and allocation

Win32 Interface

The WdmIo driver is a Plug and Play driver for Windows 98 and Windows 2000. It is based on most of the previous drivers. However, it does not support Power Management, WMI, or NT events. Its installation INF file specifies the resources that are available to a WdmIo device. A WdmIo device must have one (and only one) I/O port address range. It usually has one interrupt line resource as well. This is needed if interrupt-driven reads or writes are to be used. More than one WdmIo device can be installed, provided that separate INF files with different resources are used.

The PHDIo driver is an NT style driver, primarily for NT 3.51, NT 4, and Windows 2000. However, it works perfectly well in Windows 98. When the driver is loaded, it creates one device called \\.\PHDIo. The I/O port and interrupt resources are specified as a part of the filename used to open a handle to this device. Again, one I/O port must be given, and one interrupt can be specified, if need be.

Win32 programs must open a handle to a WdmIo or PHDIo device. WdmIo devices are accessed using the device interface identified by WDMIO_GUID. The one PHDIo device has a symbolic link name of \\.\PHDIo. However, the filename passed to CreateFile must include a definition of the resources required.

Having opened a handle to a WdmIo or PHDIo device, the Win32 program can use several IOCTLs as well as issuing read and write requests. Finally, it must close the device handle.

IOCTLs

Table 15.1 lists the supported IOCTLs, including whether they need input or output parameters.

Most of the IOCTLs pass an input buffer that contains one or more PHDIO commands and their parameters.

IOCTL_PHDIO_RUN_CMDS is used to run a set of commands straightaway. The three following IOCTLs store the commands that are used later in the processing of read and write requests. Finally, IOCTL_PHDIO_GET_RW_RESULTS retrieves the command results and output from the last read or write.

Table 15.1 WdmIo **and** PHDIo **IOCTL codes**

IOCTL	Input	Output	Description
IOCTL_PHDIO_RUN_CMDS	Yes	Optional	Run the passed commands
IOCTL_PHDIO_CMDS_FOR_READ	Yes	No	Store the commands to read a byte and store in the read buffer
IOCTL_PHDIO_CMDS_FOR_READ_START	Yes	No	Store the commands that start the read process
IOCTL_PHDIO_CMDS_FOR_WRITE	Yes	No	Store the commands to output a byte from the write buffer
IOCTL_PHDIO_GET_RW_RESULTS	No	Yes	Get the command results of the last read or write operation

Commands

A Win32 program passes a block of commands in an IOCTL input buffer. These commands are either run straightaway or are run one or more times during the processing of read or write requests.

A command is a single byte, followed by one or more parameter bytes. Table 15.2 lists all the commands, their parameters, and their output. The Ioctl.h header in the WdmIo\Sys directory contains the command definitions. Note that all operations are currently byte-sized. However, the two top bits of the command byte are reserved to indicate 16-bit word transfers and 32-bit long word transfers. If either of these bits is currently set, the command is not run and command processing is aborted. All command parameters are BYTEs. A BYTE is an unsigned 8-bit value, so you cannot use negative values.

Most of the commands are fairly self-explanatory. They are best explained using examples in the context of a sample application, *WdmIoTest*, that outputs data to the LPT printer parallel port.

Table 15.2 WdmIo **and** PHDIo **commands**

Command	Input parameters	Output	Description
PHDIO_OR	reg,Value		Read register, OR with value, and write back. Use to set bit(s)
PHDIO_AND	reg,Value		Read register, AND with value, and write back. Use to clear bit(s)
PHDIO_XOR	reg,Value		Read register, XOR with value, and write back. Use to toggle bit(s)
PHDIO_WRITE	reg,Value		Write value to a register
PHDIO_READ	reg	Value	Read value from a register

PHDIO_DELAY	delay		Delay for given microseconds. Delay must be 60μs or less
PHDIO_WRITES	reg,count, Values,delay		Write values to same register with delay (<=60μs)
PHDIO_READS	reg,count,delay	Values	Read values from same register with delay (<=60μs)
PHDIO_IRQ_CONNECT	reg,mask,Value		Connect to interrupt
PHDIO_TIMEOUT	seconds		Specify time-out for reads and writes
PHDIO_WRITE_NEXT	reg		Write next value from write buffer
PHDIO_READ_NEXT	reg		Store next value in read buffer

LPT Printer Driver Application

The *WdmIoTest* Win32 application uses the WdmIo driver to output a couple of lines of text to a printer on the old Centronics LPT1 parallel printer port. The source and executable for this program are in the book software WdmIo\Exe directory. However, you will not be able to run this program until you have installed the WdmIo driver and configured your system correctly, as described later.

Parallel Ports

A parallel port responds at three addresses in I/O space in an x86 PC. It also generates an interrupt (when enabled) on one interrupt line. The LPT1 port traditionally lives at I/O space addresses 0x378 to 0x37A and generates ISA interrupt IRQ7.

A basic parallel port has the three registers as listed in Table 15.3. Signals with # at the end of their name use negative logic. For example, if the Status BUSY# bit is read as zero (low), the printer is busy.

The printer is ready to accept data if the Status BUSY# and ONLINE bits are high.

To write data to the printer, write the byte to the Data port. Wait 1μs for the data lines to settle. Set the Control port STROBE output high for at least 0.5μs. The printer signals that it is busy as BUSY# is read as low. When it is ready for another byte, it pulses ACK# low briefly and BUSY# goes high again.

The ACK# pulse signals the parallel port electronics to interrupt the processor (provided ENABLE_INT is high). I have no clear documentation to say what is exactly supposed to happen now. One source says that bit 2 of the Status register is set low to indicate that it has caused the interrupt. However, the two printers that I tried did not set this bit low. It seems that you just have to assume that if the right interrupt arrives (e.g., IRQ7), the correct parallel port caused the interrupt. It seems that reading the Status register clears the interrupt.

A printer may sometimes also pulse ACK# low and so generate an interrupt at other times, such as when it has finished initializing itself, when it is switched off, when it goes off-line, or when it runs out of paper. A dedicated parallel port driver might well take special action for these circumstances. The code here just needs to ensure that the WdmIo driver reads the Status port to clear the interrupt.

As you may be aware, more sophisticated versions of the parallel port are available. For example, when configured in the appropriate mode, a parallel port can input information via the Data port. In addition, some non-standard devices can be connected to a basic parallel port. These might allow input of 4-bits at a time using the Status port. Dongles are hardware devices that are used to verify that some software is licensed to run on a computer. They are designed to respond when their manufacturer's software talks to them in some special way. At all other times, they are designed to pass all parallel port signals to the printer and vice versa.

Table 15.3   Parallel port registers

Offset	Access	Register
0	Read/Write	Data
1	Read only Bit 3 Bit 5 Bit 6 Bit 7	Status ONLINE OUT_OF_PAPER ACK# BUSY#
2	Read/Write Bit 0 Bit 2 Bit 3 Bit 4 Bit 6 Bit 7	Control STROBE INIT# SELECT ENABLE_INT 1 1

WdmIoTest

The *WdmIoTest* application uses the WdmIo driver to write a brief message to the printer. It outputs information to the console screen to indicate its progress as it performs the following steps. Steps 2–8 all involve DeviceIoControl or WriteFile calls to the WdmIo driver.

1. Open a handle to the WdmIo device. The GetDeviceViaInterface routine is used as before to open a handle to the first device with a WDMIO_GUID device interface.
2. Disable the interrupts and connect to an interrupt.
3. Initialize the printer.
4. Send commands for writing each byte.
5. Read the status every second until the printer is ready. Time-out after 20 seconds.
6. Write a message to the printer.
7. Get the write results.
8. Disable the interrupt.
9. Close the handle.

PHDIoTest

PHDIoTest does exactly the same job as *WdmIoTest*, except that it obtains a handle to the `PHDIo` device in a different way. It calls `CreateFile` directly, passing the required resources in the filename, as follows.

```
HANDLE hPhdIo = CreateFile("\\\\.\\PHDIo\\isa\\io378,3\\irq7\\override",
    GENERIC_READ|GENERIC_WRITE, 0,
    NULL, OPEN_EXISTING, FILE_ATTRIBUTE_NORMAL, NULL);
```

Table 15.4 shows the different resource specifiers that may be used as a part of the `PHDIo` filename. The first specifier must be `\isa`. The other specifiers can be in any order. An I/O port specifier must be given. Use the `\override` specifier with caution and never use it in a commercial release. All letters in the filename must be in lower case.

Table 15.4   `PHDIo` **filename resource elements**

Element	Required	Description
`\isa`	Mandatory	Initial string: Isa bus
`\io<base>,<length>`	Mandatory	I/O ports <base> and <length> in hex
`\irq<number>`	Optional	IRQ <number> in decimal
`\override`	Optional	Use these resources even if they cannot be allocated

Issuing Commands

Listing 15.1 shows the commands that initialize the printer. It also shows how `DeviceIoControl` is called with the `IOCTL_PHDIO_RUN_CMDS` IOCTL code to run these commands.

First, three constants are defined to represent the offsets to each register in the parallel port electronics.

The `InitPrinter` BYTE array stores the commands to initialize the printer. Each line has one command and its parameters. First, the Control port INIT# line is set low by writing 0xC8 using the `PHDIO_WRITE` command. The `PHDIO_DELAY` command then waits for 60µs. The INIT# signal is set high. The write value of 0xDC also selects the printer and enables interrupts. A further delay of 60µs completes the operation.

The `DeviceIoControl` Win32 function is used to issue the `IOCTL_PHDIO_RUN_CMDS` IOCTL. This IOCTL runs the given commands straightaway. The `InitPrinter` array is passed as the input to the IOCTL. `IOCTL_PHDIO_RUN_CMDS` can optionally be passed an output buffer.

If there is an output buffer that is big enough, the first two 16-bit words in the output indicate any problems that the `WdmIo` driver found. The first word is an error code. The second word is the zero-based index into the command buffer in which the problem was found. Both are zero if there were no problems. The possible error codes are also found in `Ioctl.h` in the `WdmIo\Sys` directory.

The *WdmIoTest* code rather sloppily does not bother to check the returned error code. If using this driver for real, make sure that you check all error codes.

Listing 15.1 *WdmIoTest* issuing commands to run straightaway

```
const BYTE PARPORT_DATA    = 0;
const BYTE PARPORT_STATUS  = 1;
const BYTE PARPORT_CONTROL = 2;


BYTE InitPrinter[] =
{
    PHDIO_WRITE, PARPORT_CONTROL, 0xC8,  // Take INIT# low
    PHDIO_DELAY, 60,                     // Delay 60us
    PHDIO_WRITE, PARPORT_CONTROL, 0xDC,  // INIT# high, select printer,
                                         // enable interrupts
    PHDIO_DELAY, 60,                     // Delay 60us
};


int main(int argc, char* argv[])
{
    // ...

    DWORD BytesReturned;
    WORD rv[3];

    if( DeviceIoControl( hWdmIo, IOCTL_PHDIO_RUN_CMDS,
            InitPrinter, length(InitPrinter),    // Input
            rv, sizeof(rv),                      // Output
            &BytesReturned, NULL))
    {
        printf("    InitPrinter OK.  rv=%d at %d\n", rv[0], rv[1]);
    }
    else
    {
        printf("XXX  InitPrinter failed %d\n",GetLastError());
        goto fail;
    }

    // ...
```

Reading Data

If you use the PHDIO_READ or PHDIO_READS commands, you must provide an output buffer that is big enough to receive the read data. Remember that the first four bytes of the output buffer are always used for the error code and location.

Listing 15.2 shows how the ReadStatus commands are issued. It simply reads a byte value from the Status port. After DeviceIoControl has returned, the fifth byte of the output buffer contains the Status register contents. *WdmIoTest* checks that the BUSY# and ONLINE signals are 1 before continuing.

Listing 15.2 Reading data

```
BYTE ReadStatus[] =
{
    PHDIO_READ, PARPORT_STATUS,  // Read status
};

int main(int argc, char* argv[])
{
    // ...

    DWORD BytesReturned;
    WORD rv[3];

    if( DeviceIoControl( hWdmIo, IOCTL_PHDIO_RUN_CMDS,
                    ReadStatus, length(ReadStatus),  // Input
                    rv, sizeof(rv),                   // Output
                    &BytesReturned, NULL))
    {
        PBYTE pbrv = (PBYTE)&rv[2];
        printf(" ReadStatus OK.  rv=%d at %d  status=%02X\n", rv[0], rv[1],
            *pbrv);
        if( (*pbrv&0x88)==0x88)
        {
            busy = false;
            break;
        }
    }
}
```

Writing Data Using Interrupt Driven I/O

The WriteFile Win32 call is used to pass output data to the WdmIo driver. However, it needs to know how to process the data and handle interrupts. Two steps are required before Write-File is called. First, connect to the interrupt. Second, pass WdmIo a series of commands that it will run to send the first byte and process each write interrupt.

Connecting to an Interrupt

When the WdmIo device is started, it is told which interrupt to use. However, it does not connect to the interrupt (i.e., install its interrupt handler), as it does not yet know how to handle the interrupt.

The ConnectToInterrupts commands shown here are used to initialize WdmIo's interrupt handling. The first write command tells the parallel port hardware not to generate interrupts. The second command tells the WdmIo driver to use a time-out of 10 seconds when processing subsequent WriteFile (and ReadFile) requests. WdmIo must have a time-out; a default of 10 seconds is used if no PHDIO_TIMEOUT command is given.

```
BYTE ConnectToInterrupts[] =
{
    PHDIO_WRITE, PARPORT_CONTROL, 0xCC,              // Disable interrupts
    PHDIO_TIMEOUT, 10,                              // Write time-out in seconds
    PHDIO_IRQ_CONNECT, PARPORT_STATUS, 0x00, 0x00,  // Connect to interrupt
};
```

The last command, PHDIO_IRQ_CONNECT, connects the WdmIo driver to its interrupt. As mentioned before, the actual interrupt number is passed as a resource when the WdmIo device is started. WdmIo starts servicing a hardware interrupt by reading a hardware register; in this case, the Status register is read. It must determine whether the interrupt was caused by its hardware or not.

The two final parameters to the PHDIO_IRQ_CONNECT command are a mask and a value. The register contents are ANDed with the mask and compared to the value, as shown in the following WdmIo code snippet. If they are not equal, the interrupt must be intended for another driver.

```
// See if interrupt is ours
UCHAR StatusReg = ReadByte( dx, dx->InterruptReg);
if( (StatusReg&dx->InterruptRegMask) != dx->InterruptRegValue)
    return FALSE;  // Not ours
```

Suppose the Status register really did reset its bit 2 to 0 when it generated an interrupt. Specifying 0x04, as the mask would isolate bit 2. If the ANDed result is equal to 0x00, the interrupt is ours. Therefore, specifying a PHDIO_IRQ_CONNECT mask of 0x04 and value of 0x00 would have correctly detected when the parallel port interrupted.

However, as stated earlier, I found that there is no interrupt indication in the Status register. I simply have to assume that if an interrupt arrives, it came from the correct parallel port. To persuade the WdmIo code to continue regardless, a mask of 0x00 and a value of 0x00 was specified in the ConnectToInterrupts code.

Storing the Write Byte Commands

The WdmIo driver needs to be told a series of commands that write a single byte of data. These commands are used both to write the first output byte and to process an interrupt to send another byte.

IOCTL_PHDIO_CMDS_FOR_WRITE is used to store the commands used to write data. This IOCTL is issued using DeviceIoControl in the normal way. The commands are in the input buffer. No output buffer need be specified.

Listing 15.3 shows the WriteByte commands that *WdmIoTest* tells WdmIo to use to write a data byte. The first PHDIO_WRITE command ensures that the STROBE output signal in bit 0 of the Control register is off. The PHDIO_WRITE_NEXT command is used to write the next byte in the output buffer to the Data port. PHDIO_DELAY is used to delay for 1μs while the output signals settle. The STROBE signal is then set. A delay of 1μs is used before turning STROBE off again. A last delay of 1μs is introduced just to be on the safe side. Finally, the Status register is read; I shall show later on how to access this value.

Listing 15.3 Stored write byte commands

```
BYTE WriteByte[] =
{
    PHDIO_WRITE, PARPORT_CONTROL, 0xDC,     // Ensure STROBE off
    PHDIO_WRITE_NEXT, PARPORT_DATA,         // Write next byte
    PHDIO_DELAY, 1,                         // Delay 1us
    PHDIO_WRITE, PARPORT_CONTROL, 0xDD,     // STROBE on
    PHDIO_DELAY, 1,                         // Delay 1us
    PHDIO_WRITE, PARPORT_CONTROL, 0xDC,     // STROBE off
    PHDIO_DELAY, 1,                         // Delay 1us

    PHDIO_READ, PARPORT_STATUS,             // Read status
};
```

The PHDIO_WRITE_NEXT command is crucial to these data transfer commands. The WdmIo driver keeps track of the current position in the output buffer passed by WriteFile. WdmIo correctly runs the WriteByte commands until all the bytes have been transferred.

Writing Data

The *WdmIoTest* program is finally ready to output a message to the printer. The following code snippet shows that WriteFile is used in the normal way to output data.

```
char* Msg = "Hello from WdmIo example\r\nChris Cant,
    PHD Computer Consultants Ltd\r\n";
DWORD len = strlen(Msg);
if( !WriteFile( hWdmIo, Msg, len, &BytesReturned, NULL))
    printf("XXX  Could not write message %d\n",GetLastError());
else if( BytesReturned==len)
    printf("    Write succeeded\n");
else
    printf("XXX  Wrong number of bytes written: %d\n",BytesReturned);
```

The WdmIo driver uses the WriteByte commands to output the first byte, H. It then expects an interrupt when each byte has been printed. The WriteByte commands are run again by the interrupt handler to output each following byte. Assuming all goes well, when an interrupt is received after the last byte has been sent, WdmIo completes the WriteFile call successfully.

If there is a problem, WriteFile returns an error. The most likely error is ERROR_NOT_READY, which is returned if the write times out. If there is a problem running the write byte commands, ERROR_GEN_FAILURE is returned. Retrieve the WriteFile results to find the source of the problem.

Getting WriteFile **Results**

Each time the write commands are run, WdmIo gives its command processor a 5-byte output buffer. The first two words (4 bytes) are used for the error code and location. The fifth byte is filled with any output data that the commands produce. In *WdmIoTest*, the WriteByte commands read the Status register just after each byte is output. If the write commands attempt to output more than one byte, the command run will be aborted with error code PHDIO_NO_OUTPUT_ROOM.

However, what is really wanted is the value of the Status register after each byte is processed by the printer. At this point, the Status register contains some useful information, such as whether the printer is off-line or has run out of paper.

While *WdmIoTest* can just issue the ReadStatus commands again, it is more convenient if the Status register value is returned along with the command output data. Therefore, the register that the interrupt handler read is stored as a sixth byte in the WriteFile results buffer.

IOCTL_PHDIO_GET_RW_RESULTS is used to obtain the write (and read) results. Table 15.5 recaps the contents of the 6-byte buffer that is returned. Listing 15.4 shows how DeviceIoControl is used to read and display the results. When *WdmIoTest* is run successfully, the cmd status is 0x5F and the int status is 0xDF. This is expected. When a byte has just been output to the printer, BUSY# (the top bit of the Status register) goes low. When it has been printed, BUSY# goes high and an interrupt is generated.

Note that the command output buffer and the last interrupt register value locations are reused each time an interrupt occurs and the commands are run. This is not a problem. As soon as any fault occurs, processing stops with the most useful information in the results buffer.

Table 15.5   Read/Write results

Bytes	Description
2	Command error code
2	Command error offset
1	Command output value
1	Last interrupt register

Listing 15.4 Getting Write/Read command results

```
if( DeviceIoControl( hWdmIo, IOCTL_PHDIO_GET_RW_RESULTS,
                NULL, 0,       // Input
                rv, sizeof(rv), // Output
                &BytesReturned, NULL))
{
    printf("    Get RW Results OK.  rv=%d at %d\n", rv[0], rv[1]);
    BYTE* pbuf = (BYTE*)(&rv[2]);
    printf("                        cmd status=%02x\n", pbuf[0]);
    printf("                        int status=%02x\n", pbuf[1]);
}
```

WdmIoTest finally disables interrupts by running the DisableInterrupts commands and closes its handle to the WdmIo device.

Reading Data Using Interrupt Driven I/O

Reading data is a process similar to writing data. This time, two sets of commands need to be passed to WdmIo before the actual ReadFile is issued. The first set of commands starts the read process. The second set is used to process interrupts. Two sets are needed, as it is highly likely that different commands will be needed.

Use IOCTL_PHDIO_CMDS_FOR_READ_START to tell WdmIo which commands to use to start the read process. Typically, these commands might simply enable input interrupts from the device. IOCTL_PHDIO_CMDS_FOR_READ passes WdmIo the commands used to process each interrupt.

The ReadFile call passes the read buffer. The WdmIo driver can only handle fixed length transfers. However, a time-out of only one or two seconds could be used to ensure that errors are caught reasonably quickly.

As before, IOCTL_PHDIO_GET_RW_RESULTS is used to get the read results.

Reading data is not illustrated in *WdmIoTest*.

Testing WdmIo

Installing WdmIo

Installing a WdmIo device for a parallel port is a bit trickier than all the previous drivers. This is because Windows should already have installed its own driver to service the port. It is possible to run WdmIo for a parallel port in Windows 98. However, I found it was not possible in Windows 2000. The PHDIo driver can be forced to run in W2000 to talk to a parallel port.

The WdmIoLpt1Checked.inf and WdmIoLpt1Free.inf installation files are designed for a parallel port that, by default, lives at address 0x0378 and generates ISA interrupt IRQ7. Other configurations are made available for the user to select.

The best way to install WdmIo in Windows 98 is to use Hardware profiles. A Hardware profile is the combination of drivers that Windows loads. The trick is to have one profile in which the standard Windows parallel port driver is available. Another Hardware profile loads the WdmIo driver. When Windows starts, select the profile that you want to use.

In the Control Panel System applet, select the Hardware profiles tab. Copy the "Original Configuration" profile into a new profile called "WdmIo". Reboot into this new profile. In the Device Manager, select the Parallel port (LPT1) driver. First, remember which resources the LPT1 driver uses. Now remove the Parallel port driver from this profile. Simply disabling the driver is not good enough, as a disabled driver still keeps its resource assignments. Do not restart yet.

Now install a WdmIo device in the Other devices category as usual. You will have to reboot for the WdmIo device to be activated. I found that removing the Parallel port driver may mean that the parallel port was disabled in the BIOS setup. As you do the last reboot, you should go into your BIOS setup and reenable the LPT1 port again.

If you ever reboot and choose the "Original Configuration" hardware profile, you will find that the WdmIo device has been removed from the WdmIo profile. You will have to reinstall it and reboot. Sigh.

If you use WdmIo to talk to some of your own custom hardware, you will almost certainly not need to go through all of the previous shenanigans. Similarly, for a one off test of WdmIo talking to a parallel port, you should be able to install it without using different hardware profiles.

In Windows 2000 beta 2, I found that it was impossible to install a WdmIo device in place of the standard parallel port driver. If a parallel port driver was removed to make the resources available, Windows 2000 always reinstalled the driver on reboot before the WdmIo device could be loaded.

LogConfig **Sections**

The WdmIo INF files must be set up for each device to which it will talk. A LogConfig section or sections must be added to give the resource assignments. Listing 15.5 shows the amendments that have been made to the WdmIoLpt1Checked.inf and WdmIoLpt1Free.inf installation files.

Both the Windows 98 WdmIo.Install and the Windows 2000 WdmIo.Install.NTx86 sections have a LogConfig directive that refers to the three LogConfig sections, each of which have different possible resource assignments. These eventually appear to users in the Device properties Resources tab as "Basic configuration 0", "Basic configuration 1", and "Basic configuration 2". Each section has a configuration priority. WdmIo.LogConfig1 has priority DESIRED that is higher than the NORMAL priority of the other LogConfig sections.

The WdmIo.LogConfig1 resource assignments are chosen by preference. There is only one I/O port range, 0x378 to 0x37A, so this address range is used. However, two IRQ numbers are listed, 5 and 7. Only one of these is used. In some computers, IRQ5 will already be used, so IRQ7 will be selected automatically. If both IRQs are available, the user will be prompted to make a choice.

The user, therefore, has to confirm which resource assignments to make. For the WdmIo device, Windows does not detect the parallel port and load the appropriate drivers. It is up to the device installer to select the correct resource assignments. The WdmIo driver does not care which port address or IRQ it is assigned. To do a useful job, the resources must correspond to a real device, and the controlling Win32 application must use the registers and interrupts in the correct way.

If you try to install a second WdmIo device for a parallel port, Windows should choose one of the remaining configurations. Windows 98 displays the chosen resource assignments and asks the user to confirm that these are satisfactory. It then tells you to insert the card. If the system's resource assignments have to be juggled to accommodate the new device, you may have to reboot.

Listing 15.5 WdmIoLpt1Xxx.INF LogConfig **sections**

```
[WdmIo.Install]
CopyFiles=WdmIo.Files.Driver
AddReg=WdmIo.AddReg
LogConfig=WdmIo.LogConfig1,WdmIo.LogConfig2,WdmIo.LogConfig3
```

Listing 15.5 WdmIoLpt1Xxx.INF LogConfig **sections (continued)**

```
[WdmIo.LogConfig1]
ConfigPriority=DESIRED
IOConfig=378-37a
IRQConfig=7,5


[WdmIo.LogConfig2]
ConfigPriority=NORMAL
IOConfig=278-27a
IRQConfig=7,5


[WdmIo.LogConfig3]
ConfigPriority=NORMAL
IOConfig=3bc-3be
IRQConfig=7,5


[WdmIo.Install.NTx86]
CopyFiles=WdmIo.Files.Driver.NTx86
LogConfig=WdmIo.LogConfig1,WdmIo.LogConfig2,WdmIo.LogConfig3
```

A large number of entries can be put in LogConfig sections. For a start, further IOConfig and IRQConfig entries can be given if a device can have more than one of these resources simultaneously. The IOConfig and IRQConfig entries can be specified in a variety of ways. DMAConfig, PcCardConfig, MemConfig, and MfCardConfig entries can also be included. I found that the Windows 2000 DDK was the best source of information on all these entries.

Running *WdmIoTest*

With a WdmIo device finally installed for one of your parallel ports, it is now possible to run the *WdmIoTest* program. Do not forget to plug in a printer. I found that a clattery old dot matrix printer was best for testing, as it produced audible output straightaway. With a page-oriented ink jet or laser printer, you might have to press Formfeed to see the printed information.

When you run *WdmIoTest*, the printer should initialize itself and print out a two-line short message.

If the printer is switched off, *WdmIoTest* will keep checking to see if it is ready for 20 seconds. It will give up if it is not available by then. To test the time-out behavior of WriteFile, you must comment out the busy check code. If you now run *WdmIoTest* again, it will attempt to write information to the printer. It should time-out after 10 seconds with error 21 (ERROR_NOT_READY).

The WdmIo driver handles the user pressing Ctrl+C correctly. I shall show in the next chapter how this IRP cancelling and cleanup occurs.

Testing PHDIo

Installing PHDIo

The PHDIo driver was initially aimed just at the NT and Windows 2000 platforms. However, I found that it works fine in Windows 98. These installation instructions are for both types of platform. You do not need to remove or disable existing drivers yet. Installing and running PHDIo does not make it use any resources. It only asks for resources when a Win32 program, such as *PHDIoTest*, runs.

These instructions describe how to install PHDIo by hand, by copying files and making registry entries. See Chapter 11 for details of how to automate this process.

Start by copying the driver executable PHDIo.sys to the Windows System32\Drivers directory. Use the checked build version of PHDIo.sys if you want to view its *DebugPrint* output. The version of *DebugPrint* supplied with this book does not run under Windows NT 3.51 or NT 4.

Make a registry key for PHDIo at HKLM\SYSTEM\CurrentControlSet\Services\PHDIo. Make the values specified in Table 15.6. The Start value of 2 indicates that the PHDIo driver should be loaded once the system is up and running. In Windows 98, also add an ImagePath REG_SZ value with \SystemRoot\System32\Drivers\PHDIo.sys.

Table 15.6   PHDIo **driver registry values**

Name	Type	Contents
Type	DWORD	1
Start	DWORD	2
DisplayName	REG_SZ	"PHDIo"
ErrorControl	DWORD	1

Now reboot your system. The PHDIo driver should now be running. Check the *DebugPrint* output to see that it has generated a few start-up trace messages.

If you change the driver, the actions you must take to use the new version vary from platform to platform. In Windows 98, you must reboot the system. In NT 3.51 and NT 4, use the Control Panel Devices applet to stop the old driver and start the new one. In Windows 2000, use the book software *Servicer* utility to do the same job.

The PHDIo driver is started in all Hardware profiles.

Running *PHDIoTest*

The *PHDIoTest* software is in subdirectory PHDIo\Exe of the book software. As mentioned previously, PHDIoText opens the PHDIo device with this filename: \\.\PHDIo\isa\io378,3\ irq7\override. This filename tells PHDIo which resources to use. In this case, an I/O port of 0x0378-0x037A (length 3) and IRQ7 are specified. Change these values if your parallel port is configured differently. The \override specifier means that PHDIo should go ahead and use these resources even if they conflict with another device.

In a similar way to *WdmIoTest*, you must disable any existing devices that talk to the parallel port. In Windows 98, use the WdmIo hardware profile, with the WdmIo driver removed. In W2000, NT 3.51, and NT 4, you must change "parport" driver startup option to Manual (3) and reboot. The \override specifier is usually needed to make *PHDIoTest* work in Windows 2000.

You should now be able to run *PHDIoTest* successfully on all these Windows platforms.

As with WdmIo, if you use PHDIo to talk to some of your own custom hardware, you should not need to do any clever hardware configuration, as only you will know where your device lives.

Analyzing WdmIo **and** PHDIo

Which to Use

From the user perspective, WdmIo and PHDIo differ in these respects: the platforms on which they run, the installation method, the method by which their resources are specified, and the number of devices that can be controlled.

WdmIo runs in Windows 98 and Windows 2000. PHDIo also runs in NT 3.51 and NT 4.

WdmIo devices can be installed by the user without a special installation program. PHDIo needs an installation program, but this should be a much easier option for users.

WdmIo receives its resources from the installation INF file. Different configurations can be chosen by the user in the Device Manager. The PHDIo resources are specified by the program that uses it. I can imagine situations in which each technique has its advantages.

Finally, more than one WdmIo device can be installed and they can all be used simultaneously by different Win32 applications[2]. Only one PHDIo device can be installed. This device can be used to control different hardware at different times, but not by two Win32 applications simultaneously. PHDIo could be enhanced fairly easily to make a series of devices (e.g., called \\.\PHDIo1, \\.\PHDIo2, etc.). If an application found that \\.\PHDIo1 was busy, it could try \\.\PHDIo2, etc.

In most cases, PHDIo is the better bet, as it covers more platforms and can be installed fairly easily.

Deficiencies

As you might have guessed by now, WdmIo and PHDIo do have some deficiencies. However, they do let you do simple I/O.

The most fundamental criticism of WdmIo and PHDIo is that they are too dangerous. A device driver should not let mere Win32 programmers control what happens in the kernel, even in the limited way that WdmIo and PHDIo allow. Use these drivers at your own risk.

The command set deliberately does not include conditional or loop commands. Such functionality can easily be implemented in user mode. The commands that are run to handle an interrupt could possibly benefit from such commands. However, a desire for simplicity rules these commands out for now. If you need to add such facilities for your drivers, do so.

---

2.   You might want to add another IOCTL that tells you which resources a device is using, so applications know which WdmIo device to use.

The WdmIo and PHDIo drivers will correctly refuse to read or write to ports that are outside the range of allocated addresses. However, in most cases they will not report a command error if an out-of-range read or write is attempted. Only the PHDIO_IRQ_CONNECT command checks its register parameter.

Another possible improvement is to implement shadow output registers. Quite often, hardware registers are write-only. If you want to set a bit in an output port, reading the register, setting the bit, and writing out the register will not work. A shadow output register is a memory copy of the value that was last written. If the shadow value is read, the set bit operation will work successfully.

Both WdmIo and PHDIo disconnect from their interrupt when a file handle is closed. Make sure that your device does not generate any further interrupts.

The WdmIo driver copes fairly well if a WdmIo device's resources are reassigned. It will not allow resources to be reassigned if there are any open handles. When it receives a Plug and Play stop device message, it waits until all IRPs have completed.

As it stands, each WdmIo device is created as being shareable. It is probably a good idea to alter this characteristic to require exclusive access, as competing IRPs from different processes may result in some confusing results. The PHDIo driver's one \\.\PHDIo device does require exclusive access, as only one open handle can be allowed to specify the resources.

I am sure that you can come up with other ideas for ways in which WdmIo and PHDIo can be enhanced. Go for it.

Conclusion

This chapter has introduced the WdmIo and PHDIo drivers. It has described the facilities that they make available to Win32 applications. WdmIo has its resources specified in its INF file. The resources are given to PHDIo in the Win32 CreateFile call. The *WdmIoTest* example user mode application uses WdmIo to drive a parallel port printer and output a short message. The similar *PHDIoTest* application does similar tests using the PHDIo driver.

The next chapters look at the construction of these drivers, how they queue IRPs to serialize hardware access and they perform interrupt driven I/O.

16

Chapter 16

Hardware I/O IRP Queuing

This chapter starts looking at the WdmIo and PHDIo drivers. Although the text only refers to the WdmIo driver, the techniques used in PHDIo are identical.

Any driver that talks to a real device will need to control access to the hardware. Chapter 9 mentioned that Critical section routines should be used to run routines that cannot be interrupted. Critical section routines are now used for real.

However, the main new requirement for a driver that accesses hardware is that it serializes all the IRPs that it processes. If more than one Win32 process has opened a device and is issuing IRPs, each IRP must not try to talk to the hardware at the same time[1]. If the IRPs can be processed very quickly, critical sections should be used. However, in most cases, the relevant IRPs need to be put into a queue and processed serially, one by one.

Processing IRPs serially will, of course, reduce performance. If jobs can be carried out in parallel, try to do it that way.

You may wish to impose further strictures on users of your device. You could insist that a certain IOCTL be used just before a read request. Although it is best to keep these restrictions to a minimum, do put in any relevant checks. You could use the *FileObject* pointer on the IRP stack to ensure that it is the same application that is issuing the IOCTL and read requests.

Processing IRPs serially is such a common requirement that the Windows Driver Model includes special support for this technique. Each device object contains a device queue of IRPs. The standard StartIo driver callback is used to process the queued IRPs one at a time. The device object and IRP structures have fields that are used by the I/O Manager to manage the queue, and cancel and cleanup queued IRPs safely.

---

1. Or for that matter, an application must not fire off several overlapped requests at the same time.

This chapter, therefore, looks at the device IRP queue, StartIo routines, and how to cancel and cleanup IRPs. The next chapter looks at interrupt handling and all its associated topics.

Figure 16.1 illustrates the queuing of two Read IRPs and one IOCTL IRP. The main Read and IOCTL dispatch routines perform some initial checking of the IRPs before they are queued for processing by the StartIo routine. The initial processing of IRPs 2 and 3 can happen simultaneously. However, the StartIo routine definitely processes the IRPs serially, one after the other.

Figure 16.1   IRP queuing

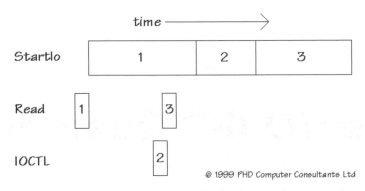

@ 1999 PHD Computer Consultants Ltd

Hardware Access

Before I start in earnest, note that the WdmIo driver is a standard WDM driver in a device stack. It is layered over the Unknown bus driver. It does not use the Unknown driver to access hardware. Instead, it talks to hardware directly.

The StartDevice, RetrieveResources, and StopDevice routines in DeviceIo.cpp have been altered slightly from their originals in Wdm2. They insist that one port or memory resource be allocated. If necessary a memory-mapped set of registers is mapped into memory, and unmapped when the device is stopped. Similarly, if WdmIo has connected to an interrupt, StopDevice disconnects and StartDevice connects again.

The WriteByte and ReadByte routines are shown in Listing 16.1. They both fail silently if the register offset is out of range. The *DebugPrint* trace calls should not be used in these routines, as they may be called at device IRQL (DIRQL) (i.e., above the maximum IRQL suitable for *DebugPrint* calls, DISPATCH_LEVEL).

If you need to delay for a short period, use the KeStallExecutionProcessor routine, specifying the stall period in microseconds. The DDK recommends that you keep the delay as short as possible, and definitely no more than 50µs. Longer delays usually mean writing code in a very different way. For example, you could use custom timers, as described in the next chapter. Alternatively, a system thread could use the KeDelayExecutionThread function for longer delays.

Listing 16.1 WriteByte and ReadByte

```
void WriteByte( IN PWDMIO_DEVICE_EXTENSION dx, IN ULONG offset, IN UCHAR byte)
{
    if( offset>=dx->PortLength) return;
    PUCHAR Port = dx->PortBase+offset;
    if( dx->PortInIOSpace)
        WRITE_PORT_UCHAR( Port, byte);
    else
        WRITE_REGISTER_UCHAR( Port, byte);
}


UCHAR ReadByte( IN PWDMIO_DEVICE_EXTENSION dx, IN ULONG offset)
{
    if( offset>=dx->PortLength) return 0;
    PUCHAR Port = dx->PortBase+offset;
    UCHAR b;
    if( dx->PortInIOSpace)
        b = READ_PORT_UCHAR(Port);
    else
        b = READ_REGISTER_UCHAR(Port);
    return b;
}
```

Finally, note that the WdmIo driver read and write dispatch routines and the IOCTLs all use Buffered I/O.

IRP Queuing

Device Queues

The *DebugPrint* in Chapter 14 used doubly-linked lists as a means of storing blocks of memory for later processing. The WdmIo driver uses the built-in IRP device queue to serialize the processing of its main IRPs. A device queue is a doubly-linked list with special features that tailor it for IRP processing. One of its special features is that it has a built-in spin lock. Thus, a driver does not need to provide its own spin lock to guard access to the queue, as *Debug-Print* had to do for its doubly-linked lists.

The WdmIo Functional Device Object (FDO) contains a device queue field that stores the queue's linked list head. Each IRP also contains a linked list entry that is used to store the pointers to its linked IRPs. For the moment, the names of these fields do not matter, as most ordinary queue actions are handled internally by the I/O Manager. However, to cleanup the device queue, these entries must be manipulated directly. The I/O Manager also initializes the device queue.

Inserting IRPs into the device queue is actually fairly straightforward. This except from the Read IRP dispatch handler in Dispatch.cpp shows the necessary steps. First, the IRP must be marked as pending using IoMarkIrpPending. Then, IoStartPacket is called to insert the IRP into the queue. If there are no queued IRPs, the IRP is sent for processing in the StartIo routine straightaway. Finally, the Read IRP returns STATUS_PENDING to confirm that the IRP is indeed pending.

```
IoMarkIrpPending(Irp);
IoStartPacket( fdo, Irp, 0, WdmIoCancelIrp);
return STATUS_PENDING;
```

The call to IoStartPacket passes a pointer to the cancel routine. I shall show later how this works. The third parameter to IoStartPacket is a key that can be used to sort the IRPs in the device queue. This feature is not used by WdmIo, so the key is always zero.

The WdmIo driver still uses the Plug and Play device locking technique to ensure that IRP processing is not interrupted by Plug and Play stop device requests. Note that the UnlockDevice routine is not called when an IRP is queued. This is deliberate, as the IRP has not been completed. I ensure that UnlockDevice is called on all paths that will later complete the IRP.

The main Read, Write, and IOCTL dispatch routines in WdmIo perform device locking and parameter checking. All valid IRPs of these types are put in the device queue for processing later serially in the WdmIo StartIo routine.

StartIo **Routines**

The WdmIoStartIo routine shown in Listing 16.2 processes IRPs for each WdmIo device one by one[2]. All queued IRPs, whatever their major function code, come through this same routine. Therefore, StartIo routines usually have a big switch statement at their heart. Listing 16.2 shows the complete code for IOCTL IRPs with major function code IRP_MJ_DEVICE_CONTROL. However, the code for Read and Write IRPs is not shown for now. This is explained in the next chapter.

When an IRP is passed to the StartIo routine, it has just been removed from the device queue. The I/O Manager has also put the IRP pointer in the FDO *CurrentIrp* field. This is particularly useful for interrupt handlers as will be seen in the chapter, but is also used by IRP cancel and cleanup routines. The IRP *Cancel* field is set to TRUE when the IRP is cancelled. The IRP *Cancel* field may even be set before StartIo is called. Both WdmIoStartIo and the interrupt handling routines check the *Cancel* field at various points, as shown later.

StartIo routines are always called at DISPATCH_LEVEL IRQL. This means that all the code and the variables it accesses must be in nonpaged memory. It restricts the set of kernel calls that can be made. However, calls to the *DebugPrint* trace output routines can be made safely.

WdmIoStartIo processes IRPs in two different ways. All the IOCTL IRPs are handled straightaway; the IRP is processed in the relevant way and is completed at the end of WdmIoStartIo. However, Read and Write IRPs use interrupt driven I/O to transfer one byte at a time. These IRPs are usually completed later, as described in the next chapter. In this case, WdmIoStartIo simply returns; the current IRP is still pending and is still being processed.

---

2. If there is more than one WdmIo device, each device has a device queue that can be processed simultaneously by WdmIoStartIo.

WdmIoStartIo is only called to process another IRP in the device queue when IoStartNextPacket is called (or IoStartNextPacketByKey). For IRPs that are processed entirely by WdmIoStartIo, IoStartNextPacket is called after the IRP is completed[3]. The TRUE second parameter to IoStartNextPacket indicates that cancel routines are being used.

WdmIoStartIo begins by zeroing the *CmdOutputCount* field in the device extension. This field stores the count of bytes transferred during immediate IRP processing. *CmdOutputCount* is passed to CompleteIrp at the end of WdmIoStartIo.

WdmIoStartIo then stops the device timer if the device extension *StopTimer* field is set to true. This timer is used to detect read and write time-outs and its use is described in the next chapter.

WdmIoStartIo now contains the obligatory huge switch statement, switching on the IRP stack major function code. As stated earlier, only IOCTL, read, and write requests should reach WdmIoStartIo. The IOCTL handler has a subsidiary large switch statement, this time switching on the IOCTL control code. All the IOCTLs are processed straightaway and eventually fall through to complete the IRP at the end of WdmIoStartIo.

WdmIoStartIo ends by completing IRPs that have finished their processing. The *Cancel* flag is checked and the cancel routine is removed. Finally, the device is unlocked using UnlockDevice, the IRP is completed, and IoStartNextPacket is called.

Listing 16.2 WdmIoStartIo **routine**

```
VOID WdmIoStartIo( IN PDEVICE_OBJECT fdo, IN PIRP Irp)
{
    PWDMIO_DEVICE_EXTENSION dx = (PWDMIO_DEVICE_EXTENSION)fdo->DeviceExtension;
    PIO_STACK_LOCATION IrpStack = IoGetCurrentIrpStackLocation(Irp);
    PUCHAR Buffer = (PUCHAR)Irp->AssociatedIrp.SystemBuffer;

    // Zero the output count
    dx->CmdOutputCount = 0;

    DebugPrint( "WdmIoStartIo: %I", Irp);

    // Stop the 1 second timer if necessary
    if( dx->StopTimer)
    {
        IoStopTimer(fdo);
        dx->StopTimer = false;
    }

    NTSTATUS status = STATUS_SUCCESS;
```

---

3. Note that the IoStartNextPacket implementation will, in fact, call WdmIoStartIo recursively (if there is another IRP in the queue). In the worst case, this recursive technique could overflow the kernel stack. One of the DDK examples shows a way of avoiding this problem by not queuing an IRP if the queue is "full".

Listing 16.2 `WdmIoStartIo` **routine (continued)**

```
// Switch on the IRP major function code
switch( IrpStack->MajorFunction)
{
//////////////////////////////////////////////////////////////////////
case IRP_MJ_DEVICE_CONTROL:
{
    ULONG ControlCode = IrpStack->Parameters.DeviceIoControl.IoControlCode;
    ULONG InputLength =
        IrpStack->Parameters.DeviceIoControl.InputBufferLength;
    ULONG OutputLength =
        IrpStack->Parameters.DeviceIoControl.OutputBufferLength;
    switch( ControlCode)
    {
    // / / / / / / / / / / / / / / / / / / / / / / / / / / / / / / / //
    case IOCTL_PHDIO_RUN_CMDS:
        DebugPrint( "WdmIoStartIo: Run Cmds %s",
            dx->ConnectedToInterrupt?"(synchronised)":"");
        // Run the commands, synchronized with interrupt if necessary
        if( dx->ConnectedToInterrupt)
        {
            if( !KeSynchronizeExecution( dx->InterruptObject,
                (PKSYNCHRONIZE_ROUTINE)RunCmdsSynch, (PVOID)fdo))
                status = STATUS_UNSUCCESSFUL;
        }
        else
            if( !RunCmds(fdo,true))
                status = STATUS_UNSUCCESSFUL;
        break;
    // / / / / / / / / / / / / / / / / / / / / / / / / / / / / / / / //
    case IOCTL_PHDIO_CMDS_FOR_READ:
        DebugPrintMsg( "WdmIoStartIo: Store cmds for read");
        status = StoreCmds( &dx->ReadCmds, &dx->ReadCmdsLen,
            InputLength, Buffer);
        break;
    // / / / / / / / / / / / / / / / / / / / / / / / / / / / / / / / //
    case IOCTL_PHDIO_CMDS_FOR_READ_START:
        DebugPrintMsg( "WdmIoStartIo: Store cmds for read start");
        status = StoreCmds( &dx->StartReadCmds, &dx->StartReadCmdsLen,
            InputLength, Buffer);
        break;
    // / / / / / / / / / / / / / / / / / / / / / / / / / / / / / / / //
    case IOCTL_PHDIO_CMDS_FOR_WRITE:
```

Listing 16.2 `WdmIoStartIo` **routine (continued)**

```
            DebugPrintMsg( "WdmIoStartIo: Store cmds for write");
            status = StoreCmds( &dx->WriteCmds, &dx->WriteCmdsLen,
                InputLength, Buffer);
            break;
        //////////////////////////////////////////////////
        case IOCTL_PHDIO_GET_RW_RESULTS:
            // Copy cmd output first
            dx->CmdOutputCount = dx->TxCmdOutputCount;
            if( dx->CmdOutputCount>OutputLength)
                dx->CmdOutputCount = OutputLength;
            RtlCopyMemory( Buffer, dx->TxResult, dx->CmdOutputCount);
            // Then add on last interrupt reg value
            if( dx->CmdOutputCount+1<=OutputLength)
                Buffer[dx->CmdOutputCount++] = dx->TxLastIntReg;

            DebugPrint( "WdmIoStartIo: Get RW Results: %d bytes"
                ,dx->CmdOutputCount);
            break;
        //////////////////////////////////////////////////
        default:
            status = STATUS_NOT_SUPPORTED;
        }
        break;
    }


    ////////////////////////////////////////////////////////////////
    case IRP_MJ_WRITE:
        // ...
    case IRP_MJ_READ:
        // ...
    default:
        status = STATUS_NOT_SUPPORTED;
        break;
    }


    ////////////////////////////////////////////////////////////////
    // Complete this IRP

    if( Irp->Cancel) status = STATUS_CANCELLED;

    // Remove cancel routine
    KIRQL OldIrql;
```

Listing 16.2 WdmIoStartIo **routine (continued)**

```
    IoAcquireCancelSpinLock( &OldIrql);
    IoSetCancelRoutine( Irp, NULL);
    IoReleaseCancelSpinLock(OldIrql);

    // Unlock device, complete IRP and start next
    UnlockDevice(dx);
    DebugPrint( "WdmIoStartIo: CmdOutputCount %d", dx->CmdOutputCount);
    CompleteIrp(Irp, status, dx->CmdOutputCount);
    IoStartNextPacket( fdo, TRUE);
}
```

Processing Commands

Critical Sections

The IOCTL_PHDIO_RUN_CMDS IOCTL is used to run a set of commands straightaway. Eventually the ProcessCmds routine is run to process the commands. ProcessCmds is fairly straightforward and so is not listed here. It is in DeviceIo.cpp on the book CD-ROM. Suffice to say, it is passed parameters to its input and output buffers, together with their length. It also has a CanTrace bool parameter that dictates whether its *DebugPrint* trace statements can be run safely.

However, there are a couple of hurdles to overcome before IOCTL_PHDIO_RUN_CMDS can call ProcessCmds.

The first hurdle is that the IOCTL input and output buffers use the same block of memory. While this is a jolly useful technique for saving memory in the first stage of IOCTL processing, it means that some nonpaged memory must be allocated for the output data. The output data is copied back to the shared buffer and the temporary memory freed. The RunCmds routine performs this task[4].

If WdmIo has not connected to a hardware interrupt, the IOCTL_PHDIO_RUN_CMDS handler in Listing 16.2 simply calls RunCmds. However, if the device extension *ConnectedToInterrupt* field is true, an interrupt may occur that could scupper any command processing. This problem is overcome by calling RunCmds in the context of a Critical Section. To recap, a Critical Section routine runs at Device IRQL (DIRQL) and so cannot be interrupted (by our interrupt at least).

In this case, the IOCTL_PHDIO_RUN_CMDS handler calls KeSynchronizeExecution to run the RunCmdsSynch routine as a Critical Section. RunCmdsSynch just calls RunCmds with the

---

4. At the last moment, I have moved this code back into WdmIoStartIo. RunCmds may not run at DIRQL and it is incorrect to allocate nonpaged memory above DISPATCH_LEVEL.

CanTrace parameter set to `false`. Both RunCmdsSynch and RunCmds return FALSE if they can not allocate enough memory for the output buffer.

Listing 16.3 RunCmdsSynch **and** RunCmds

```
BOOLEAN RunCmdsSynch( IN PDEVICE_OBJECT fdo)
{
    return RunCmds( fdo, false);
}


BOOLEAN RunCmds( IN PDEVICE_OBJECT fdo, IN bool CanTrace)
{
    PWDMIO_DEVICE_EXTENSION dx = (PWDMIO_DEVICE_EXTENSION)fdo->DeviceExtension;
    PIRP Irp = fdo->CurrentIrp;
    PIO_STACK_LOCATION IrpStack = IoGetCurrentIrpStackLocation(Irp);
    ULONG InputLength = IrpStack->Parameters.DeviceIoControl.InputBufferLength;
    ULONG OutputLength =
        IrpStack->Parameters.DeviceIoControl.OutputBufferLength;
    PUCHAR Buffer = (PUCHAR)Irp->AssociatedIrp.SystemBuffer;

    PUCHAR OutBuffer = NULL;
    if( OutputLength>0)
    {
        OutBuffer = (PUCHAR)ExAllocatePool(NonPagedPool,OutputLength);
        if( OutBuffer==NULL) return FALSE;
    }
    ProcessCmds( dx, Buffer, InputLength, OutBuffer, OutputLength, CanTrace);
    if( OutBuffer!=NULL)
    {
        RtlMoveMemory( Buffer, OutBuffer, OutputLength);
        ExFreePool(OutBuffer);
    }
    return TRUE;
}
```

Cancelling Queued IRPs

Chapter 14 explained when Windows tries to cancel IRPs using IRP cancel routines. In addition, Windows issues the Cleanup IRP to cancel IRPs in some circumstances. Chapter 14 also showed how the DebugPrint driver provides a cancel routine for its one queued IRP.

As WdmIo queues IRP, it must also provide cancel routines for these IRPs. In addition, it handles the Cleanup IRP correctly so that IRPs are cancelled if a user mode application closes its file handle with overlapped I/O requests pending.

Queued IRP Cancelling

When considering a strategy for cancelling IRPs, two cases must usually be considered. In the first case, the IRP is still being held in the device queue. The second case is when the IRP has been removed from the device queue and is being processed by StartIo.

The I/O Manager does not know whether an IRP is in the device queue[5]. It simply calls the cancel routine. The cancel routine must determine what to do. If the IRP pointer matches the FDO *CurrentIrp* field, the IRP is running in StartIo (or in the process of a transfer started by StartIo). Otherwise, the cancel routine must try to remove the IRP from the device queue.

The I/O Manager uses its *Cancel spin lock* to guard cancelling operations. An IRP's cancel routine is called at DISPATCH_LEVEL IRQL while holding the Cancel spin lock. The IRP *CancelIrql* field holds the old IRQL that should be passed to IoReleaseCancelSpinLock before the IRP is completed and the cancel routine exits.

I have already mentioned that a device queue includes a spin lock to ensure that all operations on the queue are handled safely in a multiprocessor environment. When cancel routines are involved, the Cancel spin lock must be held. This is to ensure that a Cancel routine is not called on one processor while the IRP is being dequeued on another processor.

WdmIo **IRP Cancelling Strategy**

The WdmIo driver cancels a queued IRP by removing it from the device queue and completing it with status STATUS_CANCELLED.

If the IRP is being processed by StartIo, the cancel routine in effect does nothing. Before the cancel routine is called, the I/O Manager sets the IRP's *Cancel* flag. The code called by WdmIoStartIo checks this *Cancel* flag every now and then. If it is found to be set, the current operation is abandoned and the IRP is completed with status STATUS_CANCELLED.

The DDK documentation does not say that the Cancel routine has to complete the IRP. In WdmIo, if the IRP is being processed by StartIo, the Cancel routine does not cancel the IRP. The IRP is only cancelled later. This strategy seems to work. See the following section for an alternative technique.

Listing 16.4 shows the WdmIoCancelIrp routine. If the IRP to be cancelled matched the FDO *CurrentIrp* field, the IRP is being processed by StartIo (or its interrupt driven follow on code). In this case, all WdmIoCancelIrp does is to release the Cancel spin lock and exit.

If the IRP to be cancelled is not the current IRP, KeRemoveEntryDeviceQueue is called to try to remove the IRP from the device queue. The FDO *DeviceQueue* field holds the device queue list head. The IRP *Tail.Overlay.DeviceQueueEntry* field holds the list entry. The Cancel spin lock can now be released safely. If the IRP was removed from the queue, UnlockDevice is called and the IRP is completed with status STATUS_CANCELLED.

Listing 16.4 WdmIoCancelIrp **routine**

```
VOID WdmIoCancelIrp( IN PDEVICE_OBJECT fdo, IN PIRP Irp)
{
    PWDMIO_DEVICE_EXTENSION dx = (PWDMIO_DEVICE_EXTENSION)fdo->DeviceExtension;
    DebugPrint("WdmIoCancelIrp: Cancelling %I", Irp);
    if( Irp==fdo->CurrentIrp)
```

---

5. In fact, the I/O Manager could work out if an IRP is in the device queue.

Listing 16.4 `WdmIoCancelIrp` **routine (continued)**

```
{
    DebugPrintMsg("WdmIoCancelIrp: IRP running in StartIo");
    // IRP is being processed by WdmIoStartIo.
    // Irp->Cancel flag already set.
    // WdmIoStartIo or timeout will detect Cancel flag
    // and cancel IRP in due course
    IoReleaseCancelSpinLock(Irp->CancelIrql);
}
else
{
    DebugPrintMsg("WdmIoCancelIrp: IRP in StartIo queue");
    // IRP is still in StartIo device queue.
    // Just dequeue and cancel it.  No need to start next IRP.
    BOOLEAN dequeued = KeRemoveEntryDeviceQueue( &fdo->DeviceQueue,
        &Irp->Tail.Overlay.DeviceQueueEntry);

    IoReleaseCancelSpinLock(Irp->CancelIrql);

    if( dequeued)
    {
        UnlockDevice(dx);
        CompleteIrp( Irp, STATUS_CANCELLED);
    }
}
}
```

Cancel Checking

The code in DeviceIo.cpp makes various checks to see if the I/O Manager has requested that the IRP be cancelled.

At the end of WdmIoStartIo, as shown in Listing 16.2, the status is set to cancelled if the IRP *Cancel* field has been set. The code at the end of WdmIoStartIo also removes the cancel routine before completing the IRP. It acquires the I/O Manager Cancel spin lock before calling IoSetCancelRoutine with NULL for the cancel routine parameter. Remove a cancel routine before completing an IRP.

The ProcessCmds routine also checks the IRP *Cancel* field just before it gets the next command. If it is set, processing stops with error code PHDIO_CANCELLED in the output buffer and the IRP is cancelled.

The interrupt handling code also checks the IRP *Cancel* field, as shown in the next chapter.

Alternative Cancel Strategy

An alternative IRP cancelling strategy is to remove an IRP's cancel routine as soon as it starts being processed in StartIo. The advantage is that the cancel routine can complete the IRP straightaway. The downside is that no cancel routine is available while the IRP is being processed by StartIo or its follow on code.

The IRP cancel routine is changed for the case when the IRP is the current IRP. Listing 16.5 shows that in this case, the IRP is completed and IoStartNextPacket called.

Listing 16.5 Alternative IRP cancel routine

```
if( Irp==fdo->CurrentIrp)
{
    DebugPrintMsg("WdmIoCancelIrp: IRP just dequeued for StartIo");
    // IRP has just been dequeued but StartIo has not had a chance
    // to remove this cancel routine yet.  Irp->Cancel flag already set.
    IoReleaseCancelSpinLock(Irp->CancelIrql);

    // Cancel IRP and start next one
    CompleteIrp( Irp, STATUS_CANCELLED);
    IoStartNextPacket( fdo, TRUE);
}
else
    ...
```

The start of the StartIo routine must also be changed. The main job is to remove the cancel routine using IoSetCancelRoutine. However, there is a small chance that the IRP has already been cancelled. The code in Listing 16.6 checks the IRP *Cancel* field first. If it has been cancelled, StartIo simply exits.

In my mind, this technique has a potential race condition. When the Cancel routine completes, the IRP memory may disappear straightaway. However, the StartIo routine may be just about to run on another processor. Accessing the IRP *Cancel* field might, therefore, cause an access violation (or refer to the next IRP to use this IRP structure). Moving the CompleteIrp and IoStartNextPacket calls into the StartIo routine (if Irp->Cancel is set) would solve this problem, but it would mean that the cancel routine does not complete the IRP.

Listing 16.6 Alternative StartIo initial processing

```
// Check whether cancelled already
KIRQL OldIrql;
IoAcquireCancelSpinLock( &OldIrql);
if( Irp->Cancel)
{
    IoReleaseCancelSpinLock(OldIrql);
    // IoStartNextPacket called by cancel routine
    return;
```

Listing 16.6 Alternative `StartIo` initial processing (continued)

```
}
// Remove cancel routine
IoSetCancelRoutine( Irp, NULL);
IoReleaseCancelSpinLock(OldIrql);
```

Cleanup IRP Handling

The Cleanup IRP is issued to cancel any IRPs that are outstanding when a file handle is closed. Even if the IRPs have cancel routines, they are not called.

To handle it correctly, the driver ought to cancel only IRPs that belong to the correct file handle. Only queued IRPs whose *FileObject* field matches the Cleanup IRP's stack *FileObject* field should be cancelled.

The WdmIo Cleanup IRP handler, WdmIoDispatchCleanup, shown in Listing 16.7, does not perform this *FileObject* check. This makes the code simpler.

WdmIoDispatchCleanup must hold the Cancel spin lock wherever it accesses the device queue. However, you must not be holding the Cancel spin lock when an IRP is completed.

The code keeps extracting IRPs using KeRemoveDeviceQueue until the device queue is empty. A pointer to the IRP is found using the usual CONTAINING_RECORD machinations. The IRP is marked for cancelling and its cancel routine is removed. The Cancel spin lock is released before UnlockDevice is called and the IRP is completed with status STATUS_CANCELLED.

The Cleanup IRP handler then goes on to try to cancel the IRP currently being processed by WdmIoStartIo or its follow-on code. CancelCurrentIrpSynch is called as a Critical Section routine. This returns TRUE if a transfer is in progress (i.e., if the device extension *Timeout* field is greater than or equal to zero). If a transfer is in progress, WdmIoDpcForIsr is called to complete the IRP with status STATUS_CANCELLED. The *Timeout* field and WdmIoDpcForIsr are described in the next chapter.

To handle the FileObject check correctly, you must still remove each IRP from the queue in turn. If the IRP's FileObject does not match the Cleanup FileObject, the IRP must be put in a temporary holding queue. At the end, all these IRP's must be reinserted in the main device queue.

Listing 16.7 `WdmIo` Cleanup IRP handling

```
NTSTATUS WdmIoDispatchCleanup( IN PDEVICE_OBJECT fdo, IN PIRP Irp)
{
    PWDMIO_DEVICE_EXTENSION dx = (PWDMIO_DEVICE_EXTENSION)fdo->DeviceExtension;
    DebugPrintMsg("WdmIoDispatchCleanup");
    KIRQL OldIrql;
    IoAcquireCancelSpinLock(&OldIrql);

    // Cancel all IRPs in the I/O Manager maintained queue in device object
    PKDEVICE_QUEUE_ENTRY QueueEntry;
    while( (QueueEntry=KeRemoveDeviceQueue(&fdo->DeviceQueue)) != NULL)
    {
```

Listing 16.7 `WdmIo` **Cleanup IRP handling (continued)**

```
        PIRP CancelIrp =
            CONTAINING_RECORD( QueueEntry, IRP, Tail.Overlay.DeviceQueueEntry);
        CancelIrp->Cancel = TRUE;
        CancelIrp->CancelIrql = OldIrql;
        CancelIrp->CancelRoutine = NULL;


        IoReleaseCancelSpinLock(OldIrql);
        DebugPrint("WdmIoDispatchCleanup: Cancelling %I",CancelIrp);
        UnlockDevice(dx);
        CompleteIrp( CancelIrp, STATUS_CANCELLED);
        IoAcquireCancelSpinLock(&OldIrql);
    }
    IoReleaseCancelSpinLock(OldIrql);


    // Forceably cancel any in-progress IRP
    if( dx->Timeout!=-1)
    {
        if( KeSynchronizeExecution( dx->InterruptObject,
                (PKSYNCHRONIZE_ROUTINE)CancelCurrentIrpSynch, dx))
        {
            if( fdo->CurrentIrp!=NULL)
            {
                DebugPrint("WdmIoDispatchCleanup: Cancelled in-progress IRP %I",
                    fdo->CurrentIrp);
                WdmIoDpcForIsr( NULL, fdo, fdo->CurrentIrp, dx);
            }
        }
    }


    return CompleteIrp( Irp, STATUS_SUCCESS);
}


static BOOLEAN CancelCurrentIrpSynch( IN PWDMIO_DEVICE_EXTENSION dx)
{
    if( dx->Timeout==-1)
        return FALSE;
    dx->Timeout = -1;
    dx->TxStatus = STATUS_CANCELLED;
    return TRUE;
}
```

Testing, Cancelling, and Cleanup

I amended *WdmIoTest* so that I could test the WdmIo driver's IRP cancelling and Cleanup IRP handling.

The *WdmIoCancel* application is substantially the same as that of *WdmIoTest* and is contained in the book software WdmIo\Cancel directory. You will have to uncomment some of the code and recompile to undertake some of the tests. The *PHDIoCancel* application does similar tests for the PHDIo driver.

The tests are designed to be run with the printer switched off, so that Write IRPs do not complete straightaway. WdmIoCancel opens the WdmIo device for overlapped I/O, using the GetDeviceViaInterfaceO1 function. It then issues two overlapped Write requests and does not wait for either of them to complete. It then exits straightaway. This behavior is designed to test the IRP cancelling.

Figure 16.2 shows the *DebugPrint* output that demonstrates that IRP cancelling works correctly. At points ❶ in the diagram, one of the IOCTL IRPs has been issued. They are passed to WdmIoStartIo, which processes them immediately. At point ❷, the first Write IRP is issued. At point ❸, WdmIoStartIo starts processing this first Write. At point ❹, the second Write IRP is issued. As the first Write is still in progress (the printer being off), this second Write IRP is queued.

Point ❺ is where it gets interesting. WdmIoCancel has just exited. The I/O Manager tries to cancel all pending IRPs. It calls the cancel routine of the second Write. WdmIoCancelIrp finds that it is not the current IRP. It therefore removes it from the StartIo queue and completes the IRP with a cancelled status.

The I/O Manager then calls the cancel routine of the first Write IRP at point ❻. WdmIoCancelIrp finds that this IRP is the current IRP. In this design, WdmIoCancelIrp simply exits. In due course, the Timeout1s time-out routine runs. Timeout1s finds that the IRP *Cancel* flag is set. The WdmIoDpcForIsr routine is called to complete the IRP with a cancelled status.

Finally, the Cleanup IRP is issued at point ❼. However, this finds that it has no work to do as the device queue is empty and there is no interrupt driven I/O in progress.

If you look carefully at the *DebugPrint* output, you will notice that the IRP pointer for the first four IRPs issued is the same, 0xC8548180. The I/O Manager is obviously reusing the IRP structures from its pool of available IRPs. The second write needs another IRP structure, which is at 0xC17C3E00.

Figure 16.2 Cleanup handling *DebugPrint* output

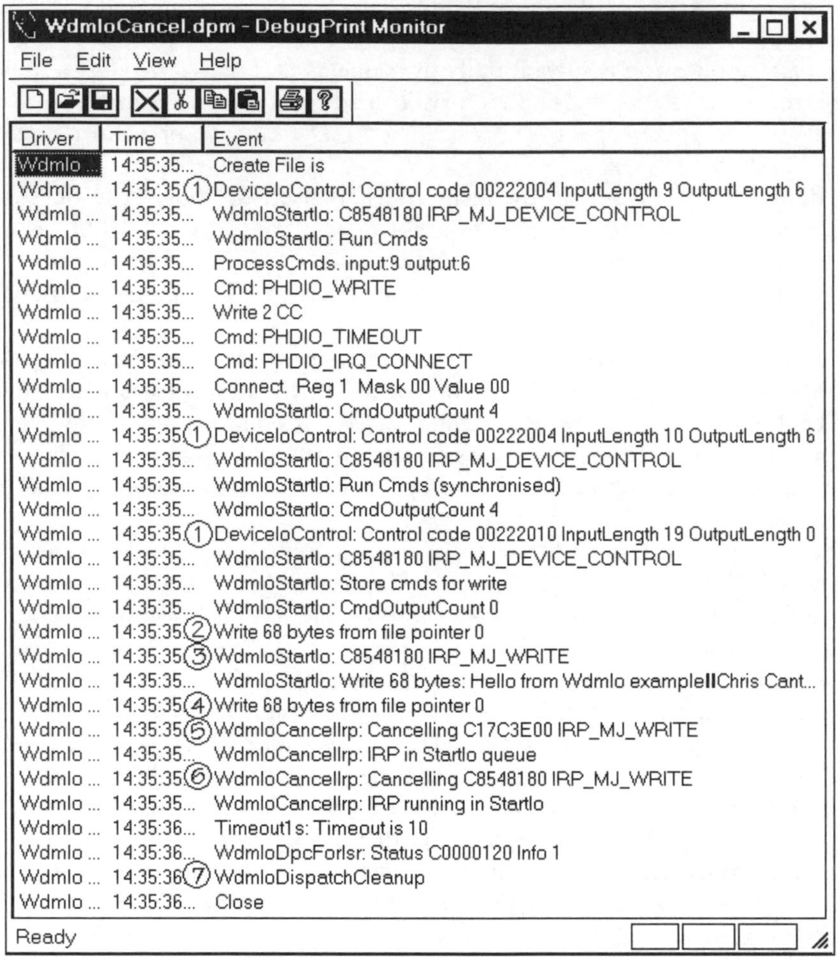

```
WdmIoCancel.dpm - DebugPrint Monitor                    _  □  ×

 File   Edit   View   Help

 [ ][ ][ ] [×][ ][ ][ ] [ ][ ]

 Driver     Time      Event
 Wdmlo ...  14:35:35...  Create File is
 Wdmlo ...  14:35:35 (1) DeviceIoControl: Control code 00222004 InputLength 9 OutputLength 6
 Wdmlo ...  14:35:35...  WdmIoStartIo: C8548180 IRP_MJ_DEVICE_CONTROL
 Wdmlo ...  14:35:35...  WdmIoStartIo: Run Cmds
 Wdmlo ...  14:35:35...  ProcessCmds. input:9 output:6
 Wdmlo ...  14:35:35...  Cmd: PHDIO_WRITE
 Wdmlo ...  14:35:35...  Write 2 CC
 Wdmlo ...  14:35:35...  Cmd: PHDIO_TIMEOUT
 Wdmlo ...  14:35:35...  Cmd: PHDIO_IRQ_CONNECT
 Wdmlo ...  14:35:35...  Connect. Reg 1 Mask 00 Value 00
 Wdmlo ...  14:35:35...  WdmIoStartIo: CmdOutputCount 4
 Wdmlo ...  14:35:35 (1) DeviceIoControl: Control code 00222004 InputLength 10 OutputLength 6
 Wdmlo ...  14:35:35...  WdmIoStartIo: C8548180 IRP_MJ_DEVICE_CONTROL
 Wdmlo ...  14:35:35...  WdmIoStartIo: Run Cmds (synchronised)
 Wdmlo ...  14:35:35...  WdmIoStartIo: CmdOutputCount 4
 Wdmlo ...  14:35:35 (1) DeviceIoControl: Control code 00222010 InputLength 19 OutputLength 0
 Wdmlo ...  14:35:35...  WdmIoStartIo: C8548180 IRP_MJ_DEVICE_CONTROL
 Wdmlo ...  14:35:35...  WdmIoStartIo: Store cmds for write
 Wdmlo ...  14:35:35...  WdmIoStartIo: CmdOutputCount 0
 Wdmlo ...  14:35:35 (2) Write 68 bytes from file pointer 0
 Wdmlo ...  14:35:35 (3) WdmIoStartIo: C8548180 IRP_MJ_WRITE
 Wdmlo ...  14:35:35...  WdmIoStartIo: Write 68 bytes: Hello from Wdmlo exampleIIChris Cant...
 Wdmlo ...  14:35:35 (4) Write 68 bytes from file pointer 0
 Wdmlo ...  14:35:35 (5) WdmIoCancelIrp: Cancelling C17C3E00 IRP_MJ_WRITE
 Wdmlo ...  14:35:35...  WdmIoCancelIrp: IRP in StartIo queue
 Wdmlo ...  14:35:35 (6) WdmIoCancelIrp: Cancelling C8548180 IRP_MJ_WRITE
 Wdmlo ...  14:35:35...  WdmIoCancelIrp: IRP running in StartIo
 Wdmlo ...  14:35:36...  Timeout1s: Timeout is 10
 Wdmlo ...  14:35:36...  WdmIoDpcForIsr: Status C0000120 Info 1
 Wdmlo ...  14:35:36 (7) WdmIoDispatchCleanup
 Wdmlo ...  14:35:36...  Close

 Ready                                          [   ][   ][   ]
```

@ 1999 PHD Computer Consultants Ltd

If you remove the comments around the CloseHandle call at the end of WdmIoCancel, you will be able to test the Cleanup IRP handling correctly. In this case, the IRP cancel routines will not be called. Instead, the Cleanup IRP handler removes the second Write IRP from the device queue. It then goes on to cancel the current IRP (i.e., the first Write).

Finally, the WdmIoCancel code contains a commented out CancelIo call after the second Write is issued. Uncomment this and recompile to check that this cancels the Write IRPs correctly.

Supplemental Device Queues

You can set up your own device queues, if need be. They are usually termed *supplemental device queues* to differentiate them from the standard device queue for StartIo requests.

Supplemental device queues might be used in the following situations.

1. A full duplex driver will need a separate queue of IRPs for each direction of transfer. This lets a Read IRP and a Write IRP be processed at the same time. Interrupt routines need to be designed carefully to ensure that interrupts are handled correctly. The standard device queue could be used for Write IRPs. A supplemental device queue could then be used for the Read IRPs.

2. The I/O Manager Cancel spin lock must be acquired whenever there is an operation on any standard device queue. This is a considerable bottleneck for the whole system. Supplemental device queues can be designed to operate with a much-reduced use of the Cancel spin lock. The example in the SRC\GENERAL\CANCEL directory of the Windows 2000 DDK shows this technique in action.

3. When a Plug and Play device is stopped for resource reassignment, all I/O IRPs should be queued[6]. The IRPs should be run when the device is restarted with new resource assignments. All the Plug and Play drivers in this book avoid this complication by not letting a device be stopped if there are any open handles. These IRPs can be queued in the standard device queue. However, you may deem that a supplemental device queue is needed instead of, or in addition to, the standard device queue.

   These queued IRPs may not need to be processed serially. When the device is started, all the IRPs are eligible for running straightaway. To achieve this, system worker thread work items may be scheduled to process all the IRPs. System worker threads are covered briefly in Chapter 14.

   Do not forget that any queued IRPs need to be cancellable. Adding a queue is not necessarily a quick operation.

Implementing a Supplemental Device Queue

Let's see how to set up and use a supplemental device queue. This queue will perform exactly the same function as the kernel-managed device queue for StartIo requests and so shows what must be happening behind the scenes in the kernel IoStartPacket call, etc. The aim is to serialize processing of IRPs in the AbcStartIo2 function. An IRP is put in the queue for processing using the AbcStartPacket2 routine. AbcStartNextPacket2 starts the processing of the next packet. Listing 16.8 shows the code for these routines.

These routines also handle the cancelling of IRPs correctly using the Cancel spin lock and cancel routines.

Declare a KDEVICE_QUEUE field somewhere in nonpaged memory. In this case, it is in the device extension in a field called *AbcIrpQueue*. Initialize it when the device object has been created by calling KeInitializeDeviceQueue at PASSIVE_LEVEL.

```
KeInitializeDeviceQueue( &dx->AbcIrpQueue);
```

The device extension also needs a current IRP pointer, here in a field called *AbcCurrentIrp*.

---

6. You may also decide to queue IRPs while a device is asleep. The example drivers in this book opt to wake up the device when an I/O request arrives.

Inserting IRPs into the Queue

AbcStartPacket2 uses KeInsertDeviceQueue to insert the IRP into the device queue. Note that KeInsertDeviceQueue must be called at DISPATCH_LEVEL IRQL. The call to IoAcquire-CancelSpinLock in AbcStartPacket2 does this. Subsequent calls to IoReleaseCancelSpinLock return to the old IRQL.

If the device queue is empty, KeInsertDeviceQueue returns FALSE, but it sets the queue into a busy state. Subsequent insertion attempts return TRUE. In AbcStartPacket2, if KeInsertDe-viceQueue returns FALSE, then the IRP is passed for processing in AbcStartIo2 straightaway. First, the IRP pointer is stored as the current IRP in the *AbcCurrentIrp* field in the device extension. Then the Cancel spin lock is released, returning the IRQL to its old level. StartIo routines must run at DISPATCH_LEVEL, so calls to KeRaiseIrql and KeLowerIrql are put around the call to AbcStartIo2.

If KeInsertDeviceQueue returns TRUE, the IRP has been put in the device queue. In this case, no further processing is required in AbcStartPacket2, apart from releasing the Cancel spin lock.

IRP Processing

The AbcStartIo2 function initially does some IRP cancel checks. Then, it sets about process-ing the IRP. Eventually, it completes the IRP and calls AbcStartNextPacket2 to begin process-ing the next IRP. The call to AbcStartNextPacket2 can occur in a different routine (e.g., after a hardware interaction).

Starting the Next IRP

AbcStartNextPacket2's job is to see if there are any more IRPs queued for processing. If there are, it dequeues one and sends it for processing in AbcStartIo2. Note that the calls to AbcStartNextPacket2 and AbcStartIo2 are recursive, which I trust will not be a problem. AbcStartNextPacket2 must be called at DISPATCH_LEVEL IRQL.

AbcStartNextPacket2 calls KeRemoveDeviceQueue to try to remove an IRP from the device queue. KeRemoveDeviceQueue must be run at DISPATCH_LEVEL IRQL. When running an IRP device queue, it is usual to call KeRemoveDeviceQueue while holding the Cancel spin lock. The call to IoAcquireCancelSpinLock raises the IRQL to the correct level, as before.

If KeRemoveDeviceQueue returns NULL, no IRPs are queued. The *AbcCurrentIrp* field is reset to NULL and the Cancel spin lock is released.

If KeRemoveDeviceQueue returns non-NULL then the return value represents the device queue entry in the IRP structure. Get the actual IRP pointer using the CONTAINING_RECORD macro. Store this in *AbcCurrentIrp*, release the Cancel spin lock, and call AbcStartIo2.

Listing 16.8 Supplemental device queue routines

```
VOID AbcStartPacket2( IN PDEVICE_OBJECT DeviceObject, IN PIRP Irp)
{
    PABC_DEVICE_EXTENSION dx = (PABC_DEVICE_EXTENSION)fdo->DeviceExtension;

    KIRQL OldIrql;
    IoAcquireCancelSpinLock(&OldIrql);
    IoSetCancelRoutine( Irp, AbcCancelIrp);
```

Listing 16.8 Supplemental device queue routines (continued)

```
        if( KeInsertDeviceQueue( &dx->AbcIrpQueue,
                            &Irp->Tail.Overlay.DeviceQueueEntry))
            IoReleaseCancelSpinLock(OldIrql);
        else
        {
            DeviceExtension->AbcCurrentIrp = Irp;
            IoReleaseCancelSpinLock(OldIrql);

            KeRaiseIrql( DISPATCH_LEVEL, &OldIrql);
            AbcStartIo2( DeviceObject, Irp);
            KeLowerIrql(OldIrql);
        }
}


VOID AbcStartNextPacket2( IN PDEVICE_OBJECT DeviceObject)
{
    PABC_DEVICE_EXTENSION dx = (PABC_DEVICE_EXTENSION)fdo->DeviceExtension;

    KIRQL OldIrql;
    IoAcquireCancelSpinLock(&OldIrql);

    PKDEVICE_QUEUE_ENTRY QueueEntry = KeRemoveDeviceQueue(&dx->AbcIrpQueue);

    if( QueueEntry!=NULL)
    {
        PIRP Irp =
            CONTAINING_RECORD( QueueEntry, IRP, Tail.Overlay.DeviceQueueEntry);
        dx->AbcCurrentIrp = Irp;
        IoReleaseCancelSpinLock(OldIrql);
        AbcStartIo2( DeviceObject, Irp);
    }
    else
    {
        dx->AbcCurrentIrp = NULL;
        IoReleaseCancelSpinLock(OldIrql);
    }
}


VOID AbcStartIo2( IN PDEVICE_OBJECT DeviceObject, IN PIRP Irp)
{
```

Listing 16.8 Supplemental device queue routines (continued)

```
    PABC_DEVICE_EXTENSION dx = (PABC_DEVICE_EXTENSION)fdo->DeviceExtension;
    PIO_STACK_LOCATION IrpStack =  IoGetCurrentIrpStackLocation(Irp);

    KIRQL OldIrql;
    IoAcquireCancelSpinLock(&OldIrql);

    if( Irp->Cancel)
    {
        IoReleaseCancelSpinLock(OldIrql);
        return;
    }
    IoSetCancelRoutine( Irp, NULL);
    IoReleaseCancelSpinLock(OldIrql);

    // Process and complete request
    // ...

    // Start next packet
    AbcStartNextPacket2(DeviceObject);
}
```

IRP Cancelling and Cleanup

Cancelling IRPs in supplemental device queues can use exactly the same techniques as the standard device queue. In this example, the cancel routine is removed when AbcStartIo2 is called. As shown in Listing 16.9, if the IRP to cancel is the current IRP, it is completed with a cancelled status in AbcCancelIrp. When AbcStartIo2 starts, it checks the *Cancel* flag and removes the cancel routine.

The Cleanup IRP handling should use exactly the same techniques as the WdmIo driver. The WdmIo driver uses a *Timeout* variable to stop an IRP that is being processed. You will need to use some similar technique if any of your IRPs may take a long time to be processed.

Listing 16.9 Supplemental device queue IRP cancel routine

```
VOID AbcCancelIrp( IN PDEVICE_OBJECT DeviceObject, IN PIRP Irp)
{
    PABC_DEVICE_EXTENSION dx = (PABC_DEVICE_EXTENSION)fdo->DeviceExtension;
    if( Irp==dx->AbcCurrentIrp)
    {
        IoReleaseCancelSpinLock(Irp->CancelIrql);
        CompleteIrp(Irp, STATUS_CANCELLED);
        AbcStartNextPacket2( DeviceObject, TRUE);
    }
    else
```

Listing 16.9 Supplemental device queue IRP cancel

```
    {
        KeRemoveEntryDeviceQueue(
            &dx->AbcIrpQueue,
            &Irp->Tail.Overlay.DeviceQueueEntry);


        IoReleaseCancelSpinLock(Irp->CancelIrql);
        CompleteIrp(Irp, STATUS_CANCELLED);
    }
}
```

Conclusion

This chapter has shown how to queue IRPs for serial processing in a driver's StartIo routine. It has then shown how to cope with the necessary evil of cancelling IRPs. Make sure you clean up any IRPs that are still not completed when the device handle is being closed. Use techniques similar to that used in the example *WdmIoCancel* application to test that cancelling and cleanup happen correctly.

The next chapter inspects the next part of the WdmIo and PHDIo drivers, how to handle interrupts and do interrupt-driven programmed I/O. It also looks at timers, both for IRP time-outs in the order of seconds, and custom timers for finer grain intervals.

Chapter 17

Interrupt-Driven I/O

This chapter shows how the WdmIo and PHDIo drivers handle interrupts, use Deferred Procedure Calls (DPCs), and catch time-outs. It shows how Read and Write interrupt-driven I/O is started and demonstrates how interrupt handling routines should work.

A driver's interrupt handler must service a hardware device's immediate needs. Any major processing must wait until a driver's DPC routine runs. Eventually, the DPC completes the read or write request and starts the next queued IRP.

The chapter also covers two types of timer. A basic "one-second interval" timer is best used to implement a time-out for interrupt-driven operations. However, custom timers and their associated Custom DPCs can be used for finer grain timing.

Interrupt Handling

First, let me be clear exactly what a hardware interrupt is. Devices that generate interrupts use an electrical signal to indicate that some special condition has occurred. The interrupt system is designed to stop the processor (or one of the processors) in its tracks — as soon as possible — and start some code to service the interrupt.

When an interrupt occurs, the processor saves a small amount of information on the kernel stack: the processor registers and the instruction pointer before the interrupt. This is just enough to restore the context when the interrupt service routine has completed.

The Nature of the Beast

Interrupts are usually used when something important has happened, or when a driver ought to do something important to its hardware device. Following are some example interrupt events.

- From the modem: I have just received a character. Come and get it! Another might arrive soon and overwrite this one.
- From the disk controller: I have just finished a DMA transfer of a sector of data. What shall I do now?
- From the printer: I have just printed a character. Give me another!
- From the printer: I have just run out of paper. Please tell the user to buy some more!

As you can see, not all interrupts are equally important. Of the previous four situations, the modem and disk controller interrupts are the most important.

x86 processors have a NonMaskable Interrupt (NMI) pin input and an Interrupt (INTR) pin input, with all normal device interrupts sharing the INTR input. External 8259A controllers are used to provide several interrupt lines to devices (i.e., IRQ0–IRQ15). IRQ0–IRQ15 share the INTR pin. An interrupt vector specifies the memory location that contains the address of the interrupt service routine. The 8259A controllers provide a different vector for each IRQ number. Therefore, IRQ0–IRQ15 each have their own interrupt service routines inside the Windows kernel.

x86 processors prioritize their interrupts in a simple way. Processor exceptions have the highest priority. NMI interrupts come next and then INTR interrupts. The INTR interrupts are maskable, as they can be disabled by setting a bit in the processor status register.

If an INTR interrupt occurs, its service routine will run until completion, stopping other INTR interrupts from occurring. However, an NMI interrupt could still butt in half-way through the INTR service routine. An interrupt service routine should do its job quickly, as it could stop other equally important INTR interrupts from being serviced. The interrupt latency is the time between a device asserting its interrupt signal and its service routine starting. Help keep the interrupt latency down for all drivers by making your interrupt service routine run quickly.

Interrupts

As described previously, Windows provides the initial interrupt handler for IRQ0–IRQ15. If your driver has connected to one of these interrupts (IRQ7 in the default case for WdmIo), it calls your interrupt handling routine to service the interrupt. More than one device can share the same interrupt line, so your first job is to determine if it really was your device that interrupted the processor.

Most hardware has a mind of its own, especially when fiddled with by users, and so is likely to generate interrupts at the most awkward time. In particular, you can guarantee that the interrupts will not occur in the context of the user mode thread that should receive its data. Note that an interrupt may occur at any point in your driver code, so you have to be especially careful that you can cope.

Naturally, the Windows kernel helps considerably. In particular, Critical Section routines forcibly increase the IRQ level (IRQL) to the appropriate Device IRQL (DIRQL) so that an interrupt cannot intervene while your Critical Section routine is running. In an x86 system, it sets the appropriate bit in the processor status register that disables INTR interrupts.

Most devices that interrupt have some similar sort of interrupt disabling facility. At power up, a device's interrupts are usually disabled. However, it is best to disable your device's interrupts at the first possible opportunity. Once your driver has installed its interrupt handler, you can enable interrupts in the device. Even now, you should note that interrupts may not conveniently arrive after you have started processing your read request. From the word go, a modem might start interrupting with incoming data. Your driver may well not have received a Read IRP yet, so you will have to decide what to do with these incoming bytes. You could buffer them somewhere in the expectation of a Read IRP coming round the corner. Or you could just dump them on the floor.

Connecting to Interrupts

A WDM driver receives its interrupt resource assignments in its Plug and Play IRP *Start Device* handler. Each interrupt has these details:

- Interrupt level[1]
- Interrupt vector
- Interrupt affinity: a set of bits indicating which processors it can use in a multiprocessor system
- Interrupt mode: latched or level sensitive

Usually, you just need to store these details and pass them on to IoConnectInterrupt when you are ready to handle interrupts.

First, declare a PKINTERRUPT object that will be initialized by IoConnectInterrupt. Like most drivers, WdmIo declares this in its device extension. However, if all your devices use the same interrupt, this object may well need to be a driver global; you would need to connect to the interrupt only once (e.g., in your DriverEntry routine).

Table 17.1 shows the parameters to pass to IoConnectInterrupt. Apart from the various assigned interrupt values, you must pass the name of your interrupt handling routine and a context to pass to it. The WdmIo driver uses the device extension pointer as this context parameter. If all your devices share an interrupt, you will need to use a different context pointer; the interrupt handler will have to work out for itself to which device the interrupt refers.

The only complication to IoConnectInterrupt is what happens when you use more than one interrupt. You need to call IoConnectInterrupt once for each interrupt, using the same

---

1.   Note that the interrupt level is not the same as the IRQ number. IRQ7 has interrupt level 20.

initialized spin lock pointer in each call. This is used to resolve tensions between the various handlers so that only one is called at once. Pass the highest DIRQL value in the `SynchronizeIrql` parameter in each call.

Table 17.1   `IoConnectInterrupt` **function**

NTSTATUS IoConnectInterrupt	(IRQL==PASSIVE_LEVEL)
Parameter	**Description**
OUT PKINTERRUPT *InterruptObject	Interrupt object pointer
IN PKSERVICE_ROUTINE ServiceRoutine	Name of interrupt handling routine
IN PVOID ServiceContext	Context to pass to the interrupt handler, usually the device extension pointer
IN PKSPIN_LOCK SpinLock	Optional spin lock parameter used when a driver uses more than one interrupt; NULL, otherwise
IN ULONG Vector	Assigned interrupt vector
IN KIRQL Irql	Assigned interrupt IRQL
IN KIRQL SynchronizeIrql	The highest IRQL of the interrupts that a driver uses (i.e., usually the same as IRQL)
IN KINTERRUPT_MODE InterruptMode	Assigned LevelSensitive or Latched value
IN BOOLEAN ShareVector	TRUE if the interrupt vector is shareable
IN KAFFINITY ProcessorEnableMask	Assigned interrupt affinity
IN BOOLEAN FloatingSave	TRUE if the floating-point stack should be saved. For x86 systems, this value must be FALSE

Do not forget to disconnect your interrupt handler using `IoDisconnectInterrupt` before your device disappears. And you must disconnect from the interrupt if the Plug and Play system stops your device. The `WdmIo` driver always stops its device when the device is about to be removed, so the `StopDevice` routine in `DeviceIo.cpp` is the only place where `WdmIo` disconnects its interrupt handler.

The `WdmIo` driver connects to its interrupt handler when it processes a `PHDIO_IRQ_CONNECT` command in its `ProcessCmds` routine. It remembers if it connected successfully in the device extension *ConnectedToInterrupt* field. If the device is stopped for Plug and Play resource reallocation, this flag is left `true` after `IoDisconnectInterrupt` is called. When the device is restarted, `WdmIo` reconnects to the new interrupt if *ConnectedToInterrupt* is `true`.

The actual call to `IoConnectInterrupt` is done in a system worker thread as this function must be called at PASSIVE_LEVEL IRQL. In this case, it is the work item that completes the IRP and starts the next IRP using `IoStartNextPacket`.

WdmIo **Reads and Writes**

Before I look at its interrupt handler, let's see how `WdmIo` starts and runs its interrupt-driven reads and writes.

Listing 17.1 shows the mass of extra fields that are needed in the device extension. The *ConnectedToInterrupt* field, as mentioned before, indicates whether `WdmIo` has connected to the interrupt. *InterruptReg*, *InterruptRegMask*, and *InterruptRegValue* are the values given by the controlling Win32 application in its PHDIO_IRQ_CONNECT command.

Remember that the controlling application uses three IOCTLs, such as IOCTL_PHDIO_ CMDS_FOR_WRITE, to store the commands that are used to process read and write transfers. WdmIoStartIo calls StoreCmds in each case to allocate some memory from the nonpaged pool to store a copy of the passed commands. The device extension *WriteCmds*, *StartReadCmds*, and *ReadCmds* fields store the pointers to this memory, with *WriteCmdsLen*, *StartReadCmdsLen*, and *ReadCmdsLen* storing the lengths. WdmIo frees the allocated memory when the device is removed.

The three time-out variables are used to detect time-outs, as described later. If *Timeout* is -1, no read or write is in progress. The final new group of variables are used to keep track of where WdmIo is in the read or write transfer. The next sections describe how these fields are used.

Listing 17.1 Interrupt handling fields in the device extension

```
// Interrupt handling support
bool ConnectedToInterrupt;
UCHAR InterruptReg;
UCHAR InterruptRegMask;
UCHAR InterruptRegValue;

ULONG CmdOutputCount;      // Count of bytes output from commands

PUCHAR WriteCmds;          // Stored commands for write IRP
ULONG WriteCmdsLen;        //                          length
PUCHAR StartReadCmds;      // Stored commands for start read IRP
ULONG StartReadCmdsLen;    //                          length
PUCHAR ReadCmds;           // Stored commands for read IRP
ULONG ReadCmdsLen;         //                          length

UCHAR SetTimeout;          // Timeout stored from script
int Timeout;               // Seconds left to go.  -1 if not in force
bool StopTimer;            // Set to stop timer

ULONG TxTotal;             // R/W total transfer size in bytes
ULONG TxLeft;              // R/W bytes left to transfer
PUCHAR TxBuffer;           // R/W buffer.  Moves through current IRP SystemBuffer
bool TxIsWrite;            // R/W direction
NTSTATUS TxStatus;         // R/W status return
UCHAR TxResult[5];         // R/W output buffer (2 Failcode, 2 Offset, 1 user)
UCHAR TxLastIntReg;        // R/W last interrupt register value
ULONG TxCmdOutputCount;    // R/W Copy of last CmdOutputCount
```

Starting Requests

Listing 17.2 shows how `WdmIoStartIo` initiates a write request. The device extension *TxIsWrite* field is set `true` for writes, and `false` for reads. *TxTotal* contains the total number of bytes to transfer. *TxLeft* stores the number of bytes left to transfer, and so counts from *TxTotal* down to zero. *TxBuffer* points to the next byte to transfer in the IRP buffer, so it moves through the buffer as each byte is written. The IRP buffer is always accessible, as long as the IRP is being processed, so there is no need to make a copy of it. The *TxStatus* field contains the IRP's eventual completion status, which is initially assumed to be successful.

The *TxResult* array is used to contain the output from running the stored write commands. This 5-byte array is, therefore, zeroed before the write begins in earnest. *TxLastIntReg* stores the last value read by the interrupt handler from its status register. The contents of *TxResult* and *TxLastIntReg* can eventually be obtained using the `IOCTL_WDMIO_GET_RW_RESULTS` call, as shown in Listing 15.4.

The "one-second interval" timer is now started to detect time-outs.

`WdmIo` is now finally ready to output the first data byte by running the stored write data commands. As interrupts have been enabled, they must be run in the context of a Critical Section routine to avoid being interrupted. Listing 17.2 shows how `RunWriteCmdsSynch` does this job, calling `ProcessCmds` to run the commands in `dx->WriteCmds` with the output going to `dx->TxResult`. If `ProcessCmds` fails or if there are bytes left to transfer, `RunWriteCmdsSynch` returns `TRUE` and `WdmIoStartIo` completes the Write IRP straight away with status `STATUS_UNSUCCESSFUL`.

Listing 17.2 also shows the code in `ProcessCmds` that handles the `PHDIO_WRITE_NEXT` command. Basically, it retrieves the next byte from *TxBuffer* and writes it to the Data register. It increments the *TxBuffer* pointer and decrements the count of bytes left to process, *TxLeft*.

Listing 17.2 How `WdmIoStartIo` starts write requests

```
case IRP_MJ_WRITE:
    if( dx->WriteCmds==NULL || !dx->ConnectedToInterrupt)
    {
        status = STATUS_INVALID_DEVICE_REQUEST;
        break;
    }


    // Store transfer details
    dx->TxIsWrite = true;
    dx->TxTotal = IrpStack->Parameters.Write.Length;
    dx->TxLeft = dx->TxTotal;
    dx->TxBuffer = (PUCHAR)Buffer;
    dx->TxStatus = STATUS_SUCCESS;
    RtlZeroMemory( dx->TxResult, sizeof(dx->TxResult));
    DebugPrint( "WdmIoStartIo: Write %d bytes: %*s",
        dx->TxTotal,dx->TxTotal,dx->TxBuffer);
```

Listing 17.2 How WdmIoStartIo **starts write requests (continued)**

```
    // Start timeout timer
    dx->Timeout = dx->SetTimeout+1;
    IoStartTimer(fdo);

    // Send first value
    if( KeSynchronizeExecution( dx->InterruptObject,
            (PKSYNCHRONIZE_ROUTINE)RunWriteCmdsSynch, (PVOID)dx))
    {
        status = STATUS_UNSUCCESSFUL;
        break;
    }
    return;

// ...



BOOLEAN RunWriteCmdsSynch( IN PWDMIO_DEVICE_EXTENSION dx)
{
    if( dx->TxLeft==0) return TRUE;

    dx->CmdOutputCount = 0;
    BOOLEAN rv = ProcessCmds( dx, dx->WriteCmds, dx->WriteCmdsLen,
                        dx->TxResult, sizeof(dx->TxResult), false);
    dx->TxCmdOutputCount = dx->CmdOutputCount;
    if( !rv)
    {
        dx->TxStatus = STATUS_UNSUCCESSFUL;
        return TRUE;
    }
    return FALSE;
}



// In ProcessCmds...

case PHDIO_WRITE_NEXT:
{
    if( dx->Timeout==-1) { FailCode = PHDIO_CANNOT_RW_NEXT; goto fail; }
    if( dx->TxLeft==0) { FailCode = PHDIO_NO_DATA_LEFT_TO_TRANSFER; goto fail; }
    GetUChar(reg);
```

Listing 17.2 How WdmIoStartIo starts write requests (continued)

```
    WriteByte( dx, reg, *dx->TxBuffer++);
    dx->TxLeft--;
    break;
}
```

Read requests are processed in a very similar way. As Chapter 15 mentioned, one set of stored commands, in *StartReadCmds*, is used to start read requests. A different set, in *ReadCmds*, is used to process read interrupts. RunStartReadCmdsSynch and RunReadCmdsSynch are run as Critical Section routines to process these two sets of commands. The PHDIO_READ_NEXT command is run in a very similar way to PHDIO_WRITE_NEXT, except that it stores the byte that it reads in the next location in the IRP buffer.

Interrupt Handler

Listing 17.3 shows the interrupt service routine for WdmIo, InterruptHandler. Remember that this is a general-purpose service routine. If the user mode controlling application has set up its control fields or commands wrongly, things could go horribly wrong.

An interrupt handler should complete its job as quickly as possible. It is run at Device IRQL (DIRQL), so do not make *DebugPrint* calls. Remember that your device could interrupt at any time, not just when you have started a write request.

The first job of an interrupt handler is to see if the interrupt was generated by the correct device. If it was not, the routine should return FALSE as quickly as possible to let any other chained interrupt service routines have their go. Otherwise it should process the interrupt (at the very least, stop the device from interrupting) and return TRUE.

InterruptHandler reads the device register at *InterruptReg*. As described in Chapter 15, it then ANDs this value with *InterruptRegMask* and compares it with *InterruptRegValue*. If they are equal, the interrupt is valid and the handler can continue.

If the device extension *Timeout* flag is -1, no transfer is in progress and so TRUE is returned straight away. If your device requires further processing to cancel such "spurious" interrupts, you need to amend WdmIo so that it does what you want.

Next, WdmIo's interrupt handler gets the current IRP for this device. I double-check that there is a current IRP and then see if the I/O Manager has signalled that it should be cancelled by setting its *Cancel* flag.

If the IRP is still in progress, the *TxIsWrite* flag indicates whether the Read or Write stored commands should be run. The RunWriteCmdsSynch and RunReadCmdsSynch routines return TRUE if the transfer is now complete (i.e., if all the bytes have been transferred or there has been some error).

If the transfer is now complete, this is remembered by setting *Timeout* to -1. Interrupt routines run at Device IRQL and so cannot complete IRPs. Completing an IRP also takes some time, so this job ought not to be done in the interrupt handler, anyway. The driver model uses

Deferred Procedure Calls (DPCs) to solve these problems, as described in the next section. `InterruptHandler` calls `IoRequestDpc` to request that the `WdmIo` DPC be run.

Listing 17.3 `WdmIo` **Interrupt Handler**

```
BOOLEAN InterruptHandler(IN PKINTERRUPT Interrupt,
    IN PWDMIO_DEVICE_EXTENSION dx)
{
    // See if interrupt is ours
    dx->TxLastIntReg = ReadByte( dx, dx->InterruptReg);
    if( (dx->TxLastIntReg&dx->InterruptRegMask) != dx->InterruptRegValue)
        return FALSE;

    // If no transfer in progress then no further processing required
    if( dx->Timeout==-1) return TRUE;

    // See if current IRP being cancelled
    PDEVICE_OBJECT fdo = dx->fdo;
    PIRP Irp = fdo->CurrentIrp;
    if( Irp==NULL) return TRUE;
    BOOLEAN TxComplete = Irp->Cancel;
    if( !TxComplete)
    {
        // Run relevant set of commands
        if( dx->TxIsWrite)
            TxComplete = RunWriteCmdsSynch(dx);
        else
            TxComplete = RunReadCmdsSynch(dx);
    }
    // If all done, in error or being cancelled then call DPC to complete IRP
    if( TxComplete)
    {
        dx->Timeout = -1;
        IoRequestDpc( fdo, Irp, dx);
    }
    return TRUE;
}
```

Deferred Procedure Calls

Code that runs at an elevated IRQL needs to run as quickly as possible. An elevated IRQL is any IRQL above `DISPATCH_LEVEL` (e.g., at Device IRQL in an interrupt service routine). Code that runs at an elevated IRQL cannot make most useful kernel calls.

The Windows kernel helps solve both these problems with Deferred Procedure Call (DPC) routines, that run at `DISPATCH_LEVEL`. When an interrupt service routine has done all the jobs that must be performed, it should request that its DPC routine be run. This DPC routine should continue where the interrupt service routine left off.

A DPC typically either starts another transfer or completes an IRP.

In WdmIo, all the bytes are transferred in the interrupt handler. When the transfer is complete, WdmIo asks that its DPC be run to complete the IRP.

Other drivers may use their DPC routine to do data transfers. While WdmIo could use this technique, it would be slower. Processing the read or write commands in WdmIo is usually fairly quick, so it is simplest to get it over and done with[2].

Direct Memory Access (DMA) is a hardware facility for transferring many bytes of data from place to place without having to be handled by the processor. The DMA controller or the relevant device usually interrupts when the whole transfer has finished. The interrupt service routine usually runs a DPC routine to start its next transfer or complete the IRP. DMA transfers cannot be started at elevated IRQL, so the next transfer must be set up in a DPC.

An indeterminate amount of time elapses between a DPC being requested and when it is run. Do not defer any interrupt servicing to the DPC if data might be lost (e.g., read data being overwritten). If necessary, store any data in the device extension or some other preallocated nonpaged memory.

Using Basic DPCs

The Windows kernel makes it easy to use one DPC routine within a driver. If you want to use more than one DPC routine, check out the next section.

A standard device object contains the necessary KDPC object. However, you need to initialize it using IoInitializeDpcRequest, passing the name of your DPC routine. WdmIo makes this call in Pnp.cpp just after its FDO is created in its AddDevice routine, passing the name of its DPC routine, WdmIoDpcForIsr.

```
// Initialize our DPC for IRQ completion processing
IoInitializeDpcRequest( fdo, WdmIoDpcForIsr);
```

Asking for your DPC to be run is very easy — you just call IoRequestDpc, passing an IRP pointer and a context that you want passed to the DPC routine. As shown previously, WdmIo only asks for its DPC routine to be run when it has transferred all bytes or when an error has occurred. In both cases, the current IRP must be completed and the next queued IRP started.

As shown in Listing 17.4, WdmIoDpcForIsr starts by indicating that the transfer has stopped by setting the device extension *Timeout* field to -1. It then works out how many bytes have been transferred by subtracting the *TxLeft* field from *TxTotal*. If the I/O Manager wants the IRP cancelled, store STATUS_CANCELLED in the *TxStatus* return status field.

---

2.   Actually, the commands could take a long time to run if the Win32 application includes large delay commands. However, moving the command processing to the DPC would not help, as a Critical section routine would have to be run, bringing the IRQL back up the DIRQL.

WdmIoDpcForIsr then removes the cancel routine, unlocks the device, completes the IRP, and starts the next queued IRP.

Listing 17.4 WdmIo **Interrupt Deferred Procedure Call handler**

```
VOID WdmIoDpcForIsr(IN PKDPC Dpc, IN PDEVICE_OBJECT fdo,
                    IN PIRP Irp, IN PWDMIO_DEVICE_EXTENSION dx)
{
    dx->Timeout = -1;
    ULONG BytesTxd = dx->TxTotal-dx->TxLeft;
    if( Irp->Cancel) dx->TxStatus = STATUS_CANCELLED;

    DebugPrint("WdmIoDpcForIsr: Status %x Info %d", dx->TxStatus, BytesTxd);

    // Remove cancel routine
    KIRQL OldIrql;
    IoAcquireCancelSpinLock( &OldIrql);
    IoSetCancelRoutine( Irp, NULL);
    IoReleaseCancelSpinLock(OldIrql);

    // Unlock device and complete IRP
    UnlockDevice(dx);
    CompleteIrp(Irp, dx->TxStatus, BytesTxd);
    IoStartNextPacket( fdo, TRUE);

    // Stop timer calls
    dx->StopTimer = true;
}
```

DPC Gotchas

Even with this simple DPC for deferred interrupt processing, there are some potential problems that you have to look out for.

The first point to note is that if two or more calls to IoRequestDpc are made before the DPC can be run, the DPC is only run once. Suppose two interrupts occur in quick succession. If each interrupt handler calls IoRequestDpc, you might expect that the DPC is run twice. However, if there is a long time before the DPC is run, then it is only run once. You must cope with this. In this situation, it might be that one interrupt wants a read call completed, while the next wants a write call completed. Use a separate flag in the device extension to indicate each condition. The DPC routine should be prepared to handle both situations. An easier alternative might be to use a different DPC routine for each condition.

For the WdmIo driver, this problem should almost never arise. The only situation I can envision is one in which an IRP is cancelled just after the interrupt handler calls IoRequestDpc. However, in this case, the late-running DPC routine will find that the IRP has been cancelled, which is correct.

The other potential problems regarding DPCs will only occur in multiprocessor systems.

- A DPC routine is running on one processor as a device interrupt is handled on another.
- When an interrupt handler asks for a DPC to be run, it is run straight away on another processor before the interrupt service routine exits.
- Two or more DPC routines may be running at the same time on different processors.

The main solution to these problems is to use Critical Section routines whenever a DPC routine needs to access fields that an interrupt handler or another DPC use.

Custom DPCs

If you need to use more than one DPC, this is fairly straightforward. These "custom DPCs" are also used for Custom Timers with fine grain time-outs.

Declare a KDPC object in nonpaged memory (e.g., in the device extension). Initialize it using KeInitializeDpc. The DeferredContext parameter is eventually passed to the custom DPC routine.

To ask that your DPC routine be run, call KeInsertQueueDpc from within your interrupt handler. KeInsertQueueDpc returns TRUE if the request was successfully queued. It returns FALSE if the DPC is already in the queue. The SystemArgument1 and SystemArgument2 parameters to KeInsertQueueDpc are eventually passed to your DPC routine. You can use KeRemoveQueueDpc to remove your DPC request from the queue.

Table 17.2 shows the function prototype for your custom DPC routine. There are three context parameters. To be compatible with the basic DPC handler, these should be your FDO, the IRP, and usually the device extension. However, use these as you wish.

Table 17.2   CustomDpc **prototype**

VOID CustomDpc	(IRQL==DISPATCH_LEVEL)
Parameter	Description
IN PKDPC Dpc	DPC
IN PVOID Context	DeferredContext parameter given to KeInitializeDpc
IN PVOID SystemArg1	SystemArgument1 passed to KeInsertQueueDpc (NULL for custom timers)
IN PVOID SystemArg2	SystemArgument2 passed to KeInsertQueueDpc (NULL for custom timers)

Timers

Two different types of timer can be used. A basic timer is called once every second; WdmIo uses this to detect device time-outs. Custom timers may be set up with resolutions starting from 100ns.

One-Second Interval Timers

The kernel provides easy access to a device timer that calls you back every second. The timer must be initialized at PASSIVE_LEVEL IRQL using IoInitializeTimer. WdmIo calls IoInitializeTimer as follows in its AddDevice routine just after the FDO has been created. The final parameter to IoInitializeTimer is passed to the timer callback.

```
status = IoInitializeTimer( fdo, (PIO_TIMER_ROUTINE)Timeout1s, dx);
if( !NT_SUCCESS(status))
{
    IoDeleteDevice(fdo);
    return status;
}
```

Use IoStartTimer to start timer calls and IoStopTimer to stop them. Do not call IoStopTimer from within your timer routine. The first timer call may occur after less than one second.

The timer routine is called at DISPATCH_LEVEL. You will usually need to use a Critical Section routine if you wish to coordinate your activities with interrupt handling routines.

WdmIo Time-Outs

Some drivers start their timer calls when a device is created and stop them when it is removed. WdmIo reduces the number of timer callbacks. It starts the timer whenever an interrupt driven I/O starts. When the transfer is completed, it sets a *StopTimer* flag to indicate that its timer should be stopped. The next call to WdmIoStartIo checks this flag and calls IoStopTimer, if necessary. The timer is also stopped when the device is removed.

Listing 17.5 shows the WdmIo timer callback, Timeout1s, and the Critical Section routine that it uses, called Timeout1sSynch.

The device extension *Timeout* field serves two purposes. If -1, it indicates that no interrupt-driven transfer is in progress. The WdmIo code in DeviceIo.cpp checks this value in several places to ensure that a read or write is indeed in progress. If *Timeout* is zero or more, it indicates the number of seconds left before the current transfer times out. The first timer callback may occur after less than one second, so the code that starts reads and writes adds one to the given time-out to be on the safe side.

Timeout1s, therefore, first checks whether there is a transfer in progress. If not (i.e., if *Timeout* is -1), it returns immediately. If the IRP *Cancel* flag is set, it calls the DPC routine directly. Otherwise, it calls Timeout1sSynch as a Critical Section routine. Timeout1sSynch checks and decrements *Timeout*. If *Timeout* has reached zero, the read or write must be stopped. *Timeout* is set to -1, the IRP return status is set appropriately and TRUE is returned.

If Timeout1sSynch returns TRUE, Timeout1s calls the DPC routine directly. This call to WdmIoDpcForIsr uses NULL as the Dpc parameter. This fact could be used in DPC processing, if necessary. For example, for IRP cancelling and time-outs, there is no need to provide a priority boost for the IRP.

Listing 17.5 WdmIo **time-out routines**

```
VOID Timeout1s( IN PDEVICE_OBJECT fdo, IN PWDMIO_DEVICE_EXTENSION dx)
{
    if( dx->Timeout==-1) return;

    DebugPrint("Timeout1s: Timeout is %d",dx->Timeout);
    PIRP Irp = fdo->CurrentIrp;
    if( Irp->Cancel || KeSynchronizeExecution( dx->InterruptObject,
                        (PKSYNCHRONIZE_ROUTINE)Timeout1sSynch, dx))
        WdmIoDpcForIsr( NULL, fdo, fdo->CurrentIrp, dx);
}


static BOOLEAN Timeout1sSynch( IN PWDMIO_DEVICE_EXTENSION dx)
{
    if( dx->Timeout==-1 || --dx->Timeout>0)
        return FALSE;
    dx->Timeout = -1;
    dx->TxStatus = STATUS_NO_MEDIA_IN_DEVICE;  // Win32: ERROR_NOT_READY
    return TRUE;
}
```

Custom Timers

Custom timers may be used if you want timer resolutions other than one second. You can detect when the timer goes off in two ways. Either use a Custom DPC callback, or wait for the timer object to become signalled.

NT 3.51 timers are one-shot only. NT 4, W2000, and WDM drivers can use periodic timers. Declare a KTIMER field in nonpaged memory (e.g., in your device extension), and initialize it with KeInitializeTimer or KeInitializeTimerEx. To start a one-shot timer, call KeSetTimer. If the DueTime LARGE_INTEGER parameter is positive, it represents an absolute time. If it is negative, it is a relative time in units of 100ns.

Use KeSetTimerEx if you want to specify a periodic timer. The DueTime parameter is used to specify the first time-out. The Period parameter specifies the period in milliseconds for subsequent time-outs.

You can cancel a timer using KeCancelTimer, and use KeReadStateTimer to find out if the timer has gone off.

A custom DPC routine may used as the timer callback. Initialize a custom DPC as described previously. Pass the KDPC pointer to KeSetTimer or KeSetTimerEx.

System threads can wait for a timer to go off using the KeWaitForSingleObject and KeWaitForMultipleObjects calls. KeInitializeTimerEx lets you specify whether a timer is a NotificationTimer or a SynchronizationTimer. A SynchronizationTimer timer releases only one waiting thread, while a NotificationTimer timer releases all waiting threads.

Conclusion

This chapter has looked at how to write interrupt service routines. Any nonessential processing is best done in a Deferred Procedure Call (DPC) routine at a lower IRQL. A basic one-second interval timer can be used to detect device time-outs. Custom timers can be used with Custom DPCs to receive notification of other time-out periods.

Chapter 18

NT Hardware

This chapter looks at how to find, allocate, and use hardware resources in non-WDM drivers that do not support Plug and Play. I call these "NT style" drivers because they run in NT 3.51, NT 4, and Windows 2000. These drivers sometimes also work in Windows 98, slightly to my surprise.

I have already shown how to use the translated hardware resource assignments in the last two chapters. The WdmIo and PHDIo drivers use MmMapIoSpace to map I/O ports into memory and IoConnectInterrupt to install an interrupt handler.

An NT style driver has three jobs to do to obtain its translated hardware resources.

1. Find its device's raw resource requirements.
2. Allocate these raw resources: check for conflicts with existing devices and reserve the resources so that other new devices cannot use them.
3. Translate the raw resource information.

The PHDIo driver receives its resource requirements from the Win32 application. The filename passed in the Create IRP specifies the I/O port details and optionally the Interrupt IRQ number. PHDIo can currently only accept ISA resource details.

This chapter shows how PHDIo allocates its raw resources and translates them. At the end of the chapter, I look at how NT and W2000 find various devices and make the resource information available to drivers.

First, I look at how to build and structure an NT style driver.

NT Style Driver Construction

An NT style driver is built in a slightly different way from a WDM driver. It also has none of the Plug and Play infrastructure to support.

DDK Issues

You must build an NT style driver in Windows 2000 or NT 4. The Windows 2000 DDK or NT 4 DDK must be installed as appropriate. If you alter an NT style driver, you must reboot Windows 98 to use the changed driver.

It is possible that crucial kernel structures have been changed between NT 4 and W2000, so it is safest to have one NT 4/NT 3.51 version of your driver and one W2000 version. In practice, it seems as though a driver compiled in W2000 using the W2000 DDK works in the NT platforms and Windows 98. I compared the free build driver made by the Windows 2000 Beta DDK with the same build using the NT 4 DDK. Although the files were both the same size, they were not identical.

An NT style driver can support Power Management and Windows Management Instrumentation (WMI), as long as it is only run on Windows 2000 or Windows 98. The major function code for the Power Management IRP, IRP_MJ_POWER, is defined as 0x16 in the W2000 DDK NTDDK.H. IRP_MJ_POWER is not defined in the NT 4 DDK. Instead, the IRP major function code 0x16 is defined as IRP_MJ_QUERY_POWER in NT 4. I am not sure whether this IRP is issued in NT 4. However, it is best if you do not handle IRP_MJ_POWER or IRP_MJ_SYSTEM_CONTROL in a driver installed in NT 4.

Compile Environment

An NT style driver must use NTDDK.H as its main header file, rather than WDM.H. In general, this gives the driver access to more facilities than would be available to a WDM device driver.

In the SOURCES build file, remove the line that says DRIVERTYPE=WDM.

NT Style Driver Structure

An NT style driver like PHDIo creates devices in a different way from WDM device drivers. A WDM device driver has an AddDevice routine and receives Plug and Play IRP notifications. PHDIo does not handle the PnP IRP and, therefore, loses its Pnp.cpp file[1]. It also does not handle Power Management and WMI IRPs.

As a consequence of not using Plug and Play, PHDIo does not deal with PnP device stacks. The device extension does not need to have fields for the Physical Device Object (PDO) or *NextStackDevice*. Similarly, there is no need for the *GotResources*, *Paused*, *IoDisabled*, and *OpenHandleCount* fields. The StartDevice and RetrieveResources routines in DeviceIo.cpp have been removed, as there are no PnP Start Device IRPs to process.

Instead, PHDIo creates one device in its DriverEntry routine. This is not a Functional Device Object (FDO) in the PnP sense. Instead, it is just called a "device object". The PHDIo device extension now has a *phddo* field that contains a pointer this device object. *phddo* is used in exactly the same way as the *fdo* field in all the previous drivers.

---

1.   The LockDevice and UnlockDevice routines are still used and so have been moved from Pnp.cpp into Init.cpp.

Most NT style drivers that need resources find and allocate them in the DriverEntry call. However, PHDIo only receives its resource requirements when a handle is opened to its device. Later in the chapter, I will look at various techniques for finding a driver's resource requirements.

The one PHDIo device is deleted using PHDIoUnload when the driver is removed.

Device Creation and Deletion

The PHDIo DriverEntry routine delegates its device object creation to PHDIoCreateDevice, also in Init.cpp.

Listing 18.1 shows how PHDIo creates its device in PHDIoCreateDevice. Its kernel device name is \Device\PHDIo and the Win32 symbolic link name is \??\PHDIo. This is used by Win32 programs with a CreateFile filename of \\.\PHDIo.

PHDIoCreateDevice first sets up UNICODE_STRINGs for the kernel and symbolic link names. It then calls IoCreateDevice in the same way as before. This creates the phddo device object and its associated device extension. The IoCreateDevice Exclusive parameter is TRUE indicating that only one Win32 application can open this device at a time.

All the appropriate fields are set up in the device extension, including the *phddo* field. The device timer is initialized with IoInitializeTimer and the device Deferred Procedure Call is initialized with IoInitializeDpcRequest. Finally, an explicit symbolic link is set up using IoCreateSymbolicLink. Device interfaces are not used.

Listing 18.1 NT style device creation

```
PDEVICE_OBJECT phddo = NULL;

#define NT_DEVICE_NAME    L"\\Device\\PHDIo"
#define SYM_LINK_NAME     L"\\DosDevices\\PHDIo"

NTSTATUS PHDIoCreateDevice( IN PDRIVER_OBJECT DriverObject)
{
    NTSTATUS status = STATUS_SUCCESS;

    // Initialise NT and Symbolic link names
    UNICODE_STRING deviceName, linkName;
    RtlInitUnicodeString( &deviceName, NT_DEVICE_NAME);
    RtlInitUnicodeString( &linkName, SYM_LINK_NAME);

    // Create our device
    DebugPrint("Creating device %T",&deviceName);
    status = IoCreateDevice(
                DriverObject,
                sizeof(PHDIO_DEVICE_EXTENSION),
                &deviceName,
                FILE_DEVICE_UNKNOWN,
                0,
```

Listing 18.1 NT style device creation (continued)

```
                TRUE,    // Exclusive
                &phddo);
if( !NT_SUCCESS(status))
{
    DebugPrintMsg("Could not create device");
    return status;
}


phddo->Flags |= DO_BUFFERED_IO;

// Initialise device extension
PPHDIO_DEVICE_EXTENSION dx =
    (PPHDIO_DEVICE_EXTENSION)phddo->DeviceExtension;
dx->phddo = phddo;
dx->UsageCount = 1;
KeInitializeEvent( &dx->StoppingEvent, NotificationEvent, FALSE);
dx->Stopping = false;
dx->GotPortOrMemory = false;
dx->GotInterrupt = false;
dx->ConnectedToInterrupt = false;
dx->SetTimeout = 10;
dx->Timeout = -1;
dx->StopTimer = false;
dx->WriteCmds = NULL;
dx->ReadCmds = NULL;
dx->StartReadCmds = NULL;

// Initialise timer for this device (but do not start)
status = IoInitializeTimer( phddo, (PIO_TIMER_ROUTINE)Timeout1s, dx);
if( !NT_SUCCESS(status))
{
    DebugPrintMsg("Could not initialise timer");
    IoDeleteDevice(phddo);
    return status;
}

// Create a symbolic link so our device is visible to Win32...
DebugPrint("Creating symbolic link %T",&linkName);
status = IoCreateSymbolicLink( &linkName, &deviceName);
if( !NT_SUCCESS(status))
{
```

Listing 18.1 NT style device creation (continued)

```
        DebugPrintMsg("Could not create symbolic link");
        IoDeleteDevice(phddo);
        return status;
    }


    // Initialise our DPC for IRQ completion processing
    IoInitializeDpcRequest( phddo, PHDIoDpcForIsr);


    return status;
}
```

When PHDIo is unloaded, its PHDIoUnload routine is called. This runs PHDIoDeleteDevice to stop the PHDIo device, remove its symbolic link, and delete its device object.

If a driver makes a variable number of devices, the unload routine could use the following technique to find all the devices to remove. The kernel sets the driver object *DeviceObject* field to point to the first device that belongs to the driver. The *NextDevice* field in each device object points to the next device. It is, therefore, a simple task to traverse this chain of device objects and delete them. The only catch is that you still have to generate the correct symbolic link name for each device so the symbolic link can be removed.

Claiming Resources

This section looks at how to allocate resources. The crucial kernel function is IoReportResourceUsage. This checks to see if any other devices are using the resources. If the resources are free, they are reserved so no other device can use them.

The PHDIo driver only finds out which resources are needed when the user calls Create-File, passing the resource description in the filename string. The Create IRP arrives in the PHDIoCreate routine. All the previous Create IRP handlers have not done very much. However, PHDIoCreate has these jobs to do

1. Get the resource details from the filename.
2. Check for resource conflicts and reserve the resources.
3. Translate and map the resources.

Getting the resource details is handled by the GetResourcesFromFilename routine. I will not go into the details of this code here, apart from saying that it uses three support routines that work with the UNICODE_STRING structure: usStrCmpN, usGetHex, and usGetDec. In the end, the *GotPortOrMemory* device extension field is true if an I/O port specifier has been found, *GotInterrupt* is true if an interrupt has been found and *ResourceOverride* is true if the \override specifier was used.

PHDIoCreate checks that an I/O port was specified. It then calls ClaimResources to check for resource conflicts and reserve the resources. Finally, TranslateAndMapResources is used to translate resource information and map memory. PHDIoCreate carefully ensures that all the resource bool fields are reset to false at the fail label if any error occurs.

Listing 18.2 PHDIoCreate **routine**

```
NTSTATUS PHDIoCreate( IN PDEVICE_OBJECT phddo, IN PIRP Irp)
{
    PPHDIO_DEVICE_EXTENSION dx =
        (PPHDIO_DEVICE_EXTENSION)phddo->DeviceExtension;
    PIO_STACK_LOCATION IrpStack = IoGetCurrentIrpStackLocation(Irp);
    DebugPrint( "Create File is %T", &(IrpStack->FileObject->FileName));

    dx->GotPortOrMemory = false;
    dx->GotInterrupt = false;
    dx->PortNeedsMapping = false;
    dx->ConnectedToInterrupt = false;
    dx->ResourceOverride = FALSE;

    // Get resources from filename string
    PUNICODE_STRING usfilename = &(IrpStack->FileObject->FileName);
    NTSTATUS status = *(usfilename,dx);
    if( !NT_SUCCESS(status)) goto fail;

    // We must have IO port resource
    if( !dx->GotPortOrMemory)
    {
        DebugPrintMsg("No IO Port resource in filename");
        status = STATUS_INVALID_PARAMETER;
        goto fail;
    }

    // Claim resources
    status = ClaimResources(phddo);
    if( !NT_SUCCESS(status))
    {
        DebugPrintMsg("Could not ClaimResources");
        goto fail;
    }

    // Translate and map resources
    status = TranslateAndMapResources(phddo);
    if( !NT_SUCCESS(status))
    {
        UnclaimResources(phddo);
```

Listing 18.2 `PHDIoCreate` **routine (continued)**

```
        goto fail;
    }


    // Complete
    return CompleteIrp(Irp,status);

    // On error, make sure everything's off
fail:
    dx->GotPortOrMemory = false;
    dx->GotInterrupt = false;
    dx->PortNeedsMapping = false;
    dx->ConnectedToInterrupt = false;
    return CompleteIrp(Irp,status);
}
```

Claiming resources means working with Full and Partial Resource Descriptors. Chapter 9 showed that a device's resource assignments are given in these structures when the Plug and Play *Start Device* is received. The WdmIo driver obtains its resource assignments in this way.

NT style drivers have to build a resource list of these descriptors to pass to the IoReportResourceUsage routine. Note carefully that the raw resource details must be passed to IoReportResourceUsage, not the translated values. The kernel resource list structures are sufficiently intricate that it is worth showing them in Listing 18.3.

A *resource list* consists of one *Full Resource Descriptor* for each bus instance. Note carefully that this is an expandable structure. Although it is declared with only one Full Resource Descriptor, it may in fact contain one or more such descriptors.

A *Full Resource Descriptor* specifies the bus type and instance number. It also contains a *Partial Resource List* structure.

A *Partial Resource List* primarily contains an array of *Partial Resource Descriptors*. Again, note that a Partial Resource List is a structure that expands as more Partial Resource Descriptors are used.

A *Partial Resource Descriptor* finally contains the details of an individual resource. Table 9.2 in Chapter 9 gives full details of Partial Resource Descriptors.

Listing 18.3 Kernel resource list structures

```
typedef struct _CM_RESOURCE_LIST {
    ULONG Count;
    CM_FULL_RESOURCE_DESCRIPTOR List[1];
} CM_RESOURCE_LIST, *PCM_RESOURCE_LIST;


typedef struct _CM_FULL_RESOURCE_DESCRIPTOR {
    INTERFACE_TYPE InterfaceType;      // unused for WDM
    ULONG BusNumber;                   // unused for WDM
    CM_PARTIAL_RESOURCE_LIST PartialResourceList;
} CM_FULL_RESOURCE_DESCRIPTOR, *PCM_FULL_RESOURCE_DESCRIPTOR;
```

Listing 18.3 Kernel resource list structures (continued)

```
typedef struct _CM_PARTIAL_RESOURCE_LIST {
    USHORT Version;
    USHORT Revision;
    ULONG Count;
    CM_PARTIAL_RESOURCE_DESCRIPTOR PartialDescriptors[1];
} CM_PARTIAL_RESOURCE_LIST, *PCM_PARTIAL_RESOURCE_LIST;

typedef struct _CM_PARTIAL_RESOURCE_DESCRIPTOR {
    UCHAR Type;
    UCHAR ShareDisposition;
    USHORT Flags;
    union {
    // ...
    } u;
} CM_PARTIAL_RESOURCE_DESCRIPTOR, *PCM_PARTIAL_RESOURCE_DESCRIPTOR;
```

The PHDIo driver deals with only the first instance of the ISA bus, so it needs only one Full Resource Descriptor. It always has one Partial Resource Descriptor for the I/O port. It can also have a second Partial Resource Descriptor for the interrupt, if one was specified.

ClaimResources, in Listing 18.4, builds a resource list structure and passes it to IoReportResourceUsage.

As you can guess, it is quite a job building the resource list correctly. The correct size for the whole structure must be determined first. A suitably sized block of paged memory is allocated and zeroed.

ClaimResources gradually fills the resource list. The resource list *Count* is set to one, as there is only one Full Resource Descriptor. The Full Resource Descriptor *InterfaceType* field is set to Isa and the *BusNumber* is set to 0. The Partial Resource List *Count* is set to 1 or 2, depending on how many resources are declared. The Partial Resource Descriptor for the I/O port is generated, then the one for the interrupt, if required. Whew!

There is a final complication to calling IoReportResourceUsage. You must either specify the resource list as belonging to the whole driver, or associate the resource list with an individual device. PHDIo says that the resources belong to the whole driver. If PHDIo were enhanced to provide more than one device, it would make sense to allocate resources on a per-device basis. One call to IoReportResourceUsage per device would be needed in this case.

Listing 18.4 ClaimResources **routine**

```
NTSTATUS ClaimResources( IN PDEVICE_OBJECT phddo)
{
    PPHDIO_DEVICE_EXTENSION dx =
        (PPHDIO_DEVICE_EXTENSION)phddo->DeviceExtension;

    // Get resource count: either 1 (IOport) or 2 (IOport&IRQ)
    ULONG PartialResourceCount = 1;
```

Listing 18.4 `ClaimResources` **routine (continued)**

```
    if( dx->GotInterrupt) PartialResourceCount++;

    // Get size of required CM_RESOURCE_LIST
    ULONG ListSize = FIELD_OFFSET( CM_RESOURCE_LIST, List[0]);

    ListSize += sizeof( CM_FULL_RESOURCE_DESCRIPTOR) +
        ((PartialResourceCount-1) * sizeof(CM_PARTIAL_RESOURCE_DESCRIPTOR));

    // Allocate CM_RESOURCE_LIST
    PCM_RESOURCE_LIST ResourceList =
        (PCM_RESOURCE_LIST)ExAllocatePool( PagedPool, ListSize);
    if( ResourceList==NULL)
    {
        DebugPrintMsg("Cannot allocate memory for ResourceList");
        return STATUS_INSUFFICIENT_RESOURCES;
    }
    RtlZeroMemory( ResourceList, ListSize);

    // Only one Full Resource Descriptor needed, for ISA
    ResourceList->Count = 1;

    // Initialise Full Resource Descriptor
    PCM_FULL_RESOURCE_DESCRIPTOR FullRD = &ResourceList->List[0];
    FullRD->InterfaceType = Isa;
    FullRD->BusNumber = 0;


    FullRD->PartialResourceList.Count = PartialResourceCount;

    // Initialise Partial Resource Descriptor for IO port
    PCM_PARTIAL_RESOURCE_DESCRIPTOR resource =
        &FullRD->PartialResourceList.PartialDescriptors[0];
    resource->Type = CmResourceTypePort;
    resource->ShareDisposition = CmResourceShareDriverExclusive;
    resource->Flags = CM_RESOURCE_PORT_IO;
    resource->u.Port.Start = dx->PortStartAddress;
    resource->u.Port.Length = dx->PortLength;

    // Initialise Partial Resource Descriptor for Interrupt
    if( dx->GotInterrupt)
    {
        resource++;
        resource->Type = CmResourceTypeInterrupt;
```

Listing 18.4 `ClaimResources` **routine (continued)**

```
        resource->ShareDisposition = CmResourceShareDriverExclusive;
        if( dx->Mode==Latched)
            resource->Flags = CM_RESOURCE_INTERRUPT_LATCHED;
        else
            resource->Flags = CM_RESOURCE_INTERRUPT_LEVEL_SENSITIVE;
        resource->u.Interrupt.Level = dx->Irql;
        resource->u.Interrupt.Vector = dx->Irql;
        resource->u.Interrupt.Affinity = 1;
    }


    // Ask for resources for the driver
    DebugPrint("Allocating %d resources",PartialResourceCount);
    DebugPrint("phddo->DriverObject %x",phddo->DriverObject);
    if( dx->ResourceOverride) DebugPrintMsg("Resource override conflict");
    BOOLEAN ConflictDetected;
    NTSTATUS status = IoReportResourceUsage( NULL,
        phddo->DriverObject, ResourceList, ListSize,    // Driver resources
        NULL, NULL, 0,                                  // Device resources
        dx->ResourceOverride, &ConflictDetected);
    // Cope (or override) if resource conflict found
    if( ConflictDetected)
    {
        DebugPrintMsg("ConflictDetected");
        if( dx->ResourceOverride
        {
            DebugPrintMsg("Conflict detected and overridden");
            status = STATUS_SUCCESS;
        }
    }
    // Free allocated memory
    ExFreePool(ResourceList);
    return status;
}
```

Table 18.1 shows the parameters for `IoReportResourceUsage`. If you are allocating resources for the whole driver, use the `DriverObject`, `DriverList`, and `DriverListSize` parameters; otherwise, set these to NULL. Do the same for the per-device parameters, `DeviceObject`, `DeviceList`, and `DeviceListSize`.

In the NT and Windows 2000 platforms, the resource assignments end up in the registry in the `HKLM\HARDWARE\RESOURCEMAP` key[2]. If you specify a `DriverClassName` parameter, this string is used as a subkey to hold the resource assignments. Otherwise, in NT 4 and NT 3.51,

---

2. I do not know where they go in Windows 98.

the resource assignments end up in the "OtherDrivers" subkey. In W2000, the resource assignments are in the "PnP Manager" subkey. You can use *RegEdt32* to inspect the raw and translated resource assignments. In NT 3.51 and NT 4 this is the only way to view the resources used by the system. In Windows 2000 and Windows 98, the Device Manager properties for a device shows its resource usage; in W2000, the PHDIo device can be found if you opt to show hidden devices.

IoReportResourceUsage checks for resource conflicts and, if none, assigns the new resources. The OverrideConflict parameter can be used to force the storage of the new resource list, even if there were conflicts. The output ConflictDetected BOOLEAN says whether a conflict was detected. If its *ResourceOverride* field is true, PHDIo ignores a resource conflict and forces a STATUS_SUCCESS return.

Table 18.1 IoReportResourceUsage **function**

NTSTATUS IoReportResourceUsage	(IRQL==PASSIVE_LEVEL)
Parameter	Description
IN PUNICODE_STRING DriverClassName	Optional resource class name
IN PDRIVER_OBJECT DriverObject	Driver object pointer
IN PCM_RESOURCE_LIST DriverList	Resource list for driver
IN ULONG DriverListSize	Driver resource list size
IN PDEVICE_OBJECT DeviceObject	Device object pointer
IN PCM_RESOURCE_LIST DeviceList	Resource list for device
IN ULONG DeviceListSize	Device resource list size
IN BOOLEAN OverrideConflict	If TRUE, store resource list even if a conflict was detected
OUT PBOOLEAN ConflictDetected	BOOLEAN that is set TRUE if a conflict was detected

A driver releases its claim on any resources by calling IoReportResourceUsage again; this time to report that it uses no resources. Listing 18.5 shows how UnclaimResources does this job. UnclaimResources is called when the file handle is closed.

Listing 18.5 UnclaimResources **routine**

```
void UnclaimResources( IN PDEVICE_OBJECT phddo)
{
    DebugPrintMsg("Freeing all allocated resources");
    // Release all driver's resources by declaring we have none.
    CM_RESOURCE_LIST ResourceList;
    ResourceList.Count = 0;
    BOOLEAN ConflictDetected;
    IoReportResourceUsage( NULL,
        phddo->DriverObject, &ResourceList, sizeof(ResourceList),    // Driver
```

Listing 18.5 UnclaimResources **routine (continued)**

```
            NULL, NULL, 0,                              // Device resources
            FALSE, &ConflictDetected);
    // ignore return result
}
```

Translating Resources

At this point, PHDIo has found its raw resource requirements and claimed them for itself. The final stage is to translate the raw resource information into a form that can be used by the driver. Listing 18.6 shows the TranslateAndMapResources routine that does this job.

HalTranslateBusAddress is used to translate a bus address. The bus type and number are passed as the first parameters, followed by the raw address. The AddressSpace parameter is used as both an input and an output. If set to 1, the address is in I/O space. If AddressSpace is 0 on output, the output address in memory space must be mapped using MmMapIoSpace. The *PortStartAddress* device extension field receives the translated address.

The raw interrupt information must be translated using HalGetInterruptVector. The bus type and number as passed as before. For the ISA bus, specifying the IRQ number for the raw interrupt level and raw vector parameters seems to work. The translated interrupt *Vector*, *Irql*, and *Affinity* are stored in the PHDIo device extension. Do not forget that PHDIo still needs to connect to the interrupt to install its interrupt service routine.

The final step in TranslateAndMapResources is to get a usable pointer to the I/O port. If the port needs mapping, MmMapIoSpace is called to obtain this pointer. Otherwise, PHDIo can just use the low part of the translated port address.

Listing 18.6 TranslateAndMapResources **routine**

```
NTSTATUS TranslateAndMapResources( IN PDEVICE_OBJECT phddo)
{
    PPHDIO_DEVICE_EXTENSION dx =
        (PPHDIO_DEVICE_EXTENSION)phddo->DeviceExtension;


    // Translate IO port values
    ULONG AddressSpace = 1;     // IO space
    if( !HalTranslateBusAddress( Isa, 0, dx->PortStartAddress,
        &AddressSpace, &dx->PortStartAddress))
    {
        DebugPrint( "Create file: could not translate IO %x",
            dx->PortStartAddress.LowPart);
        return STATUS_INVALID_PARAMETER;
    }
    DebugPrint( "IO trans %x,%d", dx->PortStartAddress.LowPart,
        dx->PortLength);
    dx->PortNeedsMapping = (AddressSpace==0);
    dx->PortInIOSpace = (AddressSpace==1);
```

Listing 18.6 `TranslateAndMapResources` **routine (continued)**

```
    // Translate IRQ values
    if( dx->GotInterrupt)
    {
        ULONG irq = dx->Irql;
        dx->Vector = HalGetInterruptVector( Isa, 0, irq, irq, &dx->Irql,
            &dx->Affinity);
        if( dx->Vector==NULL)
        {
            DebugPrint( "Create filename: Could not get interrupt vector
                for IRQ %d", irq);
            return STATUS_INVALID_PARAMETER;
        }
        DebugPrint("Interrupt vector %x IRQL %d Affinity %d Mode %d",
            dx->Vector, dx->Irql, dx->Affinity, dx->Mode);
    }


    // Map memory
    if( dx->PortNeedsMapping)
    {
        dx->PortBase = (PUCHAR)MmMapIoSpace( dx->PortStartAddress,
            dx->PortLength, MmNonCached);
        if( dx->PortBase==NULL)
        {
            DebugPrintMsg( "Cannot map IO port");
            return STATUS_NO_MEMORY;
        }
    }
    else
        dx->PortBase = (PUCHAR)dx->PortStartAddress.LowPart;

    return STATUS_SUCCESS;
}
```

The `FreeResources` routine is used to unmap memory and disconnect from the interrupt, if necessary.

Well, that wraps up the discussion of `PHDIo`. The rest of its functionality is the same as `WdmIo`. The rest of the chapter backtracks to revisit the subject of finding the resources that a driver needs.

Finding Resources

The `PHDIo` driver is told which resources to use in the filename passed with the Create IRP. However, most NT style drivers do not have this luxury. Instead, they must use one of the following techniques to find out what resources to use.

- Ask what resources the kernel has detected.
- Interrogate configurable buses.
- Save the resource requirements in the driver's registry key when it is installed.
- Poke around in memory to see if you can find your devices.

    Calling `IoGetConfigurationInformation` returns a count of certain types of device.

Hardware detection

Auto-Detected Hardware

NT and Windows 2000 attempt to identify all the hardware devices attached to the system when they boot. A driver can look for suitable devices using the `IoQueryDeviceDescription` kernel call. `IoQueryDeviceDescription` calls your configuration callback routine for each matching hardware element. I have not tried to find out if this call works in Windows 98.

The automatic detection process finds all the buses on the system, all recognized controllers on each bus, and, if possible, each peripheral attached to a controller. It starts by locating any standard serial and parallel ports and finds any attached mice or printers. Along the way, it finds any disk drivers, network cards, etc. The detected information is put in the registry in the `HKLM\HARDWARE\DESCRIPTION` key.

Table 18.2 shows the parameters for `IoQueryDeviceDescription`. You must supply a `BusType` parameter, and can optionally provide `ControllerType` and `PeripheralType` parameters. These parameters are pointers, so specifying NULL means that you do not want to find this type of hardware. You also pass a callback routine and a context to pass to it.

The possible bus, controller, and peripheral types are found in the `INTERFACE_TYPE` and `CONFIGURATION_TYPE` enumerations in `NTDDK.H`.

Table 18.2   `IoQueryDeviceDescription` **function**

NTSTATUS IoQueryDeviceDescription	(IRQL==PASSIVE_LEVEL)
Parameter	**Description**
IN PINTERFACE_TYPE BusType	Bus type
IN PULONG BusNumber	Bus number, zero-based
IN PCONFIGURATION_TYPE ControllerType	Controller type
IN PULONG ControllerNumber	Controller number, zero-based
IN PCONFIGURATION_TYPE PeripheralType	Peripheral type
IN PULONG PeripheralNumber	Peripheral number, zero-based
IN PIO_QUERY_DEVICE_ROUTINE CalloutRoutine	Configuration callback routine name
IN PVOID Context	Context for configuration callback
Returns	STATUS_OBJECT_NAME_NOT_FOUND if no match found, or status returned by callback

Listing 18.7 shows some code that finds any parallel ports that have been detected (i.e., where the `ControllerType` is `ParallelController`). The code can be found in the book software file `PHDIo\autodetect.cpp`. Note carefully that the code will find parallel ports that are on all the available buses, not just the ISA bus. At the moment, this code just prints out the basic raw resource details using *DebugPrint*. If you use this code, you will eventually have to claim these resources and translate them into usable values. Do not forget that you need a separate Full Resource Descriptor for each bus type in the resource list passed to `IoReportResourceUsage`.

`FindParallelPort` loops through all possible bus types. For each bus type, it keeps incrementing the bus number from zero and checking if the bus instance exists using `IoReportResourceUsage`. If the bus instance does exist, it calls `IoReportResourceUsage` again to find all printers on the bus instance.

The `AbcConfigCallback` callback is called when a bus instance is found and when a parallel port is found. In the first case, the `BusInfo` parameter is valid. `CtrlrInfo` is valid when looking for parallel ports. If looking for a peripheral, `PeripheralInfo` is valid. If the relevant parameter is non-NULL, it points to some information obtained from the registry: a device identifier, configuration data, and information about its subcomponents. The configuration data has the detected raw resource assignments. Use the code in Listing 18.7 to enumerate these resources.

Listing 18.7 Finding any autodetected parallel ports

```
NTSTATUS FindParallelPort()
{
    NTSTATUS status;

    for( int BusType=0; BusType<MaximumInterfaceType; BusType++)
    {
        INTERFACE_TYPE iBusType = (INTERFACE_TYPE)BusType;
```

Listing 18.7 Finding any autodetected parallel ports (continued)

```
            CONFIGURATION_TYPE CtrlrType = ParallelController;
            ULONG BusNumber = 0;
            while(true)
            {
                // See if this bus instance exists
                status = IoQueryDeviceDescription(
                        &iBusType, &BusNumber,
                        NULL, NULL,
                        NULL, NULL,
                        AbcConfigCallback, NULL);
                if( !NT_SUCCESS(status))
                {
                    if( status != STATUS_OBJECT_NAME_NOT_FOUND)
                        return status;
                    break;
                }

                // See what printers exist on this bus instance
                status = IoQueryDeviceDescription(
                        &iBusType, &BusNumber,
                        &CtrlrType, NULL,
                        NULL, NULL,
                        AbcConfigCallback, NULL);
                if( !NT_SUCCESS(status) &&
                    (status != STATUS_OBJECT_NAME_NOT_FOUND))
                    return status;
                BusNumber++;
            }
        }
    return status;
}


NTSTATUS AbcConfigCallback(
    IN PVOID Context, IN PUNICODE_STRING PathName,
    IN INTERFACE_TYPE BusType, IN ULONG BusNumber,
    IN PKEY_VALUE_FULL_INFORMATION *BusInfo,
    IN CONFIGURATION_TYPE CtrlrType, IN ULONG CtrlrNumber,
    IN PKEY_VALUE_FULL_INFORMATION *CtrlrInfo,
    IN CONFIGURATION_TYPE PeripheralType, IN ULONG PeripheralNumber,
    IN PKEY_VALUE_FULL_INFORMATION *PeripheralInfo
    )
```

Listing 18.7 Finding any autodetected parallel ports (continued)

```
{
    DebugPrint( "ConfigCallback: Bus: %d,%d",BusType,BusNumber);
    DebugPrint( "ConfigCallback: Controller: %d,%d",CtrlrType,CtrlrNumber);
    DebugPrint( "ConfigCallback: Peripheral:
        %d,%d",PeripheralType,PeripheralNumber);

    if( CtrlrInfo!=NULL)
    {
        PCM_FULL_RESOURCE_DESCRIPTOR frd = (PCM_FULL_RESOURCE_DESCRIPTOR)
            (((PUCHAR)CtrlrInfo[IoQueryDeviceConfigurationData])
            +CtrlrInfo[IoQueryDeviceConfigurationData]->DataOffset);
        for( ULONG i=0; i<frd->PartialResourceList.Count; i++)
        {
            PCM_PARTIAL_RESOURCE_DESCRIPTOR resource =
                &frd->PartialResourceList.PartialDescriptors[i];
            switch( resource->Type)
            {
            case CmResourceTypePort:
                DebugPrint( "ConfigCallback: I/O port %x,%d",
                        resource->u.Port.Start.LowPart, resource->u.Port.Length);
                break;
            case CmResourceTypeInterrupt:
                DebugPrint( "ConfigCallback: Interrupt level %d vector %d",
                    resource->u.Interrupt.Level, resource->u.Interrupt.Vector);
                break;
            default:
                DebugPrint( "ConfigCallback: Resource type %d",resource->Type);
            }
        }
    }
    return STATUS_SUCCESS;
}
```

Interrogating Configurable Buses

A configurable bus is one that can be configured in software. Devices on these buses must have Plug and Play WDM device drivers. The PCI bus driver will retrieve a device's details and generate an appropriate Hardware ID. This Hardware ID will be used to identify the correct driver. Its AddDevice routine is called, etc.

In NT systems, you will need to use a different technique to get configurable bus data. This method still works in Windows 2000. The HalGetBusData and HalGetBusDataByOffset calls can be made to retrieve configuration data for one slot of a specified bus instance. For example, if you ask for `PCIConfiguration` bus data, the passed buffer is filled with PCI_ COMMON_CONFIG data.

You can then claim the necessary resources using `IoAssignResources`, in a similar way to `IoReportResourceUsage`. As an alternative, drivers of PCI-type devices can call `HalAssignSlotResources`. `HalSetBusData` and `HalSetBusDataByOffset` can be used to write back configuration data.

I have not tried it, but it looks as though `HalGetBusData` and `HalSetBusData` can be used to access the BIOS CMOS data.

Final Resource Discovery Techniques

There are two final ways of obtaining a driver's resource requirements. The first is to store the relevant information in the driver's registry key, say in its `Parameters` subkey. This information will typically be set up when the device is installed.

Last and least, you can simply poke around in likely memory locations to see if your device is present. Dangerous.

Conclusion

This chapter has looked at how to write an NT style device driver such as `PHDIo`. As it does not use Plug and Play, it must use different techniques to find what hardware resources it needs, reserve them, and translate them into usable values.

The `PHDIo` driver receives its resource details for its controlling Win32 application. Other drivers must find out what devices the kernel has found or interrogate all the configurable buses itself.

19

Chapter 19

WDM System Drivers

This chapter serves as a brief introduction to the system drivers that are provided as a part of the Windows Driver Model (WDM). These bus and class drivers are a very important part of the model. You can and should use the relevant system driver to access standard types of bus. These bus drivers are sometimes called class drivers, as they let you talk to a whole class of devices.

Figure 19.1 shows the main system drivers. Chapter 2 gave a brief overview of each of these. The remaining chapters of this book look in detail at the Universal Serial Bus (USB) and Human Input Device (HID) class drivers.

Talk to your bus driver

Figure 19.1 WDM system drivers

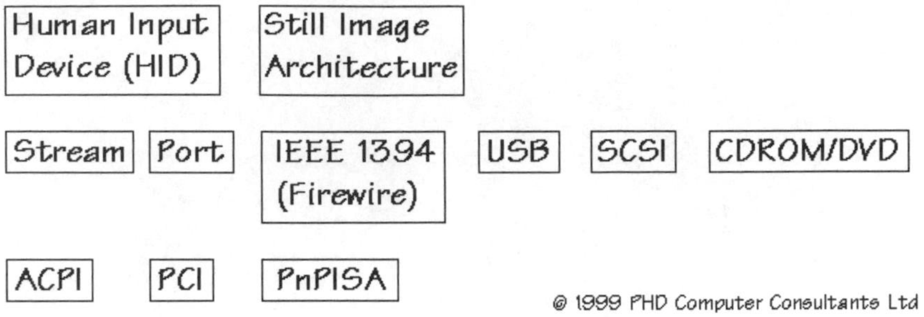

Human Input Device (HID) Still Image Architecture

Stream Port IEEE 1394 (Firewire) USB SCSI CDROM/DVD

ACPI PCI PnPISA

@ 1999 PHD Computer Consultants Ltd

Writing Client Drivers

Each type of system driver has its own documentation that you must consult. In many cases, there will be a specification provided by a major standards body and a further specification provided by Microsoft for its system driver. For example, the core USB specification is produced jointly by Compaq, Intel, Microsoft, and NEC. However, the Windows USB Driver Interface (USBDI) is solely Microsoft's responsibility.

There are two main ways to work with system drivers. The first, a *client*, is when you use the system drivers to access your device. The second driver category interfaces the system to your hardware, by writing a minidriver, a miniclass driver, or a miniport driver. This book does not cover this last type of driver.

If you are writing a client driver, you need to understand and use the core WDM driver technologies as well as all the system driver's capabilities. This usually means that you must write a Plug and Play device driver with Power Management and Windows Management Instrumentation capabilities. Although this seems like an onerous task, the provision of standard system drivers definitely makes your task easier. To help you further, for the USB and HID cases, you can base your driver on the examples given in the rest of this book.

There are two main types of client driver that you can write[1]. The first is when you use a standard Plug and Play (PnP) INF file to make yourself part of the driver stack for a particular device, as all the WDM drivers in this book have done so far. The second type of client is an NT style driver that uses PnP Notification. Such a driver is notified when a device of the right type arrives. Your client can then make its own device object to represent the underlying device. The USB and HID drivers in the rest of this book illustrate how to write both types of client.

Common Devices

Occasionally you may want to control the behavior of all your devices. Rather than find each device and send IOCTLs to each one, a useful technique is to provide an additional "common device" that can be used to control the features that are common to all the other devices.

This common device object is typically an NT style device object created in your `Driver-Entry` routine with a fixed symbolic link name. Your control application would open a handle

---

1. You can also access the HID class driver in a user-mode client application.

to this common device, and, for example, use some IOCTLs to change the general behavior of your driver.

The complication to this approach is that your IRP dispatch routines will receive requests both for your ordinary devices and for your common device. Control requests for the ordinary devices should be rejected, and vice versa. The simplest way to achieve this functionality is to have a flag in the device extension that indicates the device type. You could have two separate device extensions, with the device type flag in a common portion of each structure.

Filter Drivers

Filter drivers let you modify the behavior of standard system drivers. A WDM filter driver is inserted into the device stack when the device stack is built. Figure 19.2 shows the different types of filter driver. This book does not cover filter drivers.

Figure 19.2   WDM filter drivers

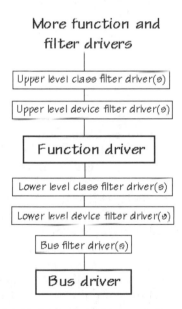

@ 1998 PHD Computer Consultants Ltd

Filter drivers can also be written to work in Windows NT 4 and NT 3.51. However, WDM device stacks are not available in these older operating systems. Instead, the `IoGetDeviceObjectPointer` or `IoAttachDevice` routines can be used to layer your device transparently above existing device objects. This book does not cover these types of filter driver, either.

NT Layering

Windows NT 4, NT 3.51, and 2000 use layers of NT style drivers. This is not the layering of Plug and Play devices seen in most of this book. Instead, each driver must be loaded in the correct order and use any existing drivers.

Parallel Port Drivers

As an example, Figure 19.3 shows the layering of the parallel port drivers. The parport driver is loaded first. parport arbitrates access to the parallel port hardware. The parallel and parvdm drivers are then loaded. When these drivers want to access the parallel port hardware, they must ask parport for access rights. Once access rights have been obtained, they can talk to the parallel port electronics directly.

New parallel port drivers should use the same technique to get access to the hardware[2]. The DDK recommends that you grab access to the port on a per-IRP basis. For most applications, I recommend taking control of the port for the duration of a larger transaction (e.g., while a handle is open to your device).

Figure 19.3 Parallel port layering

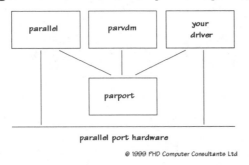

parallel port hardware

© 1999 PHD Computer Consultants Ltd

The parport driver creates a named kernel device object for each parallel port it finds. These device objects have kernel names starting with \Device\ParallelPort0. Use IoGetDeviceObjectPointer to obtain a pointer to the relevant device object[3]. Store this in your device extension for later use.

The parport driver supports several Internal IOCTLs. IOCTL_INTERNAL_GET_PARALLEL_PORT_INFO returns information about the parallel port (the base register address and port allocation routines). When the information is safely stored away, a similar (but undocumented) IOCTL_INTERNAL_RELEASE_PARALLEL_PORT_INFO call is made to release the information.

Two port allocation routines are available. TryAllocatePort tries to allocate the port and returns straightaway. If this succeeds, FreePort is used (later) to return the port. Call TryAllocatePort when you want to start accessing a parallel port and FreePort when you are finished with it.

When you install a driver that uses the parport driver, you must make sure that your driver loads after parport has started. Do this by including a DependOnService string value in your driver's registry entry that is set to parport.

Microsoft may be introducing a new WDM architecture for these parallel port drivers. A parallel port bus driver would replace parport. It would enumerate the ports available on the "bus". This is particularly important, as several IEEE 1284.3 printers may be daisy-chained

---

2. The only alternative is to remove the parport, parallel, and parvdm drivers, which would stop any existing parallel port access.

---

3. See Chapter 23 for details of IoGetDeviceObjectPointer.

on a single parallel port connection. The header file `parallel.h` in the W2000 DDK `src\kernel\inc` directory gives a hint of how the new interface will work.

Conclusion

Writing a client driver to access standard buses has been made much easier with the provision of several important system drivers. In addition, you can write minidrivers to interface these standard drivers to your own specialized hardware.

Let's now start looking at the standard system drivers by considering the Universal System Bus.

Chapter 20

The Universal Serial Bus

This chapter introduces the Universal Serial Bus (USB) by giving an overview of the technology, its low-level structure, and its device framework. The next chapter looks at how to write Windows device drivers using the USB Device Interface (USBDI). This chapter takes a brief look at the available classes of USB device that have common characteristics. The subsequent chapters look in detail at one of these classes for Human Input Devices (HID)[1].

The Universal Serial Bus (USB) is the recommended way to connect new low-speed or medium-speed external devices to PCs. It is simple for users to plug in a new USB device or remove one, even if the PC is switched on. The system configures itself automatically to accommodate the new device, prompting for the necessary driver.

The USB Specification defines all the basic physical, electrical, and logical characteristics of all USB systems. The Windows system drivers define the closely related USB Driver Interface (USBDI) for use by USB client device drivers. This chapter and the next look at both these USB models.

The USB specifications are available from the USB website at www.usb.org/developers/ in PDF format. I recommend that you also get the USB Compatibility Test software, usb-comp.exe. This includes the *USBCheck* tool to check whether a USB device meets some of the higher-level specifications. It also contains the *HIDView* program to inspect and test HID devices. The "Ignore hubs (Memphis only)" box should be checked for Windows 98. The Windows 2000 DDK also includes the *UsbView* tool that shows all the USB buses in the system and the devices that are connected to each USB bus.

---

1. Note that there are two uses of the word "class". Device drivers use the Windows USB class drivers to get access to the USB system. Each USB device can be categorized as a basic USB device, or as belonging to a class of USB devices such as HID, printer, etc.

Membership of USB Implementers Forum costs $2500 per year including a free vendor ID. Alternatively, vendor IDs can be purchased separately for $200.

Device Classes

Some USB devices just exhibit basic USB features. However, USB goes on to specify several classes of devices, with common behavior and protocols. This makes it easier to write generic device drivers.

Table 20.1 shows the main USB device classes. The USB website has the full specifications for each class. The hub device class is specified in the standard USB Specification. I shall show later how the class constant (in the Interface descriptor) is sometimes used to identify the correct drivers for the device.

Table 20.1   Main USB device classes

Device class	Example device	Class constant
Audio	Speakers	USB_DEVICE_CLASS_AUDIO
Communication	Modem	USB_DEVICE_CLASS_COMMUNICATIONS
Human Input Device (HID)	Keyboard, Mouse	USB_DEVICE_CLASS_HUMAN_INTERFACE
Display	Monitor	USB_DEVICE_CLASS_MONITOR
Physical feedback devices	Force feedback joystick	USB_DEVICE_CLASS_PHYSICAL_INTERFACE
Power	Uninterruptible power supply	USB_DEVICE_CLASS_POWER
Printer		USB_DEVICE_CLASS_PRINTER
Mass storage	Hard drive	USB_DEVICE_CLASS_STORAGE
Hub		USB_DEVICE_CLASS_HUB

Human Input Devices (HID)

One of the USB device classes is for Human Input Devices (HID), described fully in Chapter 22. HID devices do not have to run on the USB, but they fit neatly into the USB device model.

HID provides an abstract model of input devices. A program that uses HID should not care how a device is connected, as long as a suitable HID minidriver is available for the interface. HID hides the implementation details.

Windows includes system HID drivers to interact with USB devices. If a USB device has an interface descriptor that says it is in the HID class, Windows loads the system HID drivers and the USB HID minidriver and reads the HID descriptors. You just need to write a HID client to do something with the device.

The Big Picture

Figure 20.1 shows the logical structure of a USB device. The physical connection of the USB devices does not effect this logical view.

Figure 20.1   USB logical structure

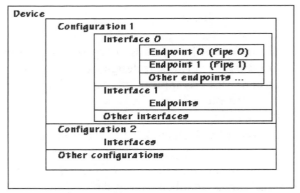

@ 1999 PHD Computer Consultants Ltd

Each device has one or more logical connection points inside it, called *endpoints*, each assigned one of the following transfer types: *control*, *interrupt*, *bulk*, and *isochronous*. All devices have an *Endpoint 0* for transfers used to configure and control the device.

A connection between the host and a device's endpoint is called a *pipe*. The connection between the host USB system software and a device's Endpoint 0 is called the *Default Pipe*.

A device presents an appropriate set of endpoints to the host. A set of related endpoints is called an *interface*. A device with more than one interface is called a *composite device*.

Finally, a device may have more than one set of interfaces. Each set is called a *configuration*. Only one configuration can be active at a time. However, all interfaces (and their endpoints) in the current configuration can be active at the same time. Most devices just have one configuration and one interface. Windows prompts you to choose the desired configuration when a device is first inserted.

The host reads various *descriptors* from a device to find out what configurations, interfaces, and endpoints are available. The host reads these descriptors when the device is first plugged in using the default pipe. Windows tries to set up a kernel device for each configuration or interface found.

Windows USB Driver Interface

The system USB drivers handle most of the hard work of connecting to a USB device. Indeed, some HID USB devices, such as keyboards, mice, and game devices are recognized automatically and do not need extra drivers.

However, most USB devices will need a new driver to talk to the device and respond to kernel or user application requests. At the kernel level, commands are issued by a client driver to the USB system using Internal IOCTLs. The most useful IOCTL allows you to issue USB Request Blocks (**URBs**) to the system USB driver. URBs let you issue a multitude of function calls to the USB system.

User mode USB utilities can also issue a few ordinary IOCTLs to a USB device, solely to get information about the connected devices.

Before I look at how to write USB device drivers using this USB Driver Interface (USBDI), it is necessary to explore the physical and logical structure of the USB system. The next chapter shows that Windows provides a fairly plain wrapper for the USB logical structure. The main USB concepts are reflected in the interface available to device driver writers.

Transfer Types

At the USB level, a device can communicate in four different transfer types: *control* transfers, *interrupt* transfers, *bulk* transfers, and *isochronous* transfers. If designing a USB device from scratch, you will have to decide what transfer types are appropriate. As a driver writer, you may simply be told what transfer types to implement.

Transfers are always initiated by the PC.

- Control transfers are used by the USB system and clients to send or receive relatively small amounts of data.
- Interrupt transfers signal input events from a device to the PC. Actually, the PC polls the device regularly (every 1ms–255ms) to get any available interrupt data.
- Bulk transfers are used — in either direction — for large amounts of data.
- Isochronous transfers happen regularly and are usually time-sensitive (e.g., for voice data coming from a telephone). USB divides its available bandwidth into *frames*, each nominally 1ms long. A device can do one isochronous transfer per frame.

The PC USB host juggles the pending transfer requests from various clients so that they fit into the available bandwidth within the frame structure. Isochronous and interrupt transfers can take up a maximum of ninety percent of the available frame bandwidth, leaving some room for control transfers. Most of the available bandwidth in each frame could be taken up by a regular isochronous transfer from one device. In this case, attaching another device that needs isochronous transfers might fail due to lack of bandwidth.

USB Low Level Structure

USB Devices

USB devices can be plugged into any USB *ports* on a PC. Some USB devices are *hubs* that allow further USB devices or hubs to be connected. A hub has one *upstream* connection towards the PC, and multiple *downstream* ports. Up to 127 devices can be connected in total. A hub counts as a device.

In USB-speak, the PC is called the *host*. Data is always transferred between the host and a device, or vice versa. Devices never talk to each other. The PC contains the *root hub* (or hubs), usually with a couple of downstream ports.

This arrangement leads to the star or branch architecture shown in Figure 20.2.

A device that does a useful job is called a *function device*. A USB *compound device* supports several functions. Internally, they consist of a hub with several downstream function devices.

Figure 20.2 shows an example USB physical layout. A USB keyboard is connected to one of a PC's USB ports. A USB hub is connected to the other PC port. The hub has two downstream ports. One is spare and the other has a compound USB device connected. Internally, the compound device has a hub and two function devices.

Figure 20.2   Example USB physical architecture

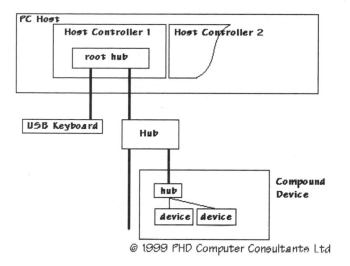

@ 1999 PHD Computer Consultants Ltd

A PC has one or more USB *host controllers* built into it (e.g., each with two ports). There are two standard types of host controller: the USB Host Controller Interface (UHCI) and the Open Host Controller Interface (OHCI). The Windows USB class drivers have a miniclass driver for each of these controller types.

USB Signals

The USB cable has four wires. Two wires carry 5V power. Some devices may use this to power themselves, while others will be self-powered (i.e., battery or mains powered).

The other two USB wires carry the serial data. Data transfer is in one direction only at a time. Serial data is either transferred at full speed, 12 megabits per second (Mbs), or at low speed, 1.5 Mbs. All the serial data is at one of these speeds, so there is no rate conversion in hubs. A device is either a full-speed device or a low-speed device.

Note that there is no intention to uprate these communication speeds, unlike IEEE 1394. Move to IEEE 1394 if you need higher speed.

The USB protocol defines what appears on the serial bus. The host starts all transactions, but data transfer direction is switched, if necessary, during a transaction. The protocol allows devices to be plugged in at runtime and configured, and for devices to be unplugged at any time. Communication occurs only between the host and one device at a time.

Hubs repeat downstream data to all downstream ports (but not full-speed data to low-speed devices). For upstream data, hubs understand the protocol so they know which port (if any) to route upstream. Hubs do not buffer data.

Bus Signalling

The bus can be *suspended* by the host to reduce power, either globally or just for a specific device. If a device detects the suspended state then it must power down. The host can *resume* signalling on the bus to wake up all devices. Alternatively, a device can issue a *remote wake-up* to the host (e.g., for a telephone to indicate an incoming call).

The host can signal to a device that it should *reset*.

In some circumstances, communications on a pipe to a device can be *stalled* (e.g., if the data is corrupted). The host has to clear the problem via another pipe before normal communication can continue.

A hub detects electrically when a device is removed or plugged in, and whether it is a full-speed or low-speed device. How USB and Windows react after this is described later.

Low Level Protocol

The USB protocol defines what happens on the serial data lines. The available data transmission time — the bandwidth — is divided up into frames, each 1ms long. At full speed a frame contains a maximum of 1500 bytes and, at low speed, 187 bytes.

Frames are primarily used as a means of allocating bandwidth to the different transfers that want to occur. The regular frame timing can also be used by devices to synchronize their activities to the USB bus. Frames are of particular interest to isochronous devices and drivers.

Packets

The smallest block of data transferred on the serial lines is called a *packet*. A packet is sent in only one direction, either from the host or to the host.

A packet consists of synchronization signals, a Packet ID (*pid*), possibly some data, and some CRC check bytes. There are 10 Packet IDs in four categories:

Token	OUT IN SOF SETUP
Data	DATA0 DATA1
Handshake	ACK NAK STALL
Special	PRE

Transactions

A *transaction* is a discrete interaction between the host and one device using one or more packets.

The host always starts a transaction with one of the *Token* type pids. A *Data* packet or packets may follow in either direction. Finally, a *Handshake* is sent back in the opposite direction to the data transfer.

The IN, OUT, and SETUP packets specify a USB *address* and endpoint, to specify the device and endpoint that should respond. One hundred and twenty-eight addresses are supported, with 0 being the default address on power-up or reset.

A SETUP packet includes eight data bytes to indicate what transaction is happening. USB defines standard control transactions, as described later. Other class and vendor specific SETUP control transactions are outside the basic USB spec.

The two different DATA pids are used to make it easier to detect if a complete data packet has been missed.

The ACK *Handshake* packet indicates that the transfer completed successfully, while NAK indicates that it did not (or that no data was available). A STALL *Handshake* return means that some serious error occurred in the device or endpoint, which the host must clear using another pipe.

When any of the previous packets are sent at low speed to a low-speed device, it is seen by a high-speed device as a PRE packet and so is ignored.

There are various limits on the packet size. DATA packets may be up to 1023 bytes long.

Start of Frame (SOF)

The SOF packet is sent by the host to indicate the start of a frame. This packet includes an 11-bit frame number, continuously going from 0 to 0x7FF (i.e., USBD_ISO_START_FRAME_RANGE-1) and rolling over to 0 again. SOF packets are seen by all high-speed devices. Note that SOF packets can be corrupted, as can any other packets.

Transaction Packet Structure

A control transfer involves the host sending the device a SETUP packet, zero or more DATA packets in either direction, and a *Handshake*.

An IN packet or an OUT packet starts all the other transfers. The definition of the device's endpoint determines whether it is an interrupt, bulk, or isochronous transfer.

The host starts Interrupt transfers regularly to see if the device endpoint has any data available. The device endpoint may return the data. Alternatively, the device might send NAK to indicate that there is no data available or that the state has not changed.

Similarly, bulk input requests may return data or NAK.

Remember that all pipes might stall.

Power

When first reset, a device can draw 100mA (milliampere) from the bus. When configured, it can draw the power it asked for, up to the maximum of 500mA. If suspended, a device may only use 500μA (microampere).

USB Device Framework

This section shows how the low-level USB packets and transactions are used. The USB protocol defines how the USB system is set up and run, as well as how it may be used by clients.

The host works with the hubs to find all the devices on the bus. It uses standard control requests to set up a device. It reads various descriptors from the device, works out which drivers to load, and enables the device. A hub reports any new device insertions or removals through its Status Change endpoint.

After this, the USB system drivers control access to the bus, scheduling all the transfer requests to fit into the available bandwidth.

Bus Enumeration

The PC host has to deal with USB device insertion and removal, both at power up and while running. Any hubs in the system help in this task. A hub is itself a device that talks to the system USB device driver usbhub.sys to inform it of any device insertions or removals.

The host knows when a device is inserted or removed. The hub tells it which port a new device is on. The host tells the hub to enable the port and issue a reset command to the device. The device initially responds at the *default address* of zero. The host reads the device's device descriptor and then allocates it a free USB address in the range 1–127.

The host can then read the rest of the device's descriptors. Eventually, the host selects one of the device configurations. This fully enables the device so that all the endpoints can be used.

The USB system will only let a device configuration be selected if there is enough bandwidth available.

Window inspects the configuration, interface, and endpoint descriptors and loads the appropriate drivers. In the unlikely case of a device with more than one configuration, Windows itself prompts the user to choose between them.

Root Hub

The PC host contains a root hub that it finds on its internal buses when the host is switched on. A USB host controller is usually found by the PCI enumerator. The host controller's PCI configuration space is accessed and set with appropriate values. The controller's registers are permanently mapped into memory. A USB host controller can usually take bus mastership to do its data transfers.

The PC talks directly to the root hub electronics embedded in the host controller. The root hub then reports any devices or further hubs that are attached. Any further hubs are enumerated in the same way until all the USB devices have been found.

If Windows finds two root hubs for example, it treats them as two separate USB buses.

Hub Functionality

A hub is a standard USB device of class HubClass, which must implement functions defined in the USB Specification. It has the necessary USB descriptors to describe its functionality.

A hub primarily consists of a signal repeater and controller. The repeater passes the USB signals in the correct direction. The controller understands the protocol and instructs the repeater. The hub controller is itself controlled by the host through two endpoints.

A hub is configured and interrogated through its standard Endpoint 0. This default pipe implements most of the standard control transactions and then goes on to implement various hub class specific control transactions (ClearHubFeature, ClearPortFeature, GetBusState, GetHubDescriptor, GetHubStatus, GetPortStatus, SetHubDescriptor, SetHubFeature, and SetPortFeature).

A hub also has a Status Change interrupt endpoint, used to inform the host of changes to each of its downstream ports. A port change bit is set if any of the following states change: device connected, port enabled, suspended, over-current indicated, reset, and powered. The host asks for the Status Change interrupt data every 255ms. The hub simply returns NAK if nothing has changed (e.g., if no devices have been added or removed).

Standard Control Transactions

The USB system uses standard control transactions to set up all devices. Further class and vendor specific requests can be defined.

A control transaction starts with a SETUP packet that includes eight data bytes in a customizable format, as shown in Table 20.2. Table 20.3 shows the bit definitions of the first byte, *bmRequestType*. The next byte is the request code.

Table 20.2   Control transfers

Field	Size	Description
bmRequestType	1 byte	Request characteristics, see Table 20.3
bRequest	1	Request code, see Table 20.4
wValue	2	word parameter
wIndex	2	index parameter
wLength	2	Number of bytes to transfer

Table 20.3   bmRequestType **bit definitions**

Bit 7	Transfer direction	0 = Host to device 1 = Device to host
Bits6..5	Type	0 = Standard 1 = Class 2 = Vendor 3 = Reserved
Bits 4..0	Recipient	0 = Device 1 = Interface 2 = Endpoint 3 = Other 4..31 reserved

The basic USB spec only defines the Standard type control transfers listed in Table 20.4. The table indicates the possible destinations, such as Device (D), Interface (I), or Endpoint (E).

Table 20.4   Standard USB SETUP **transfer requests**

Request code	Recipient	Description
CLEAR_FEATURE	DIE	Clears stall or remote wake-up and clears power states
GET_CONFIGURATION	D	Get current configuration number (or 0 if not configured)

Request code	Recipient	Description
GET_DESCRIPTOR	D	Get descriptor: Device, String or Configuration, Interface, and Endpoint.
GET_INTERFACE	I	Get alternate setting for interface
GET_STATUS	DIE	Get status, e.g., whether device is self-powered and whether remote wake-up is signalled, or whether endpoint is stalled.
SET_ADDRESS	D	Set the device's address
SET_CONFIGURATION	D	Set the configuration number
SET_DESCRIPTOR	D	Set or add a descriptor
SET_FEATURE	DIE	Sets stall or remote wake-up and sets power states
SET_INTERFACE	I	Select an alternate interface setting
SYNCH_FRAME	E	Used in isochronous transfers to indicate the start of a frame pattern.

For example, the SET_ADDRESS request is sent to assign a USB address to a device. The device returns ACK when complete. A GET_DESCRIPTOR request asks for one of the descriptors. The descriptor is returned in one or more DATA packets.

Each device class usually has various control transfer requests defined. See the appropriate specification for details.

Descriptors

Table 20.5 shows the basic types of descriptor that a device should return, along with a summary of the information in each descriptor. Figure 20.1 gives a view of the relationship between the various descriptors. String descriptors are optional. Further class specific or vendor specific descriptors can be returned. Table 20.6 lists the standard descriptor type constants.

Configuration and endpoint descriptors must not include a descriptor for Endpoint 0.

An interface may have *alternate* settings that redefine the number or characteristics of the associated endpoints.

For isochronous requests, the endpoint maximum packet size reserves bus time.

Table 20.5   USB descriptor types and fields

Descriptor	Fields
Device	Vendor ID, Product ID, and Device release number Device class, sub class, and protocol Number of configurations
Configuration	Number of interfaces Attributes: bus-powered, self-powered, supports remote wake-up Maximum power required
Interface	Interface number, class, subclass, and protocol. Number of endpoints.

Descriptor	Fields
Endpoint	Endpoint number, direction, transfer type, maximum packet size, and interrupt polling interval.
String	String index 0: Language ID supported by device Others: Unicode string and length

Table 20.6   Standard descriptor type constants

USB_DEVICE_DESCRIPTOR_TYPE	0x01	
USB_CONFIGURATION_DESCRIPTOR_TYPE	0x02	
USB_STRING_DESCRIPTOR_TYPE	0x03	
USB_INTERFACE_DESCRIPTOR_TYPE	0x04	
USB_ENDPOINT_DESCRIPTOR_TYPE	0x05	
USB_POWER_DESCRIPTOR_TYPE	0x06	W98 only
USB_CONFIG_POWER_DESCRIPTOR_TYPE	0x07	W2000 only
USB_INTERFACE_POWER_DESCRIPTOR_TYPE	0x08	W2000 only

Driver Installation

Windows uses values in the Device or Interface descriptors to choose which driver to load. Windows initially forms *Hardware IDs* using the Device descriptor vendor and product fields (*idVendor*, *idProduct*, and *bcdDevice*). If no installation INF file can be found with a model that matches the Hardware IDs, Windows forms *Compatible IDs* out of the Interface descriptor class fields (*bInterfaceClass*, *bInterfaceSubClass*, and *bInterfaceProtocol*). Windows then searches for an installation file that handles one of these Compatible IDs. If nothing can be found, it prompts the user for a new driver. The chosen installation file specifies the driver to load.

The Interface descriptor class field, *bInterfaceClass*, is zero for a basic USB device. If the interface meets one of the standard device class specifications, *bInterfaceClass* is one of the values listed earlier in Table 20.1.

Note that the Device descriptor class fields are not used in the driver selection process. Instead, the *bDeviceProtocol* field can be used to indicate which, if any, new features are supported, as described in the following text.

USB Classes

As mentioned earlier, the basic USB device framework is enhanced to provide specifications for several different classes of USB device. A *Class* constant is defined for each device class. Each class may then go on to define various *SubClass* and *Protocol* constants to further define the general type of functionality provided in the device.

These class constants are not given in the standard USB device descriptor, but set in the Interface descriptor, instead. If Windows cannot find a driver that matches the Device descriptor Hardware ID, it uses the `Interface` descriptor class constants to form a Compatible ID.

The USB class specifications define what interfaces and endpoints must appear in a particular class of device. A class may define one or more additional class-specific descriptors for the class device, interface, or endpoint. These devices still provide the standard device, configuration, interface, and endpoint descriptors.

Device classes typically also define a series of class specific request codes for control transactions. Bits 6 and 5 of the request code are set to 01 to indicate that these are class-specific requests. For example, the `Hub` class request `ClearHubFeature` has code 0x20.

Obtain the appropriate class specification from the USB website at `www.usb.org/developers/`. The HID class and its USB implementation are discussed in detail in Chapter 22. The power supply device class is defined purely as a HID device.

The class-specific descriptors are usually types of `Interface` descriptors. Class-specific descriptor IDs have the top three bits set to 001. Table 20.7 shows the first assigned class-specific descriptor IDs.

Table 20.7   Class-specific descriptor IDs

HID Class	HID descriptor	0x21
	Report descriptor	0x22
	Physical Descriptor	0x23
Communications Class	CS_INTERFACE	0x24
	CS_ENDPOINT	0x25 (not used yet)

Printer Class

As an example of a USB class, it is worth looking here at the printer class. USB printers should resemble EPP or ECP parallel port printers, as defined in the IEEE 1284 specification. The format of the print data sent to the printer is not defined by this spec.

The printer class does not define any class-specific descriptors. The standard device descriptor has zeroes for its class, subclass, and protocol. The standard interface descriptor has constant USB_DEVICE_CLASS_PRINTER (0x07) in its *bInterfaceClass* field. *bInterfaceSubClass* is always set to 0x01 indicating a Printer. The *bInterfaceProtocol* field has 1 if the printer is unidirectional (i.e., it only accept incoming data) or 2 if bidirectional (i.e., it can return status information). A printer can have only one such interface, but it could perhaps switch between the two operating modes using an alternate interface setting.

A printer class device always has a Bulk OUT pipe for data sent to the printer. A bidirectional printer also has a Bulk IN pipe. These pipes are used to carry the relevant page control/description information and status.

A printer uses the default control pipe Endpoint 0 for standard USB purposes. However, it is also used for three printer class-specific requests. GET_DEVICE_ID (request 0) reads an IEEE 1284 device ID. GET_PORT_STATUS (1) gets a status byte that mimics the Centronics printer port status. SOFT_RESET (2) flushes all the buffers, resets the Bulk pipes, and clears all stall conditions.

New Features

USB continues to evolve, both to define new classes of devices and to make implementation of these devices easier. The USB Common class specification document firstly defines how new class specifications should be written. Then it goes on to describe new features that may be used to extend the basic USB functionality.

Table 20.8 lists the USB extensions. The Shared Endpoint extension is described in the next section.

Support for the USB enhancements is indicated in the device descriptor. The device *bDeviceSubClass* field is 0x00 and *bDeviceProtocol* is a bitmap of supported features. Shared endpoint support is indicated in bit 0.

Currently, there seems to be no indication that Windows supports any of these new features[2].

Table 20.8   USB new feature summary

Shared Endpoints	Allows one physical pipe to carry information from several logical pipes.
Synchronization	Defines a standard way of reporting synchronization requirements for an endpoint.
Dynamic Interfaces	Lets a device change its interface dynamically (e.g., a modem might receive either a voice call or a data call, and would want to change its interface accordingly).
Associations	The ways in which different interfaces might be related (e.g., a modem data call might have voice and video information). Each interface requires a separate pipe. However, an association defines how the interfaces are related.
Interface Power Management	An interface power descriptor might describe different: • power down states (apart from just suspended) or • methods of adjusting each interface's power.
Default Notification Pipe	A default notification pipe describes a common format for passing interrupt data to the host.

Shared Endpoints

Designers of some devices might find that they need many endpoints and pipes, which might require lots of resources. The basic USB standard says that only the default pipe Endpoint 0 can be shared between interfaces.

The *Shared Endpoint* USB extension allows several *logical pipes* to be carried on one normal "physical" pipe. Such shared pipes must have the same transfer type (control, interrupt, bulk, or isochronous) and be in the same direction. Each logical pipe defines one or more *logical packets* for the data it might send.

---

2. The Windows 2000 DDK defines the Interface power descriptor structure. However, it is not apparent whether this information is used by the system USB class drivers.

Hosts and devices that handle shared pipes multiplex the data from each logical source on the physical pipe. For isochronous pipes, the combined maximum data packet size must fit into a frame. The software must ensure that a stall on one logical pipe does halt data flow on another logical pipe.

Flow control can optionally be implemented using a pipe in the opposite direction to data transfer. A source must not send data on a logical pipe until the target has provided a *grant* to send: a count of the number of logical packets it can receive.

Client Design

This section briefly discusses options for both device firmware engineers and Windows device driver writers.

Endpoint Type Selection

The different types of endpoint have different characteristics. Control transfers consist of *messages* or requests sent when the host dictates. Interrupt, bulk, and isochronous transfers are said to be *stream* pipes as they keep generating application defined data.

There is usually only one control pipe, for Endpoint 0. As well as carrying the USB set up and configuration standard transfers, you can use this pipe for the class-defined or your own vendor-defined requests. If need be, define any further control pipes.

Interrupt pipes are usually used for user input data, but they can also retrieve relatively infrequent data from the device. Interrupt pipes can be used in conjunction with isochronous pipes to provide feedback so that the data rate can be varied. As mentioned previously, a device can NAK an interrupt pipe request to indicate that no data has changed. Additionally, Chapter 23 shows that HID devices support the concept of an idle rate. This technique reduces the number of interrupt reports that are generated.

Devices that generate or consume data samples regularly will want to use isochronous transfers. Isochronous transfers can vary in length, and may be used in preference to bulk transfers as they reserve USB bus bandwidth. Isochronous transfers are not error checked, but a receiver can determine if a transmission error occurred by checking the frame number. Make sure that the application or device can cope with losing data. Bulk transfers are error checked, but they have the lowest priority on the bus.

Low-speed devices can only have two endpoints after Endpoint 0. Full-speed devices can have a maximum of 16 input and 16 output endpoints. Check that maximum packet size for your pipe type is acceptable.

Do not forget that if a pipe stalls, a control pipe request must be defined to clear the stall condition.

You might want to consider some of the new USB features, discussed earlier. Check that Windows supports the relevant feature.

Isochronous Devices

Isochronous devices can represent quite a challenge to software designers. An isochronous input *source* usually generates a regular number of samples to represent their data, while an output *sink* receives samples and produces the relevant output.

A driver must firstly be able to handle the volume of data coming in and take appropriate steps if it cannot keep up. It is sometimes best if a set of samples can be left to build up for processing in one fell swoop, rather than processing each sample individually. Check that your application will survive this treatment.

Synchronizing a series of devices and processes is a more complicated issue. The clocks inside each device and inside the PC may all be running at different frequencies. Even if they are nominally at the same frequency, they may well drift or produce jitter.

A basic *asynchronous* source just sends any samples it has produced. A basic asynchronous sink might use a feedback interrupt pipe to tell the PC how many samples to send.

There are various techniques that devices and drivers can use to handle these difficulties. At the very least, some sort of rate adaptation or conversion may be necessary.

A common technique is to group the samples into service intervals and buffer the sample groups through the relevant processes. A driver might expect a set number of samples in a service interval. If not enough have arrived, a new sample or samples could be generated by interpolation. Similarly, if too many samples arrive, some should be dropped. This technique requires that you buffer up a set of samples for a service interval, one for each rate adaptation process. Check that the resultant delay through the system is acceptable, and that adding or dropping samples is satisfactory.

Another technique is to synchronize all activities to the 1kHz USB frame clock. An input source would generate the correct number of samples each frame, even if a different number had arrived. Make sure that your device copes if the USB Start of Frame (SOF) frame packet is corrupted.

An *adaptive* device interacts with the PC to tell it how many samples to send or receive. It might use an interrupt pipe to tell the host how to adjust its sample rate.

Frame Length Control

Another synchronization technique is for a device to obtain control of the USB frequency (i.e., take over mastership of the [SOF] packet rate). An SOF master adjusts the USB clock slightly to match its internal clock. However, only one device can take over USB bus mastership, so a device should not rely on this technique.

There are Windows URB functions to request control of the USB frame length, and to get and set the current frame length. The frame length may be adjusted by only plus or minus one bit at a time. You can also get the current frame number.

Patterns

A final useful technique for isochronous transfers is to let the size of each transfer vary in a regular pattern to match the source data rate. The host and the endpoint must agree which frame begins the repeating pattern. One of the calls to the Windows USB drivers asks for the frame number which starts a pattern.

An audio signal could be sampled at 44.1kHz. This means that 44100 16 bit samples are generated every second. This corresponds to 44.1 samples every USB frame. Sending 45 samples per frame would be an acceptable solution, provided some special flag is used to indicate which frames have 45, rather than 44, samples. The recommended solution is to use a frame pattern. In this case, a pattern that repeats every 10ms should be used, consisting of nine frames of 44 samples and one frame of 45 samples. In total, exactly the right number of samples is sent, 44100 per second.

Conclusion

This chapter has looked at the background to the Universal Serial Bus (USB). An understanding of the USB low-level frame structure helps when writing device drivers. The Windows USB system drivers read the standard descriptors in each USB device to determine which drivers to load.

The next chapter looks at how to write a device driver that can talk to USB devices through the Windows USB Device Interface (USBDI).

Chapter 21

USB Driver Interface

This chapter describes how to write a USB driver that uses the Windows USB Driver Interface (USBDI). The UsbKbd example driver talks to a USB keyboard and retrieves the raw keyboard input data. The corresponding Win32 test application displays this input data. In addition, you finally get the opportunity to flash some LEDs (the LEDs on keyboard).

A USB client is a device driver that uses the standard Windows system USB class drivers to talk to USB devices using USBDI. USBD.sys is the USB class driver. It uses either UHCD.sys to talk to Universal Host Controller Interface devices, or OpenHCI.sys for Open Host Controller Interface devices. USBHUB.sys is the USB driver for root hubs and external hubs. The relevant drivers are loaded when the PCI Enumerator finds each USB host controller.

Windows 2000 runs USBDI version 2.00, while Windows 98 uses version 1.01. There is no documentation on the differences. USBDI version 2.0 might return some Windows Management Instrumentation (WMI) data. Use the USBD_GetUSBDIVersion function to determine what version of the USB class drivers you are using. Only one very minor aspect of this chapter's driver is affected by the USBDI version.

As far as a USB client driver is concerned, the host computer (the Windows PC) talks directly to any function device that is plugged in. The details of bus enumeration and how information is transferred are not directly of concern to a USB client.

However, a client does need to be aware of the types of transfer, when they occur, and their timing, particularly in the case of isochronous transfers.

A client talks to a function device through one or more *pipes*. A pipe is a unidirectional or bidirectional channel for data transfer between the host and a device. A pipe may be one of four types. *Control* pipes transfer commands to the device (including USB set up and configuration information), with data transfer in either direction. *Interrupt* pipes transfer device-specific information to the host (e.g., when a keyboard key is pressed). *Bulk* pipes transfer larger amounts of data. *Isochronous* pipes carry time-sensitive data, such as voice.

Control endpoints are bidirectional. Interrupt endpoints are unidirectional into the host. Bulk and isochronous endpoints are unidirectional, in either direction.

The *default pipe* is used for control transfers. The USB specification defines several standard requests on the default pipe, such as "get descriptor". In addition, a device may respond to class-defined or vendor-defined requests on the default pipe.

Figure 21.1 shows the logical entities that a device and USBDI present to the client. A device exposes a series of connection points for pipes called *endpoints*. Endpoints may be grouped together to make an *Interface*. A *configuration* consists of one or more interfaces. A device usually contains just one configuration and one interface, but can contain more.

Figure 21.1 USB logical structure

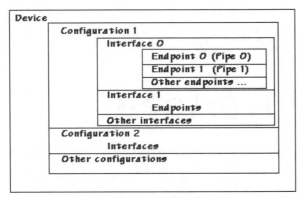

@ 1999 PHD Computer Consultants Ltd

Only one configuration is active at any one time, but all the interfaces and their endpoints within that configuration are active. Windows forces the selection of a configuration. As explained in the previous chapter, Windows forms a Hardware ID out of the USB Device descriptor, Vendor ID, and Product ID fields. This Hardware ID is used to try to find a matching device driver installation INF file. If there is no match, a Compatible ID is formed from the class fields in each interface descriptor in the selected configuration. These Compatible IDs are used to search for class-specific installation INF files; a device is created for each interface in the device.

USB Client Driver Design

A USB client driver is a standard WDM driver with Plug and Play support. As described in Chapters 11 and 20, the USB hub driver detects when a new USB device is inserted. The PnP Manager uses the vendor ID or device class information to select the driver to run. The driver's AddDevice routine is called and other PnP IRPs are issued as normal. Client USB drivers never receive any hardware resources, such ports or interrupts, as the USB class drivers handle all the low-level I/O.

USB devices are particularly prone to surprise removals, when the user accidentally (or deliberately) pulls out the USB plug. In Windows 2000, you will receive a *Surprise Removal* PnP IRP, while in Windows 98, only a *Remove Device* PnP IRP. These requests must succeed even if there are open handles to the device. Any transfers that are in progress must be aborted.

By definition, a USB client driver makes calls, at its lower edge, to the USB class drivers. However, it can implement whatever upper edge it deems necessary. The UsbKbd example driver in this chapter exposes a device interface that allows Win32 applications to read raw keyboard data and issue various IOCTLs. However, your driver could take a totally different approach, depending on what your USB device does.

The UsbKbd example driver is based on the Wdm2 driver, with Power Management support removed. The UsbKbd driver source code is in the book software UsbKbd\Sys directory. I suggest that you use UsbKbd as the basis for your USB client driver. Most of the USB handling routines are in the file Usb.cpp. You will need to alter the upper edge interface that is implemented in the dispatch routines in Dispatch.cpp.

A real USB client driver should support Power Management. An article on the Microsoft website at www.microsoft.com/hwdev entitled "OnNow requirements in the USB Core Specification" discusses this issue.

Using UsbKbd

UsbKbd devices can be found by Win32 applications using the USBKBD_GUID device interface. When a handle to a UsbKbd device is opened, the UsbKbd driver starts talking to the USB peripheral. If the peripheral is a HID USB keyboard, its configuration is selected, thereby enabling the keyboard interrupt pipe. The UsbKbd Create IRP handler also obtains some more information from the device and the USB class drivers, displaying *DebugPrint* trace output in the checked build of the driver.

Closing the file handle deselects the keyboard's configuration.

The Win32 application can then use standard ReadFile calls to obtain raw keyboard data from UsbKbd. The raw data is in the form a HID keyboard input report, sent on a USB interrupt pipe. Chapter 22 details the format of this report precisely. All you need to know for now is that the first byte has all the modifier key bits (such as Shift, Control, etc.). The third and following bytes contain the HID keypress codes for all the keys that are currently pressed.

The time-out defaults to 10 seconds for ReadFile calls. One of the UsbKbd IOCTLs can be used to alter the time-out.

The UsbKbd driver also accepts WriteFile calls to change the state of the keyboard LEDs. The first byte of the WriteFile data is sent as a HID output report in a control transfer along the default pipe to Endpoint 0. The lower five bits of this byte are defined, though most keyboards just have three LEDs.

Table 21.1 shows the IOCTLs that UsbKbd supports. These functions are provided just to let you see how to exercise the USB interface. In most cases, you would not want to expose such IOCTLs for general use, although they might be useful for debugging purposes.

Table 21.1 UsbKbd **IOCTLs**

IOCTL_USBKBD_SET_READ_TIMEOUT **Input** ULONG Time-out in seconds Specifies the time-out for subsequent read requests
IOCTL_USBKBD_GET_DEVICE_DESCRIPTOR **Output** Received device descriptor Read the device descriptor.
IOCTL_USBKBD_GET_CONFIGURATION_DESCRIPTORS Output Received descriptors Read the first configuration descriptor, along with any associated, interface, endpoint, class and vendor descriptors.
IOCTL_USBKBD_GET_SPECIFIED_DESCRIPTOR **Input** ULONG Descriptor type, ULONG Descriptor size **Output** Received descriptor Read the specified descriptor into the given sized buffer
IOCTL_USBKBD_GET_STATUSES **Output** 6-byte buffer for status values Read the device, first interface and first endpoint status values (16 bits each)
IOCTL_USBKBD_GET_FRAME_INFO **Output** 12-byte buffer for frame information Read the ULONG current frame length (in bits), the ULONG current frame number and the ULONG frame number for the next frame whose length can be altered

UsbKbd **Installation**

For your USB client driver, you will use a standard installation INF file. In the Version section, make sure that you include the following lines.

```
Class=USB
ClassGUID={36FC9E60-C465-11CF-8056-444553540000}
```

As described in Chapter 11, your driver will be loaded if it has the right Hardware ID. A Hardware ID is constructed using the vendor and product IDs in the device descriptor (e.g., USB\VID_046A&PID_0001).

However, for USB keyboards, this approach does not work. Windows 98 and Windows 2000 have built in support for USB keyboards. In both cases, they load the Human Input Device (HID) class drivers. The standard keyboard driver then gets key input through HID client drivers.

I tried fiddling with the USB keyboard installation INF files for Windows 98 and Windows 2000. However, I could not persuade Windows to load UsbKbd in place of the standard drivers. I can only assume that the standard driver names are hard-coded into the kernel and that the Windows INF files are not really used.

My brutal solution to this problem was to replace the HID USB minidriver HidUsb.sys with the UsbKbd.sys driver executable. Windows uses HidUsb.sys to process HID USB devices. Replacing HidUsb.sys seems to work. Needless to say, this approach stops you from using the USB keyboard for normal typing. In addition, it stops you from using any other USB HID devices. Therefore, you will need to have a standard PCBIOS keyboard connected to the system, as well. Do not forget to save a copy of the HidUsb.sys file (or delete it and reinstall it from the Windows CD). You need a dual-boot PC if you are going to change HidUsb.sys in W2000.

If you do not have a HID USB keyboard at hand, you obviously cannot run these tests. However, I show some of the *DebugPrint* output later in this chapter, so you can see what is going on.

While developing the UsbKbd driver, I altered makefile.inc so that it copied the final executable, UsbKbd.sys, to overwrite HidUsb.sys. However, this command line has now been commented out, so you will have to do the copy yourself.

Trying out each new version of the UsbKbd driver was very easy. Simply unplugging and plugging in the USB cable caused the old driver to be unloaded and new one loaded. Windows 2000 displays a bugcheck on shutdown when HidUsb.sys is replaced by UsbKbd.sys. The simple solution is to unplug the USB keyboard before you shutdown.

Headers and Libraries

The USB header files that you might need to include are in the DDK.

usb100.h	Various USB constants and structures
usbioctl.h	IOCTL definitions
usbdlib.h	URB building and assorted routines
usbdi.h	USBDI routines, including URB structures

These header files generate two annoying compiler warnings which you can ignore. I also found that I needed to specifically mention the USB library in the SOURCES file.

```
TARGETLIBS=C:\NTDDK\LIB\I386\FREE\Usbd.Lib
```

USBDI IOCTLs

The USB class drivers are primarily used through the USB Device Interface (USBDI) Internal IOCTLs shown in Table 21.2. As these are Internal IOCTLs, they are only available to other parts of the kernel, such as device drivers, and are not available to user mode applications.

A few ordinary IOCTLs are also available to Win32 programs. These are intended for use by diagnostic utilities, so I shall not cover them here. The DDK *UsbView* utility source shows how to use these IOCTLs.

Table 21.2　USBDI internal IOCTLs

`IOCTL_INTERNAL_USB_SUBMIT_URB`	Submit URB Block awaiting result
`IOCTL_INTERNAL_USB_RESET_PORT`	Reset and reenable a port

`IOCTL_INTERNAL_USB_GET_PORT_STATUS`	Get port status bits:
	`USBD_PORT_ENABLED`
	`USBD_PORT_CONNECTED`

`IOCTL_INTERNAL_USB_ENABLE_PORT`	Reenable a disabled port
`IOCTL_INTERNAL_USB_GET_HUB_COUNT`	Used internally by hub driver
`IOCTL_INTERNAL_USB_CYCLE_PORT`	Simulates a device unplug and replug
`IOCTL_INTERNAL_USB_GET_ROOTHUB_PDO`	Used internally by hub driver
`IOCTL_INTERNAL_USB_GET_HUB_NAME`	Get the device name of the USB hub
`IOCTL_INTERNAL_USB_GET_BUS_INFO`	Fills in a `USB_BUS_NOTIFICATION` structure (W2000 only)
`IOCTL_INTERNAL_USB_GET_CONTROLLER_NAME`	Get the host controller device name (W2000 only)

URBs

The most important Internal IOCTL is `IOCTL_INTERNAL_USB_SUBMIT_URB`, which lets you submit a USB Request Block (**URB**) for processing by the USB class drivers. There are thirty-odd different URB function codes. USB clients use URBs to do most of their hard work.

The URB structure itself is a union of some 16 different `_URB_*` structures, as shown in Listing 21.1. Each function code uses one of these other URB structures to detail its input or output parameters. All URB structures begin with a common header `_URB_HEADER` structure. The header *Length* and *Function* fields must be filled in before calling the USB Device Interface. The result of processing the URB is returned in the *Status* field.

Listing 21.1　URB structures

```
typedef struct _URB {
    union {
    struct _URB_HEADER              UrbHeader;
    struct _URB_SELECT_INTERFACE    UrbSelectInterface;
    struct _URB_SELECT_CONFIGURATION UrbSelectConfiguration;
    // ...
    };
} URB, *PURB;

struct _URB_HEADER {
    USHORT Length;
    USHORT Function;
    USBD_STATUS Status;
    // ...
};
```

The *Status* field uses its top two bits as a State code to indicate how the request completed. Table 21.3 shows the possible values along with the names of macros that can be used to detect these conditions. The rest of the *Status* field is filled with more detailed error codes. Some error codes are internal system USB errors (e.g., no memory).

Table 21.3   URB *Status* field State code bits

State code bits	Interpretation	Macro
00	Completed successfully	USBD_SUCCESS
01	Request is pending	USBD_PENDING
10	Error, endpoint not stalled	USBD_ERROR
11	Error, endpoint stalled	USBD_ERROR or USBD_HALTED

To make it easier to construct suitable URBs, various build macros are provided, such as UsbBuildGetDescriptorRequest, which fill in a preallocated URB. Other useful routines both allocate the memory for a URB and fill it in.

The reference section towards the end of this chapter lists the URB function codes. The reference section also details the other crucial USB structures. However, this chapter first illustrates how to perform most common USB actions by describing how these jobs are done in the UsbKbd driver.

Several URB structures have a *UrbLink* field. If non-NULL, this specifies a pointer to a URB that is processed if the current one completes successfully.

Calling USBDI

Listing 21.2 shows the CallUSBDI routine in Usb.cpp. This is used to issue all the Internal IOCTLs to the USB system class drivers. It has default parameters that make it easy to ask for a URB to be processed.

CallUSBDI has to create a new IRP for the Internal IOCTL, fill in the IRP, and send off the IRP down the device stack to the USB system drivers. Further, it then waits until the IRP has been processed. CallUSBDI can only be called at PASSIVE_LEVEL.

The USB Internal IOCTLs do not use the standard input and output IRP stack locations. Instead, the stack *Parameters.Others.Argument1* field is set to the URB pointer, etc. The *Parameters.Others.Argument2* field is used for one of the USB IOCTLs.

Allocating IRPs

If you want to call a lower driver, it is usually simplest to reuse an existing IRP. You simply fill in the next IRP stack location with the correct function codes and parameters and call the next driver down the stack.

In some cases however, you will need to build an IRP from scratch. For example, you may wish to generate an IRP in your DriverEntry routine. Alternatively, you may process a large incoming request by splitting it into several different IRPs.

UsbKbd could reuse an existing IRP. However, it is straightforward to allocate a new IOCTL IRP. The CallUSBDI routine keeps the entire IRP allocation process wrapped up in one neat location using this technique.

Building and issuing an IOCTL IRP is made particularly easy with the `IoBuildDeviceIo-ControlRequest` call. If you pass an initialized event, you can wait for the IRP to complete simply by waiting for the event to become signalled. You do not need to set up a completion routine.

`IoBuildDeviceIoControlRequest` can be used to make both IOCTL and Internal IOCTL IRPs, depending on the `InternalDeviceIoControl` parameter. In a similar way to the Win32 `DeviceIoControl` call, both input and output buffers can be used. The IOCTL control code must use buffered I/O.

Internally, it seems as though `IoBuildDeviceIoControlRequest` allocates some nonpaged memory to hold the combined input and output buffer, to make it work exactly like a standard IOCTL call. `IoBuildDeviceIoControlRequest` copies the input buffer there initially. It must set its own completion routine that copies the data back into your given output buffer (as well as setting the event and freeing the IRP memory).

The USB Internal IOCTLs do not use the standard input and output buffers[1]. Instead, you have to set up the next stack location *Parameters.Others.Argument1* field. This is more complicated than you might think.

Previous examples have used `IoSkipCurrentIrpStackLocation` to reuse the current IRP stack location, and `IoCopyCurrentIrpStackLocationToNext` to copy the current stack location to the next when setting a completion routine. We cannot use these routines, as the current stack location is not set up yet.

The `IoCallDriver` call moves onto the next IRP stack location. `CallUSBDI` wants to change the stack location that the next lower driver sees. The `IoGetNextIrpStackLocation` call returns the required IRP stack location pointer[2]. The `IoBuildDeviceIoControlRequest` call has already set up most of the correct values for this stack location. The required values are set into the *Parameters.Others.Argument1* field, etc.

In summary, `CallUSBDI` does the following jobs.

1. Initializes an IRP completion event
2. Builds an Internal IOCTL
3. Stores the URB pointer, etc., in the IRP's next stack location
4. Calls the next driver
5. If the request is still pending, wait for the completion event to become signalled. The `KeWaitForSingleObject WaitReason` parameter must be set to Suspended.

Listing 21.2 Calling USBDI

```
NTSTATUS CallUSBDI( IN PUSBKBD_DEVICE_EXTENSION dx, IN PVOID UrbEtc,
                    IN ULONG IoControlCode/*=IOCTL_INTERNAL_USB_SUBMIT_URB*/,
                    IN ULONG Arg2/*=0*/)
{
    IO_STATUS_BLOCK IoStatus;
```

---

1. Presumably to avoid unnecessary copying of large buffers.

---

2. Unlike a processor execution stack, there is nothing wrong with altering the next item down the stack.

Listing 21.2 Calling USBDI (continued)

```
    KEVENT event;

    // Initialise IRP completion event
    KeInitializeEvent(&event, NotificationEvent, FALSE);

    // Build Internal IOCTL IRP
    PIRP Irp = IoBuildDeviceIoControlRequest(
                    IoControlCode, dx->NextStackDevice,
                    NULL, 0,    // Input buffer
                    NULL, 0,    // Output buffer
                    TRUE, &event, &IoStatus);

    // Get IRP stack location for next driver down (already set up)
    PIO_STACK_LOCATION NextIrpStack = IoGetNextIrpStackLocation(Irp);
    // Store pointer to the URB etc
    NextIrpStack->Parameters.Others.Argument1 = UrbEtc;
    NextIrpStack->Parameters.Others.Argument2 = (PVOID)Arg2;

    // Call the driver and wait for completion if necessary
    NTSTATUS status = IoCallDriver( dx->NextStackDevice, Irp);
    if (status == STATUS_PENDING)
    {
        KeWaitForSingleObject( &event, Suspended, KernelMode, FALSE, NULL);
        status = IoStatus.Status;
    }

    // return IRP completion status
    return status;
}
```

Other IRP Allocations

Other kernel calls can be used to allocate IRPs. IoBuildSynchronousFsdRequest builds a Read, Write, Flush, or Shutdown IRP that uses an event to signal completion in the same way as IoBuildDeviceIoControlRequest. IoBuildSynchronousFsdRequest must be called at PASSIVE_LEVEL IRQL.

IoBuildAsynchronousFsdRequest works asynchronously, as it does not use an event to signal its completion. Consequently, it can be called at or below DISPATCH_LEVEL. Note that the IRP must be freed using IoFreeIrp. A common way to do this is to attach a completion routine. It is OK to call IoFreeIrp here and then return STATUS_MORE_PROCESSING_REQUIRED.

There are two final macho ways of building IRPs. IoAllocateIrp allocates an IRP, while IoInitializeIrp makes an IRP out of some driver allocated memory. Be very careful to set up all IRP and IRP stack locations correctly. Both of these methods allow you specify the size of the IRP stack size. Use IoGetNextIrpStackLocation to get the first IRP stack location if you need to set up a completion routine. Call IoFreeIrp to free an IRP created by IoAllocateIrp.

The old DDK documentation wrongly says that you can call IoInitializeIrp to reuse an IRP allocated using IoAllocateIrp. You can use this function as long as you preserve the IRP *AllocationFlags* field, as shown in Chapter 23. In W2000, you can reuse an IRP created with IoAllocateIrp using IoReuseIrp.

Multiple USBDI Calls

If you were reading carefully, you will have noticed that the CallUSBDI routine can only be called at PASSIVE_LEVEL. This means that it cannot be called from a StartIo routine, which runs at DISPATCH_LEVEL. UsbKbd makes its USB calls direct from its dispatch routines that run at PASSIVE_LEVEL.

In UsbKbd, it is possible for a user program to issue two overlapped read requests "simultaneously". This might easily result in the USB class drivers being sent two IRPs for processing at the same time. There is nothing in the documentation that says this is a problem. I suspect that it is not, as the USB class driver will almost certainly serialize requests.

If you feel that you ought to serialize your USBDI calls, you will have to use a StartIo routine. A single prebuilt IRP can be reused for each call. Chapter 23 shows how to build, use, and free such an IRP.

In many cases, it might be useful to send off a series of IRPs to the USB class drivers. As there will usually be a spare IRP queued up for processing, no incoming data will be lost.

Talking USB

Initializing a USB Device

There are several jobs that a USB client driver must do to initialize its connection to its device. The UsbKbd driver does these jobs in its Create IRP handler. However, most USB drivers will want to initialize their device when processing the *Start Device* Plug and Play IRP.

1. Check device enabled. Reset and enable the device, if necessary.
2. Select one interface in one of the configurations.
3. Possibly read other descriptors, such as the string-, class-, or vendor-specific descriptors.
4. Talk to your device to issue whatever commands are relevant, get device status, and initialize pipes.

Device Reset

Listing 21.3 shows how UsbKbd resets its device in routine UsbResetDevice when a Win32 program opens a handle to it. UsbGetPortStatus issues IOCTL_INTERNAL_USB_GET_PORT_STATUS to retrieve the port status bits. UsbResetDevice checks the USBD_PORT_CONNECTED and USBD_PORT_ENABLED bits, and calls UsbResetPort, if necessary. UsbResetPort simply issues IOCTL_INTERNAL_USB_RESET_PORT to the USB class drivers.

You may need to reset the port if some drastic communication problem has arisen with your device (e.g., if control transfers to the default pipe keep failing). However, if another pipe stalls, do not reset the port.

If a pipe stalls or USBDI detects a timeout, the pipe becomes *halted*. The USBD_HALTED macro detects this condition. All URBs linked to the current URB are cancelled. A halted pipe cannot accept any more transfers until a *Reset Pipe* URB is issued.

However, the default pipe can never be halted. If a timeout or stall occurs here, an error is reported, which is detectable using the USBD_ERROR macro, but not USBD_HALTED. The halt condition is cleared automatically to give transfers a chance of succeeding on this pipe. If they keep failing, you will have to reset the port.

Listing 21.3 Device reset routines

```
NTSTATUS UsbGetPortStatus( IN PUSBKBD_DEVICE_EXTENSION dx,
    OUT ULONG& PortStatus)
{
    DebugPrintMsg("Getting port status");
    PortStatus = 0;
    NTSTATUS status = CallUSBDI( dx, &PortStatus,
        IOCTL_INTERNAL_USB_GET_PORT_STATUS);
    DebugPrint( "Got port status %x", PortStatus);
    return status;
}


NTSTATUS UsbResetPort( IN PUSBKBD_DEVICE_EXTENSION dx)
{
    DebugPrintMsg("Resetting port");
    NTSTATUS status = CallUSBDI( dx, NULL, IOCTL_INTERNAL_USB_RESET_PORT);
    DebugPrint( "Port reset %x", status);
    return status;
}


NTSTATUS UsbResetDevice( IN PUSBKBD_DEVICE_EXTENSION dx)
{
    ULONG PortStatus;

    NTSTATUS status = UsbGetPortStatus( dx, PortStatus);

    if( !NT_SUCCESS(status))
        return status;

    // Give up if device not connected
    if( !(PortStatus & USBD_PORT_CONNECTED))
        return STATUS_NO_SUCH_DEVICE;
```

Listing 21.3 Device reset routines (continued)

```
    // Return OK if port enabled
    if( PortStatus & USBD_PORT_ENABLED)
        return status;

    // Port disabled so attempt reset
    status = UsbResetPort(dx);
    if( !NT_SUCCESS(status))
        return status;

    // See if it is now working
    status = UsbGetPortStatus( dx, PortStatus);
    if( !NT_SUCCESS(status))
        return status;
    if( !(PortStatus & USBD_PORT_CONNECTED) ||
        !(PortStatus & USBD_PORT_ENABLED))
        return STATUS_NO_SUCH_DEVICE;
    return status;
}
```

Issuing URBs

The Usb.cpp module contains many routines that build and send off URBs for processing by the USB class drivers. UsbGetDeviceDescriptor is used to retrieve the USB device descriptor. This routine is not essential in UsbKbd, but it is used to implement one of the IOCTLs that are available to user mode programs. UsbGetDeviceDescriptor allocates nonpaged memory for the device descriptor and puts the count of bytes transferred in its Size parameter. The routine that calls UsbGetDeviceDescriptor must free the memory using ExFreePool.

UsbGetDeviceDescriptor starts by allocating some nonpaged memory for the URB. The *Get Descriptor* URB function uses the URB_FUNCTION_GET_DESCRIPTOR_FROM_DEVICE function code. For this function code, the _URB_CONTROL_DESCRIPTOR_REQUEST structure is used.

Next, UsbGetDeviceDescriptor allocates memory for the device descriptor. Then, it calls UsbBuildGetDescriptorRequest to do the hard work of formatting the URB. The URB pointer and its size are passed as the first two parameters. The next parameter specifies the descriptor type, a device descriptor in this case. The following *Index* and *LanguageId* fields are only used when requesting configuration and string descriptors.

The following parameters to UsbBuildGetDescriptorRequest specify the descriptor buffer and its length. This buffer can be specified as a plain pointer to nonpaged memory. Alternatively, an MDL can be used. The final parameter is UrbLink, an optional link to a follow on URB.

UsbGetDeviceDescriptor sends off the built URB using CallUSBDI. It checks both the CallUSBDI NTSTATUS return status and the URB completion status.

The final job for `UsbGetDeviceDescriptor` is simply to save the count of bytes transferred. This was stored in the URB *UrbControlDescriptorRequest.TransferBufferLength* field. The USB memory is freed.

Listing 21.4 Getting a USB device descriptor

```
NTSTATUS UsbGetDeviceDescriptor( IN PUSBKBD_DEVICE_EXTENSION dx,
                                 OUT PUSB_DEVICE_DESCRIPTOR& deviceDescriptor,
                                 OUT ULONG& Size)
{
    // Allocate memory for URB
    USHORT UrbSize = sizeof(struct _URB_CONTROL_DESCRIPTOR_REQUEST);
    PURB urb = (PURB)ExAllocatePool(NonPagedPool, UrbSize);
    if( urb==NULL)
    {
        DebugPrintMsg("No URB memory");
        return STATUS_INSUFFICIENT_RESOURCES;
    }


    // Allocate memory for device descriptor
    ULONG sizeDescriptor = sizeof(USB_DEVICE_DESCRIPTOR);
    deviceDescriptor = (PUSB_DEVICE_DESCRIPTOR)ExAllocatePool(NonPagedPool,
        sizeDescriptor);
    if( deviceDescriptor==NULL)
    {
        ExFreePool(urb);
        DebugPrintMsg("No descriptor memory");
        return STATUS_INSUFFICIENT_RESOURCES;
    }


    // Build the Get Descriptor URB
    UsbBuildGetDescriptorRequest(
        urb, UrbSize,
        USB_DEVICE_DESCRIPTOR_TYPE, 0, 0,          // Types, Index & LanguageId
        deviceDescriptor, NULL, sizeDescriptor,    // Transfer buffer
        NULL);    // Link URB

    // Call the USB driver
    DebugPrintMsg("Getting device descriptor");
    NTSTATUS status = CallUSBDI( dx, urb);
    // Check statuses
    if( !NT_SUCCESS(status) || !USBD_SUCCESS( urb->UrbHeader.Status))
    {
        DebugPrint("status %x URB status %x", status, urb->UrbHeader.Status);
```

Listing 21.4 Getting a USB device descriptor (continued)

```
        status = STATUS_UNSUCCESSFUL;
    }
    // Remember count of bytes actually transferred
    Size = urb->UrbControlDescriptorRequest.TransferBufferLength;


    ExFreePool(urb);
    return status;
}
```

Selecting an Interface

To start using a device, your driver must select one interface within one configuration. This is more of a job than you might expect. Routine UsbSelectConfiguration performs these steps, as shown in Listing 21.5.

1. Get the configuration, interface, endpoint descriptors, etc.
2. Find the appropriate interface descriptor.
3. Issue a select configuration URB.
4. Save the configuration handle and handles to any pipes that you use.

UsbGetConfigurationDescriptors (not listed) is used to retrieve all the descriptors that are needed. This routine issues a *Get Descriptor* request twice. The first call returns just the basic configuration descriptor. The *wTotalLength* field in there tells you how much memory to allocate to retrieve all the associated descriptors.

USBD_ParseConfigurationDescriptorEx is used to find an interface. The parameters to this routine specify the criteria that must match. In this case UsbKbd is not interested in matching the interface number or alternate settings, but is interested in the device class. The HID device class is 3 (USB_DEVICE_CLASS_HUMAN_INTERFACE). To match a HID keyboard the subclass and protocol interface descriptor fields must both be 1.

If USBD_ParseConfigurationDescriptorEx finds a matching interface, it returns the interface descriptor pointer. The next task is to build a suitable *Select Configuration* URB. The helper function USBD_CreateConfigurationRequestEx allocates and fills this URB. UsbSelectConfiguration specifies an array of USBD_INTERFACE_LIST_ENTRY structures as an input to this routine, with a NULL *InterfaceDescriptor* field indicating the end of the list. Each structure specifies the interface descriptor. When the *Select Configuration* URB has been run, the *Interface* field points to valid data.

The *Select Configuration* URB is then issued. This may fail if there is not enough USB bandwidth available. If it works, a configuration handle is returned. You can use this to select a different alternate interface setting. UsbKbd stores the configuration handle in its device extension, but does not use it.

UsbKbd does need to store the pipe handle for the HID keyboard's interrupt pipe. For a HID USB keyboard, this is the first and only pipe in the USBD_INTERFACE_INFORMATION structure. UsbKbd makes *DebugPrint* calls to display other potentially useful information. UsbKbd finally frees the URB and configuration descriptor memory.

Listing 21.5 Selecting a configuration and interface

```
NTSTATUS UsbSelectConfiguration( IN PUSBKBD_DEVICE_EXTENSION dx)
{
    dx->UsbPipeHandle = NULL;

    // Get all first configuration descriptors
    PUSB_CONFIGURATION_DESCRIPTOR Descriptors = NULL;
    ULONG size;
    NTSTATUS status = UsbGetConfigurationDescriptors( dx, Descriptors, 0, size);
    if( !NT_SUCCESS(status))
    {
        DebugPrint("UsbGetConfigurationDescriptors failed %x", status);
        FreeIfAllocated(Descriptors);
        return status;
    }

    // Search for an interface with HID keyboard device class
    PUSB_INTERFACE_DESCRIPTOR id = USBD_ParseConfigurationDescriptorEx(
        Descriptors, Descriptors,
        -1, -1,      // Do not search by InterfaceNumber or AlternateSetting
        3, 1, 1);    // Search for a HID device, boot protocol, keyboard
    if( id==NULL)
    {
        DebugPrintMsg("No matching interface found");
        FreeIfAllocated(Descriptors);
        return STATUS_NO_SUCH_DEVICE;
    }

    // Build list of interfaces we are interested in
    USBD_INTERFACE_LIST_ENTRY ilist[2];
    ilist[0].InterfaceDescriptor = id;
    ilist[0].Interface = NULL;    // Will point to
                                  // urb->UrbUsbSelectConfiguration.Interface
    ilist[1].InterfaceDescriptor = NULL;

    // Create select configuration URB
    PURB urb = USBD_CreateConfigurationRequestEx( Descriptors, ilist);

    // Call the USB driver
    DebugPrintMsg("Selecting configuration");
```

Listing 21.5 Selecting a configuration and interface (continued)

```
status = CallUSBDI( dx, urb);
// Check statuses
if( !NT_SUCCESS(status) || !USBD_SUCCESS( urb->UrbHeader.Status))
{
    DebugPrint("status %x URB status %x", status, urb->UrbHeader.Status);
    status = STATUS_UNSUCCESSFUL;
}
else
{
    // Select config worked
    DebugPrintMsg("Select configuration worked");
    dx->UsbConfigurationHandle =
        urb->UrbSelectConfiguration.ConfigurationHandle;
    // Find pipe handle of first pipe,
    // ie interrupt pipe that returns input HID reports
    PUSBD_INTERFACE_INFORMATION InterfaceInfo =
        &urb->UrbSelectConfiguration.Interface;
    DebugPrint("interface Class %d NumberOfPipes %d",
        InterfaceInfo->Class, InterfaceInfo->NumberOfPipes);
    if( InterfaceInfo->NumberOfPipes>0)
    {
        PUSBD_PIPE_INFORMATION pi = &InterfaceInfo->Pipes[0];
        dx->UsbPipeHandle = pi->PipeHandle;
        DebugPrint("PipeHandle = %x", dx->UsbPipeHandle);
        DebugPrint("Pipes[0] EndpointAddress %2x"
            "Interval %dms PipeType %d MaximumTransferSize %d",
                pi->EndpointAddress, pi->Interval, pi->PipeType,
                pi->MaximumTransferSize);
    }
    if( dx->UsbPipeHandle==NULL)
        status = STATUS_UNSUCCESSFUL;
}

FreeIfAllocated(urb);
FreeIfAllocated(Descriptors);
return status;
}
```

Other Initialization

As mentioned earlier, having selected your configuration and interface, there may be some more steps needed to initialize your USB peripheral. You may want to read other descriptors. The UsbKbd Create IRP handler shows how this might be done by getting the first string descriptor. This in fact just returns the language ID that the device supports. You will have to issue further *Get Descriptor* requests to get each String Descriptor. The UsbGetSpecifiedDescriptor routine in Usb.cpp can be used to get any descriptor. This routine is also used by the IOCTL_USBKBD_GET_SPECIFIED_DESCRIPTOR handler. The *UsbKbdTest* test program issues this IOCTL to get the HID Descriptor.

The UsbKbd Create IRP handler, UsbKbdCreate, also retrieves some general information about the USB bus, such as the amount of bandwidth used and the host controller device name. The required Internal IOCTLs are only implemented in Windows 2000, so preprocessor directives are used to remove the body of the function that does this job, UsbGetUsbInfo, when compiled for Windows 98.

Most USB device drivers will now want to talk to their devices over control or other pipes. Shortly, I describe how to perform such transfers.

Deselecting a Configuration

When a driver wants to stop accessing its device, it should deselect its configuration. In UsbKbd, the Close file handle IRP handler does this job. The UsbDeselectConfiguration routine uses UsbBuildSelectConfigurationRequest to build a *Select Configuration* URB with a NULL configuration descriptor. This disables the configuration and its interfaces.

Interrupt Transfers

The specification for UsbKbd says that it reports raw keyboard data in response to ReadFile Win32 requests. It also must implement a time-out, which defaults to 10 seconds.

Before I describe how to handle the Read IRPs, I must discuss how a USB HID keyboard produces data. The keyboard responds to USB Interrupt transfers with an 8-byte data report. The exact format of this block is discussed in the next chapter on Human Input Devices (HID).

A keyboard report is produced whenever a keypress or release occurs. A report is also produced regularly even if no keypresses occur. This is what happens on the USB bus. The USB class drivers initiate an Interrupt transfer regularly. My USB keyboard specifies that interrupt transfers should occur every 8ms. However, USB HID keyboards implement an idle rate. If there are no state changes during the idle period, the keyboard NAKs each Interrupt request. At the end of the idle period (every 500ms or so), the keyboard returns data regardless. The idle rate can be read and changed using class specific control transfers on the default pipe[3].

UsbKbd issues a *Do Bulk or Interrupt Transfer* URB to receive any keyboard reports. Keyboard reports are not "saved up" ready to fulfil any interrupt transfer requests. It is not clear whether the USB class drivers simply do not do any transfers, or whether they do the transfers but just dump the data. Any keypresses before an interrupt transfer is requested are ignored. I understand that the Microsoft keyboard drivers always try to leave two Interrupt transfer requests outstanding; the hope is that this should ensure that no keypresses are lost.

It turns out that the 8-byte keyboard report is all zeroes if no keys are pressed. The UsbKbd Read IRP handler ignores any keyboard reports that are all zero. A real driver would return read data when any key is pressed or released. A higher-level driver would have to implement an auto-repeat feature.

---

3. The UsbGetIdleRate routine shows how to read the idle rate.

I can finally describe the Interrupt transfer routine, UsbDoInterruptTransfer, shown in Listing 21.6. This takes a buffer as input that must be at least 8 bytes long. UsbDoInterruptTransfer first checks that the USB pipe handle is available and that the input buffer is suitable.

In the same way as usual, memory is allocated for the URB, a _URB_BULK_OR_INTERRUPT_ TRANSFER structure. UsbDoInterruptTransfer then loops until a non-zero keyboard report is returned, or until a time-out or another error occurs. UsbBuildInterruptOrBulkTransferRequest is used to build the URB each time[4]. UsbDoInterruptTransfer calls the USB class drivers in the usual way.

The time-out is checked using the KeQueryTickCount call[5]. This returns a ULONG count of "timer interrupts" since the operating system started. The timer interrupt interval is not the same in Windows 98 and Windows 2000. Use KeQueryTimeIncrement to find out this interval in units of 100ns. In both cases, there are just over 100 timer ticks per second.

UsbDoInterruptTransfer remembers the tick count before the first keyboard Interrupt request is generated. After each Interrupt request completes, it checks the tick count again, and gives up after the specified time-out.

UsbKbd does not handle IRP cancelling or the Cleanup routine. UsbKbd relies on the time-out to complete Read IRPs. If you do implement IRP cancelling, then you will need to use the *Abort Pipe* URB request to cancel all requests on the specified port.

Listing 21.6 Doing interrupt transfers

```
NTSTATUS UsbDoInterruptTransfer( IN PUSBKBD_DEVICE_EXTENSION dx,
                                 IN PVOID UserBuffer, ULONG& UserBufferSize)
{
    // Check we're selected
    if( dx->UsbPipeHandle==NULL)
        return STATUS_INVALID_HANDLE;


    // Check input parameters
    ULONG InputBufferSize = UserBufferSize;
    UserBufferSize = 0;
    if( UserBuffer==NULL || InputBufferSize<8)
        return STATUS_INVALID_PARAMETER;


    // Keyboard input reports are always 8 bytes long
    NTSTATUS status = STATUS_SUCCESS;
    ULONG OutputBufferSize = 8;


    // Allocate memory for URB
    USHORT UrbSize = sizeof(struct _URB_BULK_OR_INTERRUPT_TRANSFER);
```

4. Note that the call to UsbBuildInterruptOrBulkTransferRequest must be in the loop for it to work in W2000.

---

5. Using KeQuerySystemTime is less efficient.

Listing 21.6 Doing interrupt transfers (continued)

```
PURB urb = (PURB)ExAllocatePool(NonPagedPool, UrbSize);
if( urb==NULL)
{
    DebugPrintMsg("No URB memory");
    return STATUS_INSUFFICIENT_RESOURCES;
}


// Remember when we started
// Get start tick count and length of tick in 100ns units
LARGE_INTEGER StartTickCount;
KeQueryTickCount( &StartTickCount);
ULONG UnitsOf100ns = KeQueryTimeIncrement();

// Loop until non-zero report read, error, bad length, or timed out
while( true)
{
    // Build Do Bulk or Interrupt transfer request
    UsbBuildInterruptOrBulkTransferRequest(
        urb, UrbSize,
        dx->UsbPipeHandle,
        UserBuffer, NULL, OutputBufferSize,
        USBD_TRANSFER_DIRECTION_IN,
        NULL);

    // Call the USB driver
    status = CallUSBDI( dx, urb);
    // Check statuses
    if( !NT_SUCCESS(status) || !USBD_SUCCESS( urb->UrbHeader.Status))
    {
        DebugPrint("status %x URB status %x", status,
            urb->UrbHeader.Status);
        status = STATUS_UNSUCCESSFUL;
        break;
    }

    // Give up if count of bytes transferred was not 8
    if( urb->UrbBulkOrInterruptTransfer.TransferBufferLength!=
        OutputBufferSize)
        break;

    // If data non-zero then exit as we have a keypress
    __int64* pData = (__int64 *)UserBuffer;
```

Listing 21.6 Doing interrupt transfers (continued)

```
        if( *pData!=0i64)
            break;


        // Check for time-out
        LARGE_INTEGER TickCountNow;
        KeQueryTickCount( &TickCountNow);
        ULONG ticks = (ULONG)(TickCountNow.QuadPart - StartTickCount.QuadPart);
        if( ticks*UnitsOf100ns/10000000 >= dx->UsbTimeout)
        {
            DebugPrint("Time-out %d 100ns", ticks*UnitsOf100ns);
            status = STATUS_NO_MEDIA_IN_DEVICE;
            break;
        }
    }
    UserBufferSize = urb->UrbBulkOrInterruptTransfer.TransferBufferLength;

    if( NT_SUCCESS(status))
    {
        PUCHAR bd = (PUCHAR)UserBuffer;
        DebugPrint("Transfer data %2x %2x %2x %2x %2x %2x %2x %2x",
            bd[0], bd[1], bd[2], bd[3], bd[4], bd[5], bd[6], bd[7]);
    }

    ExFreePool(urb);
    return status;
}
```

Control Transfers

The UsbKbd driver lets its controlling application modify the state of the keyboard LEDs. This lets me illustrate how to do Control transfers on the default pipe to endpoint zero. The Write IRP handler sends the first byte of the write buffer to the keyboard as a HID "output report". Again, the next chapter fully defines the format of this report and the SET_REPORT command. The lower three bits of the report correspond to the NumLock, CapsLock, and ScrollLock keys.

The Write IRP handler calls UsbSendOutputReport to send the output report. As per usual, it allocates a suitable URB. The UsbBuildVendorRequest helper function is used to fill this URB. Control transfers to the default pipe can take many shapes and forms. First, there are "class" and "vendor" types. For each type, you can specify whether the destination is the "device", "interface", "endpoint", or "other".

The HID Specification tells us to send a "class" control transfer request to the "interface". The call to UsbBuildVendorRequest, therefore, uses the URB_FUNCTION_CLASS_INTERFACE URB function code. The TransferFlags parameter specifies the direction of the data transfer; set the USBD_TRANSFER_DIRECTION_IN flag for input transfers. The Request, Value, and Index parameters are set as instructed by the HID Specification. Finally, the output data buffer is given.

The URB is then sent off to the class drivers. Apart from noting the URB status and freeing the URB memory, there is nothing else to do.

The UsbGetIdleRate routine in Usb.cpp shows how to input data using a control transfer over the default pipe. This issues a GET_IDLE request. The current keyboard idle rate is printed in a *DebugPrint* trace statement.

Listing 21.7 Doing an output control transfer

```
const UCHAR SET_REPORT = 0x09;

NTSTATUS UsbSendOutputReport( IN PUSBKBD_DEVICE_EXTENSION dx,
    IN UCHAR OutputData)
{
    // Allocate memory for URB
    USHORT UrbSize = sizeof(struct _URB_CONTROL_VENDOR_OR_CLASS_REQUEST);
    PURB urb = (PURB)ExAllocatePool(NonPagedPool, UrbSize);
    if( urb==NULL)
    {
        DebugPrintMsg("No URB memory");
        return STATUS_INSUFFICIENT_RESOURCES;
    }

    // Build URB to send Class interface control request on Default pipe
    UsbBuildVendorRequest(urb,
        URB_FUNCTION_CLASS_INTERFACE, UrbSize,
        USBD_TRANSFER_DIRECTION_OUT,      // Direction out
        0,                                // Reserved bits
        SET_REPORT,                       // Request
        0x0200,                           // Output report type, Report id zero
        0,                                // interface index
        &OutputData, NULL, 1,             // Output data
        NULL);

    // Call the USB driver
    DebugPrintMsg("Sending set report");
    NTSTATUS status = CallUSBDI( dx, urb);
    // Check statuses
    if( !NT_SUCCESS(status) || !USBD_SUCCESS( urb->UrbHeader.Status))
    {
        DebugPrint("status %x URB status %x", status, urb->UrbHeader.Status);
        status = STATUS_UNSUCCESSFUL;
```

Listing 21.7 Doing an output control transfer (continued)

```
    }
    ExFreePool(urb);
    return status;
}
```

Other Issues

The UsbGetStatuses routine shows how to read status words from the device, interface, and endpoint using the *Get Status* URB. These words are bundled into a 6-byte buffer in response to the UsbKbd IOCTL_USBKBD_GET_STATUSES request.

The UsbKbd IOCTL_USBKBD_GET_FRAME_INFO request is handled in the UsbGetFrameInfo routine. UsbGetFrameInfo uses the *Get Current Frame Number* and *Get Frame Length* URBs to find out the required information. Both these URBs are built by hand, as there are no Usb-Build...Request helper macros. The frame numbers returned are 32-bit ULONGs, not the USB SOF number that goes from 0–0x7FF.

Testing UsbKbd

The Win32 *UsbKbdTest* program puts the UsbKbd driver through its paces. Obviously, you must have a USB-enabled PC and a USB keyboard to try out this code. For the benefit of those who do not, I am reproducing both the *UsbKbdTest* output and the UsbKbd *DebugPrint* trace output in Listings 21.8 and 21.9.

You will first have to "install" the UsbKbd.sys driver by copying it into the Windows System32\drivers directory to replace HidUsb.sys. Do not forget to put the old HidUsb.sys back when you are finished.

The *UsbKbdTest* program does the following jobs.

1. Open a handle to the first driver that supports the USBKBD_GUID device interface.
2. Read the device descriptor.
3. Read the configuration descriptor, interface descriptor, etc. The *UsbKbdTest* output has been amended in Listing 21.8 to show each descriptor.
4. Read the HID report descriptor, type 0x22.
5. Set the read time-out to 15 seconds.
6. Keep reading keyboard input data until the Esc key is pressed (code 0x29). The output listing has been annotated to show when I have pressed these keys: Ctr+Alt+Del, then A, B, C and Esc.
7. Write a series of bit combinations to flash the keyboard LEDs, with a 1/3-second delay between changes.
8. Read the device, interface, and endpoint status words.
9. Read the frame length and frame numbers.
10. Close the file handle.

Armed with all the relevant specifications, you can decode what all the descriptors and keyboard data mean. The UsbKbd trace output lists some more useful information.

Listing 21.8 *UsbKbdTest* output

```
Test 1
Symbolic link is \\?\usb#vid_046a&pid_0001#7&4#{c0cf0646-5f6e-11d2-b677-00c0dfe4c1f3}
    Opened OK


Test 2
    Device descriptor is 12 01 00 01 00 00 00 08 6A 04 01 00 05 03 00 00 00 01


Test 3
    Configuration descriptors are
        09 02 22 00 01 01 00 A0 32     Configuration descriptor
        09 04 00 00 01 03 01 01 00     Interface descriptor
        07 05 81 03 08 00 08           Endpoint descriptor
        09 21 00 01 00 01 22 3F 00     HID descriptor


Test 4
     HID Report descriptor is 05 01 09 06 A1 01 05 07 19 E0 29 E7 15 00 25 01 75 01 95 08
81 02 75 08 95 01 81 01 05 08 19 01 29 03 75 01 95 03 91 02 75 05 95 01 91 01 05 07 19 00
29 65 15 00 25 65 75 08 95 06 81 00 C0


Test 5
    Read time-out set


Test 6
    Kbd report  1  0  0  0  0  0  0  0     Ctrl
    Kbd report  5  0  0  0  0  0  0  0     Ctrl+Alt
    Kbd report  5  0 63  0  0  0  0  0     Ctrl+Alt+Del
    Kbd report  4  0  0  0  0  0  0  0     Alt
    Kbd report  0  0  4  0  0  0  0  0     A
    Kbd report  0  0  5  0  0  0  0  0     B
    Kbd report  0  0  6  0  0  0  0  0     C
    Kbd report  0  0 29  0  0  0  0  0     Esc


Test 7
    Wrote  1 OK
    Wrote  2 OK
    Wrote  3 OK
    Wrote  4 OK
    Wrote  5 OK
    Wrote  6 OK
    Wrote  7 OK
    Wrote ff OK
```

```
        Wrote  0 OK

Test 8
        Statuses are 0 0 0


Test 9
        FrameLength 11999 bits.  FrameNumber 1847715.  FrameAlterNumber 1847715


Test 10
        CloseHandle worked
```

Listing 21.9 *UsbKbdTest* corresponding *DebugPrint* trace output in W2000

```
CreateFile
16:34:09.567  Create File is
16:34:09.567  USBDI version 00000200 Supported version 00000100
16:34:09.567  Getting port status
16:34:09.567  Got port status 00000003
16:34:09.567  Getting basic configuration descriptor
16:34:09.567  Got basic config descr.  MaxPower 50 units of 2mA
16:34:09.567  Getting full configuration descriptors
16:34:09.577  Selecting configuration
16:34:09.577  Select configuration worked
16:34:09.577  interface Class 3 NumberOfPipes 1
16:34:09.577  PipeHandle = 80973FC0
16:34:09.577  Pipes[0] EndpointAddress 81 Interval 8ms
              PipeType 3 MaximumTransferSize 4096
16:34:09.577  Getting bus info
16:34:09.577  Bus info: TotalBandwidth 12000,
              ConsumedBandwidth 0 and ControllerNameLength 32
16:34:09.577  Controller name is \DosDevices\HCD0
16:34:09.577  Getting descriptor type 03
16:34:09.577  No string descriptors
16:34:09.577  Sending Get Idle request
16:34:09.577  Idle rate is 125 units of 4ms
Get device descriptor using IOCTL_USBKBD_GET_DEVICE_DESCRIPTOR
16:34:09.577  DeviceIoControl: Control code 00222008
              InputLength 0 OutputLength 50
16:34:09.577  Getting device descriptor
16:34:09.577  DeviceIoControl: 18 bytes written
Get all configuration descriptors using
IOCTL_USBKBD_GET_CONFIGURATION_DESCRIPTORS
16:34:09.587  DeviceIoControl: Control code 0022200C
              InputLength 0 OutputLength 500
```

Listing 21.9 *UsbKbdTest* corresponding *DebugPrint* trace output in W2000 (continued)

```
16:34:09.587   Getting basic configuration descriptor
16:34:09.587   Got basic config descr.  MaxPower 50 units of 2mA
16:34:09.587   Getting full configuration descriptors
16:34:09.587   DeviceIoControl: 34 bytes written
Get HID Report Descriptor using IOCTL_USBKBD_GET_SPECIFIED_DESCRIPTOR
16:34:09.597   DeviceIoControl: Control code 00222010
               InputLength 8 OutputLength 500
16:34:09.597   Getting descriptor type 22
16:34:09.597   DeviceIoControl: 63 bytes written
Set read timeout using IOCTL_USBKBD_SET_READ_TIMEOUT
16:34:09.617   DeviceIoControl: Control code 00222004
               InputLength 4 OutputLength 0
16:34:09.617   USB timeout set to 15
16:34:09.617   DeviceIoControl: 0 bytes written
Read Ctrl+Alt+Del key presses
16:34:09.617   Read 8 bytes from file pointer 0
16:34:13.022   Transfer data 04 00 00 00 00 00 00 00
16:34:13.022   Read: 00000000 8 bytes returned
16:34:13.022   Read 8 bytes from file pointer 0
16:34:13.112   Transfer data 05 00 00 00 00 00 00 00
16:34:13.112   Read: 00000000 8 bytes returned
16:34:13.112   Read 8 bytes from file pointer 0
16:34:13.222   Transfer data 05 00 63 00 00 00 00 00
16:34:13.222   Read: 00000000 8 bytes returned
16:34:13.292   Read 8 bytes from file pointer 0
16:34:13.993   Transfer data 04 00 00 00 00 00 00 00
16:34:13.993   Read: 00000000 8 bytes returned
Read A, B, C, Esc key presses
16:34:14.023   Read 8 bytes from file pointer 0
16:34:16.757   Transfer data 00 00 04 00 00 00 00 00
16:34:16.757   Read: 00000000 8 bytes returned
16:34:16.777   Read 8 bytes from file pointer 0
16:34:17.128   Transfer data 00 00 05 00 00 00 00 00
16:34:17.128   Read: 00000000 8 bytes returned
16:34:17.158   Read 8 bytes from file pointer 0
16:34:17.609   Transfer data 00 00 06 00 00 00 00 00
16:34:17.609   Read: 00000000 8 bytes returned
16:34:17.639   Read 8 bytes from file pointer 0
16:34:17.859   Transfer data 00 00 29 00 00 00 00 00
16:34:17.859   Read: 00000000 8 bytes returned
```

Listing 21.9 *UsbKbdTest* corresponding *DebugPrint* trace output in W2000 (continued)

```
Write different bit combinations to flash the keyboard LEDs
16:34:17.969  Write 1 bytes from file pointer 0
16:34:17.969  Sending set report
16:34:17.969  Write: 1 bytes written
...
16:34:21.124  Write 1 bytes from file pointer 0
16:34:21.124  Sending set report
16:34:21.144  Write: 1 bytes written
Get statuses using IOCTL_USBKBD_GET_STATUSES
16:34:21.524  DeviceIoControl: Control code 00222014
              InputLength 0 OutputLength 6
16:34:21.524  Getting device status
16:34:21.524  Getting interface status
16:34:21.524  Getting endpoint status
16:34:21.524  DeviceIoControl: 6 bytes written
Read frame length and numbers using IOCTL_USBKBD_GET_FRAME_INFO
16:34:21.524  DeviceIoControl: Control code 00222018
              InputLength 0 OutputLength 12
16:34:21.524  Getting current frame number
16:34:21.524  FrameNumber 912595
16:34:21.524  Getting frame info
16:34:21.524  FrameLength 11999 FrameAlterNumber 912595
16:34:21.524  DeviceIoControl: 12 bytes written
CloseHandle
16:34:21.634  Close
16:34:21.634  Deselecting configuration
```

USBDI Structure Reference

This section provides a reference for the USB Driver Interface (USBDI). The basic URB structure has been covered in enough detail in earlier sections of this chapter. Therefore, I first describe various other USBDI structures. Then, I go on to detail each of the URB function codes, the corresponding URB structures and any useful build routines.

Structures

Device Descriptor

A device descriptor is returned in a USB_DEVICE_DESCRIPTOR structure. Table 21.4 shows the fields of interest in the device descriptor. The device class, subclass, and protocol fields are not used by Windows and are often zeroes. However, they may be used to indicate which new common class specification features are used.

Table 21.4   USB_DEVICE_DESCRIPTOR **fields**

bDeviceClass bDeviceSubClass bDeviceProtocol	The USB class codes
idVendor idProduct bcdDevice	Vendor, product, and version numbers
iManufacturer iProduct iSerialNumber	Indexes into the string descriptor for the relevant strings
bMaxPacketSize0	The maximum packet size for Endpoint 0
bNumConfigurations	The number of configurations

Configuration Descriptor

A configuration descriptor is returned in a USB_CONFIGURATION_DESCRIPTOR structure, with fields as shown in Table 21.5.

A request for a configuration descriptor will also get any other descriptors associated with this configuration (interface-, endpoint-, class-, and vendor-defined descriptors), if there is enough room in your buffer. You will usually get the configuration descriptor twice. First, get the basic USB_CONFIGURATION_DESCRIPTOR structure. The *wTotalLength* field tells you how big a buffer to allocate to retrieve all the descriptors in your next get configuration request.

Table 21.5   USB_CONFIGURATION_DESCRIPTOR **fields**

wTotalLength	Total length of all data for the configuration
bNumInterfaces	Number of interfaces
iConfiguration	Configuration number
bmAttributes	bit 5 set: supports remote wake up bit 6 set: self-powered bit 7 set: powered from bus
maxPower	Maximum power required, in 2mA units

Interface Descriptor

An interface descriptor is returned in a USB_INTERFACE_DESCRIPTOR structure, with fields shown Table 21.6. Interface descriptors are returned as a part of a get configuration descriptor request.

Table 21.6 USB_INTERFACE_DESCRIPTOR **fields**

bInterfaceNumber	Interface number
bAlternateSetting	Alternate setting
bNumEndpoints	Number of endpoints
bInterfaceClass bInterfaceSubClass bInterfaceProtocol	USB class codes
iInterface	Index of string descriptor describing the interface

Interface Selection Structures

Various structures are needed when you select a configuration. An array of USB_INTERFACE_ LIST_ENTRY structures, shown in Table 21.7, is passed to USBD_CreateConfigurationRequestEx. The *InterfaceDescriptor* field points to the interface descriptor that you want enabled in the configuration. When the *Select Configuration* URB has completed, the *Interface* pointer is valid and refers to valid data.

Table 21.7 USB_INTERFACE_LIST_ENTRY **fields**

InterfaceDescriptor	Points to interface descriptor
Interface	USB_INTERFACE_INFORMATION pointer

The *Select Configuration* URB fills a USB_INTERFACE_INFORMATION structure for each interface that it enables. Table 21.8 shows this structure, whose size increases as there are more pipes.

Table 21.8 USB_INTERFACE_INFORMATION **fields**

InterfaceNumber	Interface number
AlternateSetting	Alternate setting
Class SubClass Protocol	USB class codes
NumberOfPipes	Number of pipes
PipeInformation	Array of USB_PIPE_INFORMATION structures

Table 21.9 shows the USB_PIPE_INFORMATION structure. All fields apart from *MaximumTransferSize* and *PipeFlags* fields are filled in by the *Select Configuration* URB.

Table 21.9   `USB_PIPE_INFORMATION` **fields**

MaximumPacketSize	Maximum packet size
EndpointAddress	Endpoint number The top bit is set for input pipes.
Interval	Polling interval in ms for Interrupt pipes
PipeType	UsbdPipeTypeControl UsbdPipeTypeIsochronous UsbdPipeTypeBulk UsbdPipeTypeInterrupt
PipeHandle	Handle for later USBDI calls
MaximumTransferSize	Maximum transfer size
PipeFlags	Flags

String Descriptor

A device may optionally provide string descriptors. Each string descriptor contains just one Unicode string, so the driver must request each string one by one. String index zero is special. Instead of a string, it contains a list of two-byte language ID codes. Language ID 0x0009 is usually used for generic English.

As has been seen, various other descriptors contain string index numbers. If these numbers are non-zero, there is an associated string. When you request a string descriptor using the *Get Descriptor* URB, you must specify both the string index and the language ID that you want.

A `USB_STRING_DESCRIPTOR` structure simply has one useful field, *bString*, that contains the Unicode string for which you asked. The maximum string length (`MAXIMUM_USB_STRING_LENGTH`) is 255 characters. When you request string index zero, *bString* instead contains the supported language ID.

Isochronous Packet Descriptor

The `USBD_ISO_PACKET_DESCRIPTOR` structure is used to describe an isochronous transfer packet. An isochronous transfer request URB has a single transfer buffer that may contain several individual packets.

For writes, set the *Offset* field of each packet descriptor to indicate the offset into this buffer of this packet.

For reads, the packet descriptor *Length* field is filled with the number of bytes read and the *Status* field returns the status of this transfer packet.

USBDI URB Reference

URB Setup Functions

All these URBs use standard USB control transfers over the default pipe. I will later explain the details of how to set or clear features, or do class or vendor defined control transfers over the default pipe.

Get Descriptor

Function codes	URB_FUNCTION_GET_DESCRIPTOR_FROM_DEVICE
Build routine	UsbBuildGetDescriptorRequest
URB Structure	_URB_CONTROL_DESCRIPTOR_REQUEST

Use this function to get any descriptor from a device. The *DescriptorType* URB field determines which descriptor is retrieved. A suitably sized buffer must be provided. If asking for a configuration descriptor, specify an appropriate value for the URB *Index* field. If you are getting a string descriptor, specify a *LanguageId* field, as well.

If you ask for a configuration descriptor, you also get all the associated interface, endpoint, and class- and vendor-defined descriptors.

Other Get/Set Descriptors

Function codes	URB_FUNCTION_GET_DESCRIPTOR_FROM_ENDPOINT URB_FUNCTION_SET_DESCRIPTOR_TO_ENDPOINT URB_FUNCTION_SET_DESCRIPTOR_TO_OTHER
Build routine	None available
URB Structure	_URB_CONTROL_DESCRIPTOR_REQUEST

This function gets or sets other descriptors.

Get Status

Function codes	URB_FUNCTION_GET_STATUS_FROM_DEVICE URB_FUNCTION_GET_STATUS_FROM_INTERFACE URB_FUNCTION_GET_STATUS_FROM_ENDPOINT URB_FUNCTION_GET_STATUS_FROM_OTHER
Build routine	UsbBuildGetStatusRequest
URB Structure	_URB_CONTROL_GET_STATUS_REQUEST

This function gets a status word.

The device status word has self-powered and remote wakeup bits (USB_GETSTATUS_SELF_POWERED and USB_GETSTATUS_REMOTE_WAKEUP_ENABLED are the bit masks). The interface status currently has no bits defined. The endpoint status word has its bit zero set if the Halt feature has been set.

Select Configuration

Function codes	URB_FUNCTION_SELECT_CONFIGURATION
Build routine	USBD_CreateConfigurationRequestEx UsbBuildSelectConfigurationRequest
URB Structure	_URB_SELECT_CONFIGURATION

Use USBD_CreateConfigurationRequestEx to select a configuration and an interface, as described earlier.

If deselecting a configuration, use the simpler UsbBuildSelectConfigurationRequest to pass a NULL configuration descriptor.

Get Configuration

Function codes	URB_FUNCTION_GET_CONFIGURATION
Build routine	None available
URB Structure	_URB_CONTROL_GET_CONFIGURATION_REQUEST

This function gets the current configuration descriptors.

Select Alternate Interface

Function codes	URB_FUNCTION_SELECT_INTERFACE
Build routine	UsbBuildSelectInterfaceRequest
URB Structure	_URB_SELECT_INTERFACE

Use this URB function code to select an interface's alternate setting. Pass the configuration handle, the interface number, and the alternate setting.

Changing to an alternate interface setting discards any queued data on the interface end points.

Get Interface

Function codes	URB_FUNCTION_GET_INTERFACE
Build routine	None available
URB Structure	_URB_CONTROL_GET_INTERFACE_REQUEST

This function gets the current alternate interface setting for an interface in the current configuration.

Reset Pipe

Function codes	URB_FUNCTION_RESET_PIPE
Build routine	None available
URB Structure	_URB_HEADER

This function can clear a stall condition on the pipe with the given *PipeHandle*.

URB Transfer Functions

Do Control Transfer

Function codes	URB_FUNCTION_CONTROL_TRANSFER
Build routine	None available
URB Structure	_URB_CONTROL_TRANSFER

This function can transmit or receive data on a control pipe with the given *PipeHandle*. Fill the *SetupPacket* 8-byte array with the information for the SETUP packet. The *TransferBuffer* fields specify the extra data. The USBD_TRANSFER_DIRECTION_IN bit in the *TransferFlags* field specifies the direction in which the extra data flows. Specify the USBD_SHORT_TRANSFER_OK flag bit if a short input transfer is acceptable.

Do not use this function for transfers over the default pipe.

Do Bulk or Interrupt Transfer

Function codes	URB_FUNCTION_BULK_OR_INTERRUPT_TRANSFER
Build routine	UsbBuildInterruptOrBulkTransferRequest
URB Structure	_URB_BULK_OR_INTERRUPT_TRANSFER

Use this function to transmit or receive data on a bulk pipe, or receive data on an interrupt pipe. Specify the pipe using the *PipeHandle* field. The *TransferBuffer* fields specify the data. The USBD_TRANSFER_DIRECTION_IN bit in the *TransferFlags* field specifies the direction in which the data flows. Specify the USBD_SHORT_TRANSFER_OK flag bit if a short input transfer is acceptable.

Do Isochronous Transfer

Function codes	URB_FUNCTION_ISOCH_TRANSFER
Build routine	None available
URB Structure	_URB_ISOCH_TRANSFER

Use this function to request an isochronous transfer. The *TransferBuffer* fields specify the data. The data block is split into one or more packets. The number of packets is given in

NumberOfPackets. IsoPacket is an array of USBD_ISO_PACKET_DESCRIPTOR structures. Each of these specifies where a packet is in the transfer buffer (i.e., the *Offset* into the buffer, the *Length,* and the *Status* on return of the packet).

GET_ISO_URB_SIZE returns the number of bytes required to hold an isochronous request of the given number of packets.

If the START_ISO_TRANSFER_ASAP bit is set in *TransferFlags* field, the transfer is started as soon as possible. Otherwise, specify the 32-bit ULONG *StartFrame* number on which you want the transfer to start, within 1,000 frames of the current frame. The USBD_TRANSFER_DIRECTION_IN flag bit specifies the direction of data transfer.

Cancel

Function codes	URB_FUNCTION_ABORT_PIPE
Build routine	None available
URB Structure	_URB_PIPE_REQUEST

This function cancels all requests on the pipe with the given *PipeHandle*.

URB Default Pipe Functions

Set/Clear Feature

Function codes	URB_FUNCTION_SET_FEATURE_TO_DEVICE URB_FUNCTION_SET_FEATURE_TO_INTERFACE URB_FUNCTION_SET_FEATURE_TO_ENDPOINT URB_FUNCTION_SET_FEATURE_TO_OTHER URB_FUNCTION_CLEAR_FEATURE_TO_DEVICE URB_FUNCTION_CLEAR_FEATURE_TO_INTERFACE URB_FUNCTION_CLEAR_FEATURE_TO_ENDPOINT URB_FUNCTION_CLEAR_FEATURE_TO_OTHER
Build routine	UsbBuildFeatureRequest
URB Structure	_URB_CONTROL_FEATURE_REQUEST

This function sets or clears a feature, in a control transfer over the default pipe. Specify a feature code.

Vendor/Class

Function codes	URB_FUNCTION_VENDOR_DEVICE URB_FUNCTION_VENDOR_INTERFACE URB_FUNCTION_VENDOR_ENDPOINT URB_FUNCTION_VENDOR_OTHER URB_FUNCTION_CLASS_DEVICE URB_FUNCTION_CLASS_INTERFACE URB_FUNCTION_CLASS_ENDPOINT URB_FUNCTION_CLASS_OTHER
Build routine	UsbBuildVendorRequest
URB Structure	_URB_CONTROL_VENDOR_OR_CLASS_REQUEST

Use this function to issue a vendor or class-specific command to a device, interface, endpoint, or other device-defined target, in a control transfer over the default pipe.

The *TransferBuffer* fields specify the data. The USBD_TRANSFER_DIRECTION_IN bit in the *TransferFlags* field specifies the direction in which the data flows. Specify the USBD_SHORT_TRANSFER_OK flag bit if a short input transfer is acceptable.

The *ReservedBits* field is usually zero, but if it is a value between 4 and 31 this value is used for bits 4 to 0 in the SETUP packet *bmRequestType* field (see Table 20.3 in the last chapter).

The *Request*, *Value*, *Index*, and *TransferBuffer* fields are set appropriately for the vendor- or class-defined request.

URB Isochronous Frame Functions

Request Control of Frame Length

Function codes	URB_FUNCTION_TAKE_FRAME_LENGTH_CONTROL URB_FUNCTION_RELEASE_FRAME_LENGTH_CONTROL
Build routine	None available
URB Structure	_URB_FRAME_LENGTH_CONTROL

Use this function to request control of the frame length (i.e., SOF mastership).
Only one client can have control of frame length at any time

Get Current Frame Length

Function codes	URB_FUNCTION_GET_FRAME_LENGTH
Build routine	None available
URB Structure	_URB_GET_FRAME_LENGTH

Use this function to get the current frame length in the *FrameLength* field. It also indicates in *FrameNumber* the first frame in which the frame length can be changed.

Set Current Frame Length

Function codes	URB_FUNCTION_SET_FRAME_LENGTH
Build routine	None available
URB Structure	_URB_SET_FRAME_LENGTH

Use this function to specify the amount to change the current frame length (i.e., 1 or -1 in the FrameLengthDelta field).

You must have control of frame length to issue this URB.

Get Current Frame Number

Function codes	URB_FUNCTION_GET_CURRENT_FRAME_NUMBER
Build routine	None available
URB Structure	_URB_GET_CURRENT_FRAME_NUMBER

Use this function to get the current frame number in FrameNumber.

Gets Beginning Frame of Pattern Transfer

Function codes	URB_FUNCTION_SYNC_FRAME
Build routine	None available
URB Structure	_URB_CONTROL_SYNC_FRAME_REQUEST

This function retrieves the beginning frame of a pattern transfer from an isochronous pipe.

This function is documented, but both the function code and the URB structure are not defined in the W98 and W2000 DDK USB headers.

Conclusion

The chapter has looked in detail at the Windows USB Device Interface (USBDI) and shown how to use it to talk to a USB HID keyboard. The USB class drivers make several Internal IOCTLs available to USB client device drivers. The most important of these lets you issue USB Request Blocks (URBs). Use URBs to do most interactions with your USB device.

The next chapter looks at the core theory of Human Input Devices (HID). The following chapter shows how to use a HID device in a client device driver and Win32 application.

Chapter 22

The Human Input Device Model

HID Hides

This chapter describes the Human Input Device (HID) model, a standard way of interacting with user input devices. A HID device uses various descriptors to define its capabilities. A Report descriptor details the input reports that it can generate and the output reports it can receive. The next chapter describes how to write Windows client device drivers and user mode applications that can talk to HID devices.

The HID specification is an abstract model for most types of input device that people will use to control their computers. An input device can be a plain old keyboard, for example, a vehicle simulation rudder, or the soft on/off button for the computer.

The HID specification was written originally for USB devices and closely follows the USB descriptor model. However, it is sufficiently generic for HID to be used with other types of device.

Naturally, most of the time a HID device provides *input* data to the computer. However, you can *output* to a HID device (e.g., to turn on the NumLock LED on a keyboard). You can also control a *feature* of the device, such as the font used on a display device or an LED color.

In Windows, the system HID class driver provides an abstract view of the input device. The HID device itself can be a USB device, a IEEE 1394 device or even a plain PC-compatible device, provided an appropriate HID minidriver is written to interface to the bus or device.

In this chapter and the next, quite a few of the examples are about getting input from a keyboard. It does not matter whether you are using a USB keyboard or a (hypothetical)

IEEE 1394 keyboard. The system still gets keystrokes in exactly the same way, from the HID class driver. As the next chapter shows, Windows does get input from an old PC-compatible keyboard by a different route. Perhaps eventually a minidriver will be written so that HID will provide all the keyboard input.

HID in Windows

Figure 22.1 shows how HID devices are used in Windows. A user mode or kernel mode **HID** *client* program calls the HID device stack. The stack always contains the system *HID class driver*, Hidclass.sys, optionally with HID filter drivers above it. The HID class driver uses HID minidrivers to interface to hardware. In the figure, one of the minidrivers, HidUsb.sys, uses the USB stack to talk to a device.

Figure 22.1   HID drivers and clients

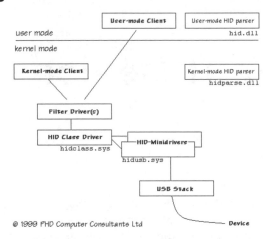

@ 1999 PHD Computer Consultants Ltd

HID provides an abstract model of input devices, so client programs are not concerned with the details of where input comes from. HID models a device as various *controls* representing the device. A keyboard, therefore, usually has single-bit control representing the state of each modifier key, such as the left Shift key, and it has an array of scan code byte controls representing the other keys that are currently pressed.

Underlying the HID model are compact, but complicated, structures for passing information. HID device engineers and minidriver writers will need to know these structures. However, clients can just use various Windows parsing routines to access the relevant data. There are parsers for HID reports available to both kernel mode, Hidparse.dll, and user mode clients, Hid.dll.

The system HID class driver does most of the hard work. It delegates its hardware interactions to HID minidrivers. Windows includes a HID minidriver HidUsb.sys for the devices on the USB bus, but new minidrivers can be written, if necessary. New USB devices that have the appropriate HID Interface class constants and HID descriptors will not need a new minidriver. The system USB HID minidriver calls the system USB driver USBD.sys, optionally through any lower-level USB filter drivers. The system USB driver then talks to the device, etc.

A HID client, therefore, just talks to the HID class driver. The Windows operating system itself is an example of a kernel client, as it can get its keyboard and mouse input via HID. You can write kernel mode client drivers to use a HID device. However, it is often easier to write a Win32 program that talks directly to the HID class driver.

The HID Model

This section is a summary of the HID Model, particularly as seen by Windows device drivers and clients users. For full details, please read the HID specifications.

The HID specification documents are found on the USB website, `www.usb.org/develop-ers/`. The "USB: Device Class Definition for HID" document `usbhid10.pdf` is referred to as the "HID Specification". The USB Compatibility Test software, also available there, includes the *HID View* program to inspect and test HID devices. The "Ignore hubs (Memphis only)" box should be checked for Windows 98. The USB website also contains a tool for generating report descriptors, the *Descriptor Tool dt*. This contains several useful example report descriptors, as well as letting you build your own and check them for errors.

Reports

A computer primarily reads *input* data from a HID device. It can also send *output* data to it, and send and receive *feature* data.

Input data might represent a mouse movement or a keyboard keypress. Output data might mean setting the NumLock LED on a keyboard. A feature is something that controls the general operation of a device, such as the display character size. A feature is usually set by a Control Panel type application and not used by ordinary HID clients.

An individual transaction with a device is called a *report*, which groups several *controls* together. An input report from a keyboard says which keys are pressed and which modifier keys are pressed (i.e., Shift, etc.).

A control is a value of one or more bits. Control values can be converted from *logical* to *physical* form (i.e., scaled and given units). (Windows refers to some controls as *buttons*, while it calls others *values*.).

I knew those MSDN CDs would come in useful.

Reports may be grouped together in *Collections*. An *Application* top-level collection usually describes all the reports that a device can produce. More than one Application collection can be given. For example, if a keyboard has a built-in pointer (i.e., a mouse), it might well have an Application collection for a keyboard report and an Application collection for a pointer report. Reports and collections may contain other collections.

Figure 22.2 shows an overview of the Report descriptor for a HID keyboard. An input report contains eight control bits for the modifier keys and six byte values for all the keys that are simultaneously pressed. An output report just contains five output control bits for the keyboard LEDs. The full keyboard report descriptor also contains a "reserved" input control byte and a 3-bit control that pads out the output report.

Figure 22.2 Keyboard report descriptor overview

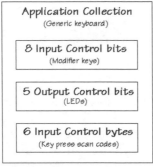

@ 1999 PHD Computer Consultants Ltd

Usages

A HID device needs a standard way to tell programmers what their device is: what Input, Output, and Feature reports it can send and receive. Discovering this information is known as getting a device's capabilities.

Each control or collection of controls is assigned a *usage* that describes its purpose. The HID Specifications list a large number of usages for a range of standard devices. (I have yet to see a device with a "Do Not Disturb" LED. Did you know that a magic carpet has standard controls?)

A usage is, in fact, represented by two bytes: a *Usage Page* and a *Usage*. Table 22.1 shows the main usage categories.

Table 22.1   Main Usage categories

Usage Page	Usage
Generic Desktop	pointer mouse pen joystick gamepad keyboard keypad
Vehicle	rudder throttle
Virtual Reality	
Sport	
Game	
Consumer	power amp video disk
Keyboard	all keys
LED	NumLock CapsLock ScrollLock power
Button	
Ordinal	
Telephony	

Note that usages are used both to specify the general characteristics of a device (i.e., of an Application collection) and the specifics of each control or group of controls.

The Application collection usage defines the overall purpose of a report. For example, a keyboard usually has an Application collection with a Usage Page of "Generic Desktop" and a Usage of "keyboard". The hidusage.h file defines these as byte constants HID_USAGE_PAGE_GENERIC and HID_USAGE_GENERIC_KEYBOARD, respectively. In most Usage Pages, Usage values in the range 1 to 0x1F usually indicate the overall purpose of a device.

The Application collection usage sometimes implies what controls should be in the report. A keyboard usually has eight input single-bit values, eight reserved bits, five output bits, three padding output bits and six input 8-bit values. Although some clients rely on devices having specific controls, most should be able to cope with different devices (e.g., with more controls) or controls in different reports.

Each control also has a usage. The eight single-bit input values in a keyboard report represent the modifier keys (i.e., whether a Shift, Ctrl, or Alt key is pressed at the same time). These modifiers are defined in the "Keyboard" Usage Page (HID_USAGE_PAGE_KEYBOARD). The first single bit value represents the state of the left control key and has a Usage of HID_USAGE_KEYBOARD_LCTRL.

The HID Specification defines Usage values in the "Keyboard" Usage Page that represent all the common keyboard key scan codes. PC-AT keyboards return usages in the range 0 to 101.

However, the left control key has a usage of 240, HID_USAGE_KEYBOARD_LCTRL. As described previously, this is usually returned as a bit value to indicate whether the key is pressed or not.

The HID Specification lets you define several reports in a report descriptor, each with different report IDs. However most Report descriptors do not use report IDs at all. For example, the keyboard Application collection just defines one report with both input and output controls in it. However, a keyboard with a built-in pointing device may have two Application collections with two reports, a keyboard report, and a pointer report.

Getting HID Capabilities

When first used, a HID device needs to be interrogated to find out what it is. The client gets the device *capabilities*, the list of reports that the device may generate or accept. The client then decides whether the device is of interest. The capabilities are in a slightly opaque format, so Windows provides a parser to help clients analyze them. The parser functions are described in the next chapter.

You may also want to read a HID device's string descriptors and physical descriptors. Both are not necessary for most HID interactions, so they are only covered briefly here. See the HID specification for more details.

Controls may optionally have a string index. You can ask the device for the corresponding string, identified by the string index and a language ID. The string descriptor for index zero is special; it provides a list of the language IDs that are supported. HID string descriptors are identical to their USB equivalent, described in the previous two chapters.

Physical descriptors describe which parts of the body can be used to activate a device's controls. Each control may optionally have a designator index that can be used to look up the appropriate physical descriptor.

Then...

After this, the device generates input reports whenever appropriate, perhaps regularly. You can send output reports to the device. It may be able to send and receive feature reports.

HID Model Representation

This section gives explicit details of how the HID model is represented in descriptors and reports. This is useful background for HID client programmers. Minidriver writers and firmware engineers will need to know these details.

HID minidrivers need to return specific data structures to the HID class driver to tell it the device's capabilities. For many new devices, these are provided by the device itself. The crucial data structures are:

- HID descriptor,
- device attributes,
- report descriptor,
- physical descriptors, and
- string descriptors.

The actual data is in little endian format, lowest byte first. A field cannot span more than four bytes in a report.

In a HID USB device, these descriptors will be hard-coded into the device ROM. If you are providing a HID minidriver for a different bus and device, your minidriver may need to synthesize these descriptors internally.

HID Descriptors

A HID descriptor provides the first information about a device. It mainly confirms that it is a HID device and gives the lengths of its report descriptor and any physical descriptors.

Table 22.2 shows the HID_DESCRIPTOR structure along with typical values. The *DescriptorList* field usually just has one element, detailing "report descriptor" and its length. However, it may also specify physical descriptor sets that provide information about the part or parts of the human body used to activate the controls on a device.

Most hardware is not localized, and so should just specify zero for the Country Code.

Table 22.2   HID descriptor

Field	Description	Typical values
bLength	length of HID descriptor	9
bDescriptorType	descriptor type	HID_HID_DESCRIPTOR_TYPE, 0x21
bcdHID	HID spec release	0x0100
bCountry	country code	0 = Not Specified
bNumDescriptors	number of HID class descriptors	1

struct _HID_DESCRIPTOR_DESC_LIST
{
   UCHAR bDescriptorType;
   USHORT wDescriptorLength;
} DescriptorList[1];
	descriptor type total length of descriptor	HID_REPORT_DESCRIPTOR_TYPE, 0x22, sizeof(MyReportDescriptor)

Device Attributes

A Device Attributes HID_DEVICE_ATTRIBUTES structure simply has *VendorID*, *ProductID*, and *VersionNumber* fields, along with a *Size*.

Report Descriptors

A Report Descriptor is a somewhat complicated (but compact) structure detailing the device's capabilities. It may describe more than one report. A parser is provided by Windows to make sense of the format.

A Report Descriptor is a variable length structure made up of *items* that provide information about a device. A group of items might describe a single input, output, feature, or collection.

Overview

A Report descriptor usually has an Application collection defining one or more reports. Each report describes the input, output, or feature controls, and collections that it contains.

Each control has a data size and a usage. Controls can be given units, ranges, and scaling exponents. The data structure allows more than one control to be given usages, units, and exponents easily.

Table 22.3 Keyboard modifier keys example

Item	Item data	Item Type	Actual Bytes
Usage Page	"keyboard"	Global	05 07
Usage Minimum	"left control key"	Local	19 E0
Usage Maximum	"left alt key"	Local	29 E2
Report Size	1	Global	75 01
Report Count	3	Global	95 03
Input	(Data,Variable,Absolute)	Main	81 02

As an example, Table 22.3 shows how three single-bit input controls are specified, the first three modifier keys on a keyboard. The last item, Input, declares the control(s) and states that they are data values (i.e., not constants). Preceding items give more details of the control(s). The preceding Report Size item says that each control has 1 bit and the Report Count item says that there are 3 controls. The preceding Usage Page item specifies the "keyboard" usage page, while the Usage Minimum and Usage Maximum items say that the first input bit corresponds to the left control key of a keyboard and that the last is the left Alt key.

Note that the Item data is actually a byte (e.g., "keyboard" is the constant 07, HID_USAGE_PAGE_KEYBOARD).

Item Parts

An item consists of one or more bytes that define an *item type*, an *item tag*, an *item size*, and possibly *item data* bytes.

An item may be in Short or Long format. Long format is not used yet, so it is not discussed here.

Figure 22.3 shows that the first byte of an item has a Tag of four bits, a Type of two bits and a Data size of two bits. Table 22.4 shows the interpretation of the size and type fields.

Figure 22.3 Short item format

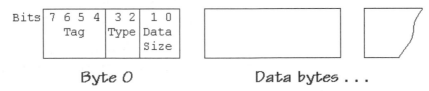

Byte 0   Data bytes . . .

© 1999 PHD Computer Consultants Ltd

Table 22.4   Short item values

Data Size	0 = 0 bytes 1 = 1 byte 2 = 2 bytes 3 = 4 bytes
Type	0 = Main 1 = Global 2 = Local 3 = Reserved

Main Item Tags

The *Main* item type is used for the Input, Output, Feature, Collection, and End Collection tags. A Main item is usually preceded by one or more Global or Local items that provide the rest of the description of the Main item.

Table 22.5 shows the set of possible item tags if the item type is Main (0). The number of data bytes and their definition is also shown. See the HID specification for full details.

Table 22.5   Main item tags

Tag Bits	Name	Data bytes	Data		
1000	Input	1-4	Bit 0: 0=Data Bit 1: 0=Array Bit 2: 0=Absolute Bit 3: 0=No Wrap Bit 4: 0=Linear Bit 5: 0=Preferred State Bit 6: 0=No Null position Bit 7: Reserved Bit 8: 0=Bit Field Bit 31-9: Reserved	1=Constant 1=Variable 1=Relative 1=Wrap 1=Non Linear 1=No Preferred 1=Null state 1=Buffered Bytes	
1001	Output	1-4	same as above, except Bit 7: 0= Non Volatile	1= Volatile	
1011	Feature	1-4	same as Output		
1010	Collection	1	0x00 Physical 0x01 Application 0x02 Logical 0x03-0x7F Reserved 0x80-0xFF Vendor-defined	group of axes mouse, keyboard interrelated data	
1100	End collection	0			
others	reserved				

The Input, Output, and Feature tags allow for up to four data bytes. However, if all the data bits in the top bytes are 0, they can be missed out. So, if bits 8-31 are 0 then the item data size can be given as 1 byte. Further, if no data bits are set at all, the data size can be zero.

For example, two bytes A1 01 (1010 0001 0000 0001) represent "Collection, Main, 1 data byte". The data byte is 0x01 (i.e., Application). Therefore, A1 01 is interpreted as "Collection (Application)".

The one byte C0 (1100 0000) is "End collection, Main, 0 data bytes", and so just means "End Collection".

The two bytes 81 02 (1000 0001 0000 0010) represent "Input, Main, 1 data byte". The data byte just has bit 1 set and so means Data, Variable, Absolute, etc. So 81 02 can be written as "Input (Data, Variable, Absolute)".

Input, Output, and Feature Controls

As shown in the earlier example, an Input, an Output, or a Feature item must be preceded by various other items that further describe the control(s). The preceding items can be Global if the setting lasts for more than one control. They can be Local if they provide just the setting for one control.

Input, Output, and Feature items must have Report Size and Report Count preceding Global items. Report Count specifies the number of controls, each of Report Size number of bits.

There must be a preceding Usage Page item. Then specify either a Usage for a single control or a Usage Minimum and Usage Maximum pair if more than one control is being defined.

Logical Minimum and Logical Maximum specify the range of values that may be returned. For example, even though a Report Size is eight bits, the actual values returned might be specified as being between Logical Minimum 0 and Logical Maximum 101 inclusive.

"Constant" bit values are often used to pad out a report so that controls are byte-aligned.

When "Array" is specified, it means that the actual report will contain an array of usage values that apply at the time of the report. Report Count sets the maximum number of usages that may be returned. If fewer usages apply then the array is padded with zeroes.

The full example below shows how various types of control are defined.

Collections

A *Physical* collection is used to group controls that represent related physical data. A mouse movement might consist of X and Y coordinates and various bit fields to indicate which buttons are pressed.

An *Application* or top-level collection groups together all the reports, controls and collections in a report descriptor. This is preceded by Usage Page and Usage items. The Application collection usage is often used by a client to determine whether the device is of any interest. For example, the Application collection usage might be "keyboard" or "mouse". A keyboard with an integrated pointing device could be defined as two different Application collections.

Collections may be nested.

Global Item Tags

Global item type tags define the attributes of any following Main items. The appropriate Global value remains in force for the rest of the report descriptor unless replaced.

Table 22.6 shows the set of possible item tags if the item type is Global (1). I have already described most of these tags.

Table 22.6   Global item tags

Tag Bits	Name	Data bytes
0000	Usage Page	1
0001	Logical Minimum	1-4
0010	Logical Maximum	1-4
0011	Physical Minimum	1-4
0100	Physical Maximum	1-4
0101	Unit Exponent	1-4
0110	Unit	1-4
0111	Report Size	1+
1000	Report ID	1
1001	Report Count	1+
1010	Push	0
1011	Pop	0
others	reserved	

While `Logical Minimum` and `Logical Maximum` bound the values returned by a device, `Physical Minimum` and `Physical Maximum` give meaning to those bounds. Physical values are logical values with units applied. For example, a thermometer might have logical extents of 0 and 999, but physical extents of 32 and 212 degrees. Logical 0 corresponds to 32 degrees and logical 999 with 212 degrees.

The `Unit` and `Unit Exponent` items further define the type and scaling of the subsequent control(s).

`Push` and `Pop` items allow the state of the item stack to be saved and restored.

Local Item Tags

Local item type tags define the attributes of the following Main item. The Local value is only in force for the following Main item.

Table 22.7 shows the set of possible item tags if the item type is Local (2).

Table 22.7   Local item tags

Tag Bits	Name	Data bytes	Data
0000	Usage	1	
0001	Usage Minimum	1	
0010	Usage Maximum	1	
0011	Designator Index	1	
0100	Designator Minimum	1	
0101	Designator Maximum	1	
0111	String Index	1	
1000	String Minimum	1	
1001	String Maximum	1	
1010	Set Delimiter	1	0 = open set 1 = close set
others	reserved		

The `Usage` or `Usage Minimum` and `Usage Maximum` give the definition of the following `Input`, `Output`, or `Feature` items.

For example, a keyboard uses eight one-bit values to specify the state of its Ctrl, Shift, Alt, etc., modifier keys. The report descriptor, therefore, has input controls in the `HID_USAGE_PAGE_KEYBOARD` Usage Page. Its `Usage Minimum` is `HID_USAGE_KEYBOARD_LCTRL` (i.e., 224) and `Usage Maximum` is `HID_USAGE_KEYBOARD_RGUI` (i.e., 231). This means that if the first input bit is set, the left Ctrl key is pressed, etc.

The `Designator` items determine the body part used for each control, and refers to a Physical Descriptor.

Strings can be associated with each control. The `String` items give the String descriptor index.

A control may have more than one usage, string, or physical descriptor associated with it. One or more alternative sets of local items may be associated with a control by simply bracketing each set with `Set Delimiter` items.

Example

A 105-key keyboard should report eight modifier keys (i.e., Shift, etc.) and up to six simultaneous key presses. It has five output LEDs (i.e., for NumLock, etc.). The complete report descriptor is shown in Table 22.8.

Table 22.8   A keyboard report descriptor

Item	Data	Data value	Actual bytes	Interpretation
Usage Page	(Generic Desktop)	HID_USAGE_PAGE_GENERIC	05 01	
Usage	(Keyboard)	HID_USAGE_GENERIC_KEYBOARD	09 06	

Item	Data	Data value	Actual bytes	Interpretation
Collection	(Application)		A1 01	"Keyboard"
Usage Page	(Key Codes)	HID_USAGE_PAGE_KEYBOARD	05 07	
Usage Minimum	left control key	HID_USAGE_KEYBOARD_LCTRL	19 E0	
Usage Maximum	right GUI key	HID_USAGE_KEYBOARD_RGUI	29 E7	
Logical Minimum	0		15 00	
Logical Maximum	1		25 01	
Report Size	1		95 01	
Report Count	8		75 08	
Input	(Data, Variable, Absolute)		81 02	Modifier key bits
Input	(Constant)		81 01	Reserved byte
Usage Page	(LEDs)	HID_USAGE_PAGE_LED	05 08	
Report Count	5		95 05	
Usage Minimum	NumLock	HID_USAGE_LED_NUM_LOCK	19 01	
Usage Maximum	kana	HID_USAGE_LED_KANA	29 05	
Output	(Data, Variable, Absolute)		91 02	LEDs
Report Size	3		75 03	
Report Count	1		95 01	
Output	(Constant)		91 01	padding
Usage Page	(Key Codes)	HID_USAGE_PAGE_KEYBOARD	05 07	
Usage Minimum	0		19 00	
Usage Maximum	101		29 65	
Report Count	6		95 06	
Report Size	8		75 08	
Logical Minimum	0		15 00	
Logical Maximum	101		25 65	
Input	(Data Array)		81 00	Key array (6 bytes)
End collection			C0	

The first three items specify the Application collection as being a keyboard in the generic usage page. A client will therefore know straightaway that this device is a keyboard.

The next items define the modifier input controls. There are eight single-bit inputs, each naturally reporting either 0 or 1. These inputs correspond to eight usage values in the keyboard usage page, the left Shift key being pressed, onwards.

The next item defines a reserved constant byte of eight single bits. Note how the Global Report Size and Report Count items persist after the previous Main item.

The next control is for the output LEDs. There are five of them, still bits needing values of 0 or 1. The LED usage page is now in force and the output bits correspond to the NumLock key, etc.

The next control is a 3-bit constant field, padding the output report to make it up to a complete byte.

The actual key presses are defined in a single Input item with its "Array" bit set. Up to 6-key scan codes may be read, each of eight bits. However, the actual values returned are in the range 0 to 101. The keyboard has 102 keys and three modifier keys, returned separately as three bits.

No Report ID items are given, and so the descriptor describes one report that contains the current state of all the input and output items. An actual input report, therefore, always contains the eight modifier bits, a reserved byte of 0x00, and one to six key scan codes padded to six bytes with zeroes. An actual output report just contains one byte with the lower five bits giving the required LED states.

Pressing Ctrl+Alt+Del might result in the six input reports shown in Table 22.9. The reserved byte values are not shown.

Table 22.9   Ctrl+Alt+Del input reports

Transition	Modifier byte	Scan codes (hex)
Left Ctrl down	00000001	00, 00, 00, 00, 00, 00
Left Alt down	00000101	00, 00, 00, 00, 00, 00
Del down	00000101	63, 00, 00, 00, 00, 00
Del up	00000101	00, 00, 00, 00, 00, 00
Left Ctrl up	00000100	00, 00, 00, 00, 00, 00
Left Alt up	00000000	00, 00, 00, 00, 00, 00

Conclusion

This chapter has introduced you to the Human Input Device (HID) model. A HID device uses various descriptors to define its functionality. The most important of these, the Report descriptor, defines what input reports it generates and the output reports it can receive. The controls within a report are given "usage" values that define their function exactly. A full example of a Report descriptor was given for a HID keyboard.

The next chapter looks at how to use the HID class drivers in Windows. Both kernel mode drivers and user mode applications can access HID devices.

Chapter 23

HID Clients

This chapter shows how to use the Human Input Device (HID) model in Windows. It is easy to write a user mode application to talk to a HID device through the Windows HID class driver. If need be, you can write a kernel mode HID client to do the same job. A HID kernel mode client can be layered over the HID class driver in the normal way. Alternatively, it can use Plug and Play Notification to find any suitable HID devices. Windows provides routines that make it much easier to analyze HID Report descriptors and send and receive reports.

This chapter initially looks at how to write a user mode application, called *HidKbdUser*, to talk to a HID keyboard. This section describes all the Windows parsing routines that are available to both user mode and kernel mode HID clients.

Later, the HidKbd kernel mode HID client driver is described. This illustrates the Plug and Play Notification technique of finding all devices with a matching device interface. HidKbd finds any devices with a HID class driver device interface. If a HID keyboard is found, it makes a device called \\.\HidKbd available to user mode applications. The *HidKbdTest* application tests the HidKbd driver.

HID Class Driver

The HID class driver is the key to using HID in Windows. It uses **HID** *minidrivers* to talk to actual devices. It handles requests from clients, directing them to the correct minidriver.

When a new HID device is found by the Plug and Play Manager, the installation INF file eventually ensures that the HID class driver and the appropriate minidriver are loaded.

The HID class driver uses the minidriver to read the HID and Report descriptors. The HID class driver then creates a device object for each of the top-level collections described in the Report descriptor. Each device object is registered as belonging to the HID class device interface.

User mode and kernel mode clients can use the HID class device interface GUID to identify all the available HID devices. The client must get the device capabilities to see if the device is of interest. If you were looking for a HID keyboard, you would ignore a HID mouse.

Each request to the HID class driver eventually ends up as a call to the appropriate HID minidriver. Each minidriver knows how to send and receive HID reports, for example. This chapter does not describe how to write minidrivers; this subject is covered in the DDK.

HID Class Driver Characteristics

The HID class driver responds to Create, Close, Read, Write, and IOCTL IRPs. Read and Write requests are used for input, output, and feature reports. IOCTLs are used to retrieve HID device capabilities. User mode applications should not use the IOCTLs directly. Instead, various HidD... routines are supplied to make the job easier.

Ring Buffer

The HID class driver has a buffer for each device for input reports. This holds input reports until a client reads them. This is a "ring buffer", so if the reports are not read quickly enough, new reports may overwrite the old reports (i.e., data may get lost).

A kernel mode client can get and set the ring buffer size using two IOCTLs.

Multiple Open Handles

In Windows 98, more than one program can access a single HID device at the same time. For example, if a HID keyboard is attached then a Windows kernel mode driver will be loaded to talk to it. This driver ensures that any key presses are routed into the normal keyboard driver. You can also open up another handle to the device object in the HID class driver that represents the HID keyboard. Therefore, in W98 at least, any input reports are received on all open handles.

However in Windows 2000, the HID keyboard driver opens a HID keyboard for exclusive access. Therefore the HidKbdUser user mode program and the HidKbd driver are unable to talk to the keyboard. I was able to develop this code in the Beta 2 version of W2000, when the keyboard was not opened for exclusive access.

Windows HID Clients

Windows 98 and Windows 2000 have built in support for various types of common HID device (e.g., keyboards, mice, game ports, audio controls, and the DirectInput system). The standard installation INF files load the required drivers automatically. If you plug in a HID USB keyboard, for example, the relevant system drivers are loaded and you should be able to start typing straightaway[1].

Figure 23.1 shows the Windows 2000 drivers that are loaded to service a "legacy" keyboard and a HID USB keyboard. The situation is identical in Windows 98, apart from the driver filenames. For mouse input and game ports, the driver diagrams are almost identical. However, in Windows 2000, all the game port input is now done using HID with a HID joystick minidriver.

---

1. In Windows 98, I found that I could not type on my HID keyboard after a reboot.

The Win32 subsystem still uses the standard keyboard driver, `KbdClass.sys` in W2000, to get all its keyboard input. This interrogates the 8042 legacy keyboard in the same way as NT 4 and earlier. However, the keyboard driver can also get its input from a HID keyboard. An intermediate driver, `KbdHid.sys` in W2000, is used. I assume that `KbdHid.sys` does all the hard work of interrogating the HID class driver, and it simply returns 8042 equivalent scan codes to the main keyboard driver.

The keyboard auto-repeat feature is not usually present in a HID keyboard. Either the `KbdClass.sys` or the `KbdHid.sys` driver must implement auto-repeat. Similarly, the Ctrl+Alt+Del key combination must be detected by one of these drivers. Quite correctly, it is not a feature of HID class driver.

I understand that `KbdHid.sys` driver always tries to keep two outstanding read requests for a keyboard: one to detect a key being depressed and one to detect a key being released. Combined with the HID class driver ring buffer, this technique should ensure that no key presses are lost.

In the future, it would be neater if Microsoft rearranged this keyboard driver layout. They would have to write a HID minidriver for the legacy 8042 port. All keyboard input would then come through the HID class driver. Although this approach might be slower, it should not matter for such a relatively slow device as a keyboard.

In addition, the figure shows that a user mode client can also access the HID keyboard as a HID device. It will receive regular HID input reports, as shown later. It will not receive 8042 equivalent scan codes. As I mentioned previously, all input reports are received by both `KbdHid.sys` and the user mode client.

Figure 23.1   Windows and a user mode client using a HID keyboard

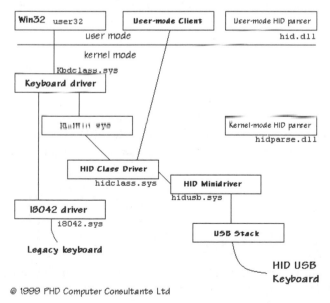

Header Files

Table 23.1 shows the HID header files that you need to include. In the Windows 98 DDK, these files are in the `src\hid\inc` directory. In the Windows 2000 DDK, most headers are in the standard `inc` directory, while the others are in `src\wdm\hid\inc`.

Table 23.1   HID header files

`hidclass.h`	IOCTL codes
`hidpddi.h`	Other parsing library routines
`hidpi.h`	Parsing routines (i.e., `HidP...` functions and structures)
`hidport.h`	Minidriver header
`hidsdi.h`	User-mode `HidD...` functions
`hidtoken.h`	Report descriptor item constants
`hidusage.h`	Standard usage values

HID USB Minidriver

HID evolved from the USB specification, but is now a general purpose input device model. Nonetheless, HID fits neatly into the USB model. The HID class is one of the classes defined in the USB Specifications.

At present, the USB bus is one of the few places where HID devices can be found. The HID class driver talks to the USB device stack through the `HidUsb.sys` driver.

A HID USB device has the standard USB device, configuration, interface, and endpoint descriptors. In addition, it has the class-specific HID, Report, and Physical descriptors described in the previous chapter.

The standard USB device descriptor has zeroes for its class, subclass, and protocol. The standard interface descriptor has constant USB_DEVICE_CLASS_HUMAN_INTERFACE (0x03) in its *bInterfaceClass* field. *iInterfaceSubClass* is set to 1 if the device supports a boot protocol, described in the following text. In this case, *bInterfaceProtocol* is 1 for a keyboard and 2 for a mouse.

A USB request for the Configuration descriptor usually returns the HID descriptor before the endpoint descriptors[2]. The report and physical descriptors must be specifically asked for, using the following codes.

HID Descriptor	`HID_HID_DESCRIPTOR_TYPE`	0x21
Report descriptor	`HID_REPORT_DESCRIPTOR_TYPE`	0x22
Physical Descriptor	`HID_PHYSICAL_DESCRIPTOR_TYPE`	0x23

A HID USB device uses the default control pipe endpoint 0 and one interrupt pipe. The control pipe is used for standard USB purposes and for sending and receiving reports, etc. The interrupt pipe is used to return new input reports.

---

2.   This is called Draft#4 compliance. An older Draft#3-compliant device returns the HID descriptor after the endpoint descriptors.

USB HID devices support the notion of an idle rate. Whenever an interrupt request is received from the host, the USB HID device can either return a report or a NAK to indicate that nothing has changed. In the former case, no input state may have changed and so host processing of the report will waste time. The idle rate determines how often the HID USB device returns a report. If at any time during the idle rate period, an input report changes, the HID USB device does return an input report at the next interrupt. The recommended idle rate for keyboards is 500ms and infinity for joysticks and mice.

The following HID-specific class requests are supported on the default pipe. GET_REPORT and SET_REPORT get or send a specific input, output, or feature report. An idle rate can be defined using SET_IDLE and read using GET_IDLE. SET_PROTOCOL and GET_PROTOCOL are used for boot devices, as described in the following text.

USB Boot Devices

A USB keyboard or mouse must be available during boot so the user can configure the BIOS settings, select the operating system, etc.

The BIOS can read the USB interface descriptor with relative ease. If it finds that *InterfaceSubClass* field is 1, the device must be a keyboard or mouse that supports the boot protocol. The Report descriptor for boot keyboards and mice must be in a standard format. Therefore, the BIOS does not have to read and decode the Report descriptor. It simply uses the SET_PROTOCOL command to enable the standard reports.

A boot keyboard input report is usually eight bytes long, in the format described in the previous chapter. Other requirements for a boot keyboard are given in the HID Specification.

A mouse boot input report is at least three bytes long. The first three bits are the mouse button states. The second and third bytes are the X and Y displacements.

User Mode HID Clients

The DDK recommends that you talk to HID devices using a user mode application, if possible. This makes sense, as it is usually far easier to write and debug a Win32 application than a device driver. See the next section if you really need to write a HID client device driver.

A user mode HID application has three main jobs to do.

1. Find all HID devices.
2. For each HID device, inspect its capabilities to see if it is of interest.
3. Read HID input reports or write HID output reports when needed. Alternatively send and receive feature reports.

The *HidUsbUser* example in HidUsb\User\HidKbdUser.cpp illustrates these tasks. *HidUsbUser* looks for a HID keyboard, reads input keys until Esc is pressed, and then — hey hey — flashes the keyboard LEDs. *HidUsbUser* does not care what sort of HID keyboard is attached. All HID keyboards ought to respond in a similar way.

HidUsbUser uses various standard routines. The HidD... routines are available only to user mode applications. The "parsing routines" HidP... can be used by kernel mode clients as well. In user mode, all these routines are provided by Hid.dll. In kernel mode, the parsing routines are in HidParse.sys.

Finding HID Devices

The HID class driver registers all its device objects as having the HID device interface. *HidUsbUser* can, therefore, use the GetDeviceViaInterface routine to find all existing HID devices. Instead of using one of the book software GUIDs, *HidUsbUser* looks for the HID class GUID. Use HidD_GetHidGuid to find this GUID.

```
GUID HidGuid;
HidD_GetHidGuid(&HidGuid);
```

Going back to Chapter 5, you will see that GetDeviceViaInterface uses the SetupDi... routines to find a device that matches the given GUID. Previously, I have always looked for the first device with the desired device interface. *HidUsbUser*, however, looks through all the available HID devices looking for a HID keyboard.

GetDeviceViaInterface opens a handle to each device. *HidUsbUser* then gets each device's capabilities. If it is a HID keyboard, the main program carries on. Otherwise it closes the device handle, and loops again so that GetDeviceViaInterface can open the next HID device. *HidUsbUser* reports an error and gives up if a HID keyboard is not found.

Note that *HidUsbUser* only looks for an appropriate HID device when it starts. As it stands, it will not discover any new HID devices that suddenly are plugged in. The Win32 RegisterDeviceNotification function can be used to listen for Plug and Play Notification events in user mode applications. The *Wdm2Notify* application described in Chapter 9 shows how to use RegisterDeviceNotification to listen for device interface events. This can easily be modified to spot new HID class devices.

If the HID USB keyboard is unceremoniously unplugged while *HidUsbUser* is running, all I/O requests fail, and GetLastError returns 1167 (ERROR_PROCESS_ABORTED).

Getting HID Capabilities

HidUsbUser uses GetCapabilities to get a HID device's capabilities, as shown in Listing 23.1. In particular, it looks for a HID keyboard. If it finds a keyboard, GetCapabilities returns a *preparsed data* pointer and the maximum input and output report sizes. The preparsed data is the HID Report descriptor parsed into a format that makes it easier for the support routines to decode and encode reports. You will need to pass the preparsed data pointer whenever you call any of the HID parsing routines.

Some applications will simply want to look for the device vendor and product IDs. If you know the capabilities of your device, there is no need to do any more checking. However, you still need to get the preparsed data. If possible, you should write code that does not rely on specific vendor and product IDs.

Get the device ID using HidD_GetAttributes. This fills in the supplied HIDD_ATTRIBUTES structure. If successful, the *VendorID*, *ProductID*, and *VersionNumber* fields are filled in. For USB devices, these values come from the USB Device descriptor.

Use HidD_GetPreparsedData to obtain a pointer to the preparsed data. This time, Windows allocates the buffer and returns a pointer to you. Make sure that you keep the preparsed data pointer handy for the rest of your HID operations. When finished with it, free the preparsed data memory using HidD_FreePreparsedData.

The first step when analyzing your device capabilities is to call HidP_GetCaps, which fills in a HIDP_CAPS structure. The *UsagePage* and *Usage* fields in there tell you the usage information for the Report descriptor top-level collection. In GetCapabilities, this is all the information that is needed to determine if the device is a keyboard. It checks that these fields are HID_ USAGE_PAGE_GENERIC and HID_USAGE_GENERIC_KEYBOARD, respectively.

The HIDP_CAPS structure also contains much other useful information. First, it contains the maximum length of input, output, and feature reports in three fields. The first two of these values are saved, because they indicate how big a buffer to allocate to send and receive reports. The other HIDP_CAPS fields are used if you check the detailed control capabilities.

Listing 23.1 Getting a HID device's capabilities

```
bool GetCapabilities( HANDLE hHidKbd, PHIDP_PREPARSED_DATA& HidPreparsedData,
                      USHORT& InputReportLen, USHORT& OutputReportLen)
{
    // Get attributes. ie find vendor and product ids
    HIDD_ATTRIBUTES HidAttributes;
    if( !HidD_GetAttributes( hHidKbd, &HidAttributes))
    {
        printf("XXX  Could not get HID attributes\n");
        return false;
    }
    printf("HID attributes: VendorID=%04X, ProductID=%04X,
        VersionNumber=%04X\n", HidAttributes.VendorID,
        HidAttributes.ProductID, HidAttributes.VersionNumber);


    // Get preparsed data
    if( !HidD_GetPreparsedData( hHidKbd, &HidPreparsedData))
    {
        printf("XXX  Could not get HID preparsed data\n");
        return false;
    }


    // Work out capabilities
    HIDP_CAPS HidCaps;
    bool found = false;
    NTSTATUS status = HidP_GetCaps( HidPreparsedData, &HidCaps);
    if( status==HIDP_STATUS_SUCCESS)
    {
        printf("Top level Usage page %d usage %d\n", HidCaps.UsagePage,
            HidCaps.Usage);
        if( HidCaps.UsagePage==HID_USAGE_PAGE_GENERIC &&
            HidCaps.Usage==HID_USAGE_GENERIC_KEYBOARD)
        {
            printf("     Found HID keyboard\n\n");
            found = true;
        }
```

Listing 23.1 Getting a HID device's capabilities (continued)

```
    // Remember max lengths of input and output reports
    InputReportLen = HidCaps.InputReportByteLength;
    OutputReportLen = HidCaps.OutputReportByteLength;
    printf("InputReportByteLength %d\n", HidCaps.InputReportByteLength);
    printf("OutputReportByteLength %d\n", HidCaps.OutputReportByteLength);
    printf("FeatureReportByteLength %d\n\n",
        HidCaps.FeatureReportByteLength);

    printf("NumberLinkCollectionNodes %d\n\n",
        HidCaps.NumberLinkCollectionNodes);

    printf("NumberInputButtonCaps %d\n", HidCaps.NumberInputButtonCaps);
    printf("NumberInputValueCaps %d\n", HidCaps.NumberInputValueCaps);
    printf("NumberOutputButtonCaps %d\n", HidCaps.NumberOutputButtonCaps);
    printf("NumberOutputValueCaps %d\n", HidCaps.NumberOutputValueCaps);
    printf("NumberFeatureButtonCaps %d\n", HidCaps.NumberFeatureButtonCaps);
    printf("NumberFeatureValueCaps %d\n\n", HidCaps.NumberFeatureValueCaps);

    ShowButtonCaps( "Input button capabilities",
        HidP_Input, HidCaps.NumberInputButtonCaps, HidPreparsedData);
    ShowButtonCaps( "Output button capabilities",
        HidP_Output, HidCaps.NumberOutputButtonCaps, HidPreparsedData);
    }
    return found;
}
```

Why Get Button and Value Capabilities?

Some programs cannot just rely on the HIDP_CAPS *UsagePage* and *Usage* fields. For example, in the future, some fancy device may use these fields to say that it is a "speech interface". A "speech interface" might still be able to generate key presses (e.g., when a user says a word). Its detailed capabilities would show that it could indeed present data that is of interest.

Each control in a HID device is seen by Windows as being either a *button* or a *value*. Anything that has a HID usage is defined as being a button. Controls that take on any other values are called *values*.

Each of the keys on a keyboard has a HID usage. All keyboard keys are in usage page 7. Within that usage page, most of the usages from 0 to 255 have key definitions. However, most Western keyboards produce usage codes in two ranges. Usages 0 to 101 contain most standard keys, while usages 224 to 231 represent the modifier keys such as Shift, Ctrl, Alt, etc. To be extra careful, check that the HID device generates usages in these two ranges.

Similarly to double check that the HID device has the correct LEDs check that its output report contains buttons in usage page 8, for LEDs. Within this usage page, usage 1 corresponds to NumLock, usage 2 with CapsLock, and usage 3 with ScrollLock.

Getting Button Capabilities

The HIDP_CAPS structure has fields that tell you how many button and value capabilities there are for each type of report. For example, *NumberInputButtonCaps* tells you how many button capabilities there are for input reports.

GetCapabilities uses the ShowButtonCaps routine shown in Listing 23.2 to show what button capabilities there are for different types of report. The NumCaps parameter is passed the number of button capabilities that are expected for the specified type of report.

ShowButtonCaps first allocates an array of HIDP_BUTTON_CAPS structures called Button-Caps. The call to HidP_GetButtonCaps fills in this array. Afterwards, ShowButtonCaps simply goes through each element in ButtonCaps and prints out the usages that can be returned. The HIDP_BUTTON_CAPS *UsagePage* field gives the usage page of all the buttons referred to in this structure. *ReportId* specifies the report in which these buttons are. If the *IsRange* BOOLEAN is TRUE, the *Range.UsageMin* and *Range.UsageMax* fields are valid. Otherwise, the *NotRange.Usage* field holds the only valid usage.

Each HIDP_BUTTON_CAPS structure has similar fields for the string descriptors and physical designators associated with controls. Finally, various link fields specify in which collection the buttons are. See the previous chapter for a description of collections.

Listing 23.2 Getting buttons capabilities

```
void ShowButtonCaps( char* Msg, HIDP_REPORT_TYPE ReportType,
                     USHORT NumCaps, PHIDP_PREPARSED_DATA HidPreparsedData)
{
    if( NumCaps==0) return;

    printf("     %s\n", Msg);

    HIDP_BUTTON_CAPS* ButtonCaps = new HIDP_BUTTON_CAPS[NumCaps];
    if( ButtonCaps==NULL) return;

    NTSTATUS status = HidP_GetButtonCaps( ReportType, ButtonCaps, &NumCaps,
        HidPreparsedData);
    if( status==HIDP_STATUS_SUCCESS)
    {
        for( USHORT i=0; i<NumCaps; i++)
        {
            printf("ButtonCaps[%d].UsagePage %d\n", i, ButtonCaps[i].UsagePage);
            if( ButtonCaps[i].IsRange)
                printf(".Usages     %d..%d\n\n",
                    ButtonCaps[i].Range.UsageMin, ButtonCaps[i].Range.UsageMax);
            else
                printf(".Usage %d\n\n", ButtonCaps[i].NotRange.Usage);
        }
    }
    delete ButtonCaps;
}
```

Here is an excerpt from the output produced by the ShowButtonCaps routine. It confirms that, in this case, the HID device has the desired controls. It could be fairly laborious if you had to check that all the right buttons were available, as the button capabilities could arrive in any order.

```
Input button capabilities
ButtonCaps[0].UsagePage 7
          .Usages    224..231

ButtonCaps[1].UsagePage 7
          .Usages    0..101

Output button capabilities
ButtonCaps[0].UsagePage 8
          .Usages    1..3
```

Use HidP_GetSpecificButtonCaps if you need to look for buttons in a different collection or search for a controls with a specific usage page or usage. This might be a better way of determining whether the controls you are interested in are supported.

Getting Value Capabilities

You can retrieve details of what control *values* are supported in a very similar way. Use HidP_GetValueCaps to get a list of all values in the top-level collection. Alternatively, HidP_GetSpecificValueCaps is used to look for values in a different collection, or search for controls with a specific usage page or usage. The information is stored in an array of HIDP_VALUE_CAPS structures.

Getting Collection Capabilities

The HidP_GetLinkCollectionNodes function is used to obtain details of all the collections in a top-level collection. It fills in an array of HIDP_LINK_COLLECTION_NODE structures. See the DDK for full details of how this array specifies the arrangement of collections within the top-level collection.

Reading Input Reports

Reading input reports from a HID device is straightforward. Listing 23.3 shows how this is done in the *HidUsbUser* main routine. It keeps reading input reports until the Esc key is pressed on the HID keyboard. See Listing 23.6 for some example output from this routine.

First, allocate and zero a buffer to receive a report, with the size given in the HIDP_CAPS structure. A keyboard input report is always eight bytes long. However, the *InputReportByteLength* field in HIDP_CAPS is one longer than this, as the first byte is used to indicate the report ID. For keyboard reports, this first byte will be zero, as report IDs are not used.

Use ReadFile to read the next available input report. If the device can return two different input reports then this call may obtain either of them. Suppose these two reports have different lengths. For the smaller input report, the returned count of bytes transferred will be less than the buffer size you passed.

Listing 23.3   Reading keyboard input reports

```
DWORD TxdBytes;
char* InputReport = new char[InputReportLen];
assert(InputReport!=NULL);
// Loop until Esc pressed on keyboard
do
{
    if( !ReadFile( hHidKbd, InputReport, InputReportLen, &TxdBytes, NULL))
    {
        printf("XXX  Could not read value %d\n", GetLastError());
        break;
    }
    else if( TxdBytes==InputReportLen)
    {
        printf("    Input report %d:", InputReport[0]);
        for( USHORT i=1; i<InputReportLen; i++)
            printf(" %02X", InputReport[i]);
        printf("\n");

        DecodeInputUsages( InputReport, InputReportLen, HidPreparsedData);
    }
    else
    {
        printf("XXX  Wrong number of bytes read: %d\n",TxdBytes);
        break;
    }
}
while( InputReport[3]!=0x29);
delete InputReport;
```

What Buttons Were Set in My Report?

You could just look directly at the received buffer and work out what it means. However, the correct HID way to analyze reports is get the HID parsing routines to tell you what usages were set in the report. DecodeInputUsages, shown in Listing 23.4, does just this job. Indeed, it goes further by telling you what changes have occurred since the last input report. It prints out the usages that have just been "made" and the ones have just been "broken".

The HidP_GetButtonsEx function analyses an input report buffer and reports which button usages were set in the report. The output is an array of USAGE_AND_PAGE structures (i.e., usage page and usage values).

DecodeInputUsages must first find out the maximum size for this output array using HidP_MaxUsageListLength, so that a suitably sized array can be allocated. Actually,

DecodeInputUsages cheats, as I know the maximum size in advance. I use the MaxPreviousUsages constant to declare fixed-size arrays; this avoids allocating and freeing buffers all over the place.

The maximum usage list length for a keyboard report is 14. This is made up of the six keys that the HID Report descriptor says can be pressed simultaneously, and the eight modifier keys, again that the Report descriptor says can be pressed simultaneously. Obviously, it is extremely unlikely that 14 keys can be pressed at the same time.

HidP_GetButtonsEx analyses the input report just received and fills in the array of USAGE_ AND_PAGE structures, called Usages. The ValidUsages variable is filled with the number of valid elements in this array. DecodeInputUsages simply prints out the usage page and usage for each of these valid key presses.

The next job, if desired, is to work out what has changed since the last input report. The HidP_UsageListDifference function does this. It is passed the previous usage list and the current usage list as input. It fills in two further arrays, one with a list of usages that have been "made" (or just arrived), and one with a list of usages that have just "broken" (or gone away). DecodeInputUsages prints out these two arrays. Note that HidP_UsageListDifference deals with arrays of USAGE values, not USAGE_AND_PAGEs. In DecodeInputUsages all the input usages are in the keyboard usage page, so this is not a problem.

Listing 23.4 Decoding input report usages

```
const ULONG MaxPreviousUsages = 14;
USAGE_AND_PAGE Usages[MaxPreviousUsages];
USAGE PreviousUsages[MaxPreviousUsages];


void DecodeInputUsages( char* KbdReport, USHORT KbdReportLen,
                        PHIDP_PREPARSED_DATA HidPreparsedData)
{
    // Get max number of USAGE_AND_PAGEs required for all input reports in
    // top-level collection
    ULONG MaxUsages = HidP_MaxUsageListLength( HidP_Input, 0, HidPreparsedData);
    if( MaxUsages==0 || MaxUsages>MaxPreviousUsages)
    {
        printf("XXX  Invalid HidP_MaxUsageListLength returned %d\n", MaxUsages);
        return;
    }


    // Get usages set in given keyboard report
    ULONG ValidUsages = MaxUsages;
    NTSTATUS status = HidP_GetButtonsEx( HidP_Input, 0, Usages, &ValidUsages,
        HidPreparsedData, KbdReport, KbdReportLen);
    if( status==HIDP_STATUS_SUCCESS)
    {
        USAGE CurrentUsages[MaxPreviousUsages];
        USAGE BreakUsages[MaxPreviousUsages];
```

Listing 23.4   Decoding input report usages (continued)

```
            USAGE MakeUsages[MaxPreviousUsages];

            // Show current usages
            memset( CurrentUsages, 0, sizeof(CurrentUsages));
            printf("    Usages set: ");
            for( ULONG i=0; i<ValidUsages; i++)
            {
                printf( " %02X:%02X", Usages[i].UsagePage, Usages[i].Usage);
                CurrentUsages[i] = Usages[i].Usage;
            }

            // Work out differences compared to previous usages
            HidP_UsageListDifference( PreviousUsages, CurrentUsages,
                                    BreakUsages, MakeUsages, MaxUsages);

            // Print out usages broken and made
            printf(" (Break: ");
            for( i=0; i<MaxUsages; i++)
            {
                if( BreakUsages[i]==0) break;
                printf( " %02X", BreakUsages[i]);
            }
            printf(") (Make: ");
            for( i=0; i<MaxUsages; i++)
            {
                if( MakeUsages[i]==0) break;
                printf( " %02X", MakeUsages[i]);
            }
            printf(")\n\n");

            // Save previous usages
            memcpy( PreviousUsages, CurrentUsages, MaxUsages*sizeof(USAGE));
        }
    }
```

The HidP_GetButtons function can be used if you are only looking for buttons in a particular usage page. Both HidP_GetButtons and HidP_GetButtonsEx have parameters that let you look for buttons in a particular collection.

What Values Were Set in My Report?

Use the HidP_GetUsageValue, HidP_GetScaledUsageValue, or HidP_GetUsageValueArray functions to retrieve control values.

Sending Output Reports

You send HID output reports using the Win32 `WriteFile` function. While you can build the output buffer by hand, it is safer to use the HID parsing routines to fill in the buffer.

The `SetLEDs` function shown in Listing 23.5 shows how a single output report is sent to set the state of the HID keyboard LEDs. It has three optional parameters, which must contain the individual LED usages that you want set on. This line in the *HidUsbUser* main routine shows how to turn the ScrollLock LED on and turn the other LEDs off.

```
SetLEDs( hHidKbd, OutputReportLen, HidPreparsedData,
    HID_USAGE_LED_SCROLL_LOCK);
```

A keyboard output report consists of just one byte. However, as before, the output buffer must have an extra byte at the beginning for the report ID. This is set to zero in this case, as keyboards do not use report IDs. `SetLEDs` allocates an output buffer and zeroes it.

`SetLEDs` must now build an array of USAGEs, filled with any of its optional parameters that are non-zero. This array and its length are passed to `HidP_SetButtons`. `HidP_SetButtons` builds the output buffer in the correct format. `SetLEDs` then simply calls `WriteFile` to send off the output report.

You can call `HidP_SetButtons` more than once for a single output report. Indeed, you may also want to call one of the set value functions, if the output report must contain values. Set values into an output report using the `HidP_SetUsageValue`, `HidP_SetScaledUsageValue`, and `HidP_SetUsageValueArray` functions.

Be careful if you are using a HID device that has more than one output report. `HidP_Set-Buttons` can only build one output report at a time. `HidP_SetButtons` sets the report ID, if appropriate. If a second call to `HidP_SetButtons` tries to set a usage that is not in the first report, it fails with a suitable error code. However, if you start afresh with a new output report and repeat the second call to `HidP_SetButtons`, it should work this time. This whole process makes it particularly laborious to set output usages in a way that is truly independent of the device that is attached.

Listing 23.5 Setting keyboard LEDs using an output HID report

```
void SetLEDs( HANDLE hHidKbd, USHORT OutputReportLen,
        PHIDP_PREPARSED_DATA HidPreparsedData,
        USAGE Usage1/*=0*/, USAGE Usage2/*=0*/, USAGE Usage3/*=0*/)
{
    // Build Output report from given usage(s)
    char* OutputReport = new char[OutputReportLen];
    assert(OutputReport!=NULL);
    memset( OutputReport, 0, OutputReportLen);

    USAGE UsageList[3];
    UsageList[0] = Usage1;
    UsageList[1] = Usage2;
    UsageList[2] = Usage3;
```

Listing 23.5 Setting keyboard LEDs using an output HID report (continued)

```
    ULONG UsageLength = 0;

    if( Usage1!=0)
    {
        UsageLength++;
        if( Usage2!=0)
        {
            UsageLength++;
            if( Usage3!=0)
            {
                UsageLength++;
            }
        }
    }


    // Convert usages into an output report
    NTSTATUS status = HidP_SetButtons( HidP_Output, HID_USAGE_PAGE_LED, 0,
        UsageList, &UsageLength, HidPreparsedData, OutputReport,
        OutputReportLen);
    if( status!=HIDP_STATUS_SUCCESS)
    {
        delete OutputReport;
        return;
    }

    printf("    Output report: ");
    for( ULONG i=1; i<OutputReportLen; i++)
        printf( " %02X", OutputReport[i]);
    printf("\n");

    // Send off output report
    DWORD TxdBytes;
    if( !WriteFile( hHidKbd, OutputReport, OutputReportLen, &TxdBytes, NULL))
        printf("XXX  Could not write value %d\n", GetLastError());
    else if( TxdBytes==OutputReportLen)
        printf("    Wrote output report OK\n");
    else
        printf("XXX  Wrong number of bytes written: %d\n",TxdBytes);
    delete OutputReport;
}
```

Other User Mode HID Client Functions

User mode HID clients can send and receive HID *feature* reports. Use HidD_GetFeature to get a feature report. The first byte of the buffer must be set to the report ID that you want to receive. The received feature report can be analyzed using HidP_GetButtonsEx, etc., as usual. Use HidD_SetFeature to send a feature report. Beforehand, use HidP_SetButtons, etc., to build up the feature report in the correct format, as usual.

HidD_FlushQueue deletes all pending information from the input queue for this HID device.

The functions HidD_GetNumInputBuffers and HidD_SetNumInputBuffers may be available in the future to let you get or set the ring buffer size used by the HID class driver for this device[3].

Running *HidKbdUser*

If you have a HID keyboard, you can run *HidKbdUser* yourself. For the benefit of those who do not, Listing 23.6 shows some example output. As mentioned earlier, HidKbdUser will not run in W2000 as the HID keyboard device cannot be shared.

There is only one HID device in the system. It is opened and *HidKbdUser* finds that it is a HID keyboard, as the top-level usage page and usage are the correct values. *HidKbdUser* then prints out the capabilities of the device.

Test 2 reads any input reports from the HID keyboard. In this example, I pressed Ctrl+Alt+Del followed by A, B, C, and Esc. You can see how an input report is produced every time a key is pressed or released. The actual input report is in exactly the same format as the raw USB interrupt transfer shown in Table 21.9.

If you remember from Chapter 21, the USB example driver, UsbKbd, kept on receiving input interrupt data even if no state changes had occurred. The HID class driver sensibly filters out these redundant input reports and only returns data when a key is pressed or released.

Test 3 sends several output reports to flash the LEDs on the keyboard. *HidKbdUser* calls the SetLEDs function several times for different LED combinations with a short delay between each call.

Listing 23.6 Example *HidKbdUser* output

```
Test 1
Symbolic link is \\.\000000000000000b#{4d1e55b2-f16f-11cf-88cb-001111000030}
    Found HID device
    HID attributes: VendorID=046A, ProductID=0001, VersionNumber=0305
    Top level Usage page 1 usage 6
    Found HID keyboard


    InputReportByteLength 9
    OutputReportByteLength 2
    FeatureReportByteLength 0
```

---

3. Kernel mode HID clients can get and set the ring buffer size using two IOCTLs.

Listing 23.6 Example *HidKbdUser* output (continued)

```
        NumberLinkCollectionNodes 1

        NumberInputButtonCaps 2
        NumberInputValueCaps 0
        NumberOutputButtonCaps 1
        NumberOutputValueCaps 0
        NumberFeatureButtonCaps 0
        NumberFeatureValueCaps 0

        Input button capabilities
        ButtonCaps[0].UsagePage 7
                    .Usages    224..231

        ButtonCaps[1].UsagePage 7
                    .Usages    0..101

        Output button capabilities
        ButtonCaps[0].UsagePage 8
                    .Usages    1..3

        Opened OK

Test 2
        Input report 0: 01 00 00 00 00 00 00 00          Left Ctrl pressed
        Usages set:  07:E0 (Break: ) (Make:  E0)

        Input report 0: 05 00 00 00 00 00 00 00          Left Alt pressed
        Usages set:  07:E0 07:E2 (Break: ) (Make:  E2)

        Input report 0: 05 00 63 00 00 00 00 00          Del pressed
        Usages set:  07:E0 07:E2 07:63 (Break: ) (Make:  63)

        Input report 0: 05 00 00 00 00 00 00 00          Del released
        Usages set:  07:E0 07:E2 (Break:  63) (Make: )

        Input report 0: 00 00 00 00 00 00 00 00          Left Ctrl & Alt
                                                         released
        Usages set:  (Break:  E0 E2) (Make: )

        Input report 0: 00 00 04 00 00 00 00 00          A pressed
        Usages set:  07:04 (Break: ) (Make:  04)
```

Listing 23.6 Example *HidKbdUser* output (continued)

```
        Input report 0: 00 00 00 00 00 00 00 00              A released
        Usages set:  (Break:  04) (Make: )


        Input report 0: 00 00 05 00 00 00 00 00              B pressed
        Usages set:  07:05 (Break: ) (Make:  05)


        Input report 0: 00 00 00 00 00 00 00 00              B released
        Usages set:  (Break:  05) (Make: )


        Input report 0: 00 00 06 00 00 00 00 00              C pressed
        Usages set:  07:06 (Break: ) (Make:  06)


        Input report 0: 00 00 00 00 00 00 00 00              C released
        Usages set:  (Break:  06) (Make: )


        Input report 0: 00 00 29 00 00 00 00 00              Esc pressed
        Usages set:  07:29 (Break: ) (Make:  29)



Test 3
        Output report:  00
        Wrote output report OK
        Output report:  04
        Wrote output report OK
        Output report:  02
        Wrote output report OK
        Output report:  01
        Wrote output report OK
        Output report:  06
        Wrote output report OK
        Output report:  05
        Wrote output report OK
        Output report:  03
        Wrote output report OK
        Output report:  07
        Wrote output report OK
        Output report:  00
        Wrote output report OK


Test 4
        CloseHandle worked
```

Kernel Mode HID Clients

A device driver can talk to a HID device using the HID class driver. As mentioned previously, it is far easier to write a user mode application to control a HID device. However, you may find that it is necessary to write a HID client driver (e.g., if you need to implement an existing device API). A kernel client should be more efficient than a user mode client, though speed ought not to be a problem for most human input devices.

Client Types

A kernel mode HID client can take one of two main forms, depending on how it relates to devices. An "AddDevice" HID client uses installation files (as usual) to layer itself above the HID class driver for each device. Alternatively, "Plug and Play Notification" HID client driver is not initially associated with any one device. Instead, it receives notifications when a HID device arrives or disappears. A PnP Notification HID client makes its own device objects if a HID device of interest arrives.

As I have not examined Plug and Play Notification before, this chapter's example driver, HidKbd, uses this technique.

"AddDevice" HID Clients

An "AddDevice" HID client will look like all the previous WDM device drivers. The driver is loaded using installation INF files in the normal way. The driver's AddDevice routine is called when a suitable device is loaded. The driver must handle Plug and Play IRPs in the same way as usual.

An "AddDevice" HID client makes calls to the HID class driver by calling the *NextStack-Device* as usual. The HID class driver responds to read and write requests as well as various IOCTL IRPs.

As is usual for WDM drivers, your driver's upper edge may be completely different. For example, the kbdhid.sys system keyboard driver has an upper edge that reports 8042 equivalent key presses. However, kbdhid.sys makes HID requests on its lower edge to find the HID keyboard data.

Things can now get complicated. Note that user mode clients will still be able to find the HID device. If they try to send HID requests to the HID device, these requests will be routed to your driver first. If your driver implements some other upper edge, it will not recognize these requests. Or, if you are being very clever, you could recognize when HID operations have been requested and route them straight through to the HID class driver. If you do this job, you are acting as a HID filter driver.

All in all, it is probably easier if you do not use the "AddDevice" device technique when writing a HID client driver. Therefore, I shall take a close look at how to write a Plug and Play Notification client. However, kernel mode PnP Notification does not seem to work in Windows 98, so you will have to write an "AddDevice" client.

"PnP Notification" HID Clients

A PnP Notification HID client is usually an NT style driver. Therefore, it is not loaded as a result of a device being plugged in. As an NT style driver, it is usually loaded when the system boots up. Installation INF files are not used. Instead, you must write a custom installation program, as described in Chapter 11.

The `DriverEntry` routine calls the `IoRegisterPlugPlayNotification` function to ask to receive any device interface change events for a particular device interface GUID. The driver `AbcUnload` routine calls `IoUnregisterPlugPlayNotification` to indicate that it no longer wants to receive such notifications.

In this case, the device interface change callback is informed whenever any HID device is plugged into the system. The `HidKbd` code then interrogates the device to see if it is a HID keyboard. If it is, it creates a `HidKbd` device.

Just to complicate matters, `HidKbd` now has two options. First, it can layer itself over the HID device so that is becomes part of the device stack for this device. However, this suffers from the same drawbacks as the "AddDevice" technique, that user mode requests to the HID device would be routed to `HidKbd`.

The `HidKbd` driver takes the second option. `HidKbd` stores a pointer to the HID device object, but does not layer itself above the HID device. Both `HidKbd` and a user mode application can therefore call the HID class driver safely.

Plug and Play Notifications

Kernel mode drivers can register to receive three different types of notification events. As far as I can tell, none of these notifications work in Windows 98, as calling `IoRegisterPlug-PlayNotification` seems to hang the system. Therefore, all the subsequent discussion here applies to Windows 2000 only. The Beta 2 version of W2000 let the `HidKbd` driver access the HID keyboard. Later versions do not, so most of the code in `HidKbd` will now not run.

Table 23.2 shows how to call `IoRegisterPlugPlayNotification`. The `EventCategory` parameter specifies what type of event you want to receive. `HidKbd` only asks for device interface change events, `EventCategoryDeviceInterfaceChange`, and so passes the relevant GUID as the `EventCategoryData` parameter. It also specifies the `PNPNOTIFY_DEVICE_INTERFACE_INCLUDE_EXISTING_INTERFACES` flag as the `EventCategoryFlags` parameter. This means that it receives notifications straightaway for any existing devices that support the given interface.

If you register to receive `EventCategoryHardwareProfileChange` events, you are supposed to receive hardware profile change events. The callback is told whether a *Query Change*, *Change Complete,* or *Change Cancelled* event occurred.

Registering for `EventCategoryTargetDeviceChange` events asks for notifications when a target device is removed. You must pass a `PFILE_OBJECT` as the `EventCategoryFlags`. The callback is told whether a *Query Remove, Remove Complete,* or *Remove Cancelled* event occurred. In my mind, there is a fundamental flaw to this notification. You must pass a file object to `IoRegisterPlugPlayNotification`. To have a file object pointer, you must have opened a file. If a file is open on a device, Windows 2000 automatically stops any PnP remove request for the device. When I tried it, my target device callback received a *Query Remove* event followed straightaway by a *Remove Cancelled* event. It seems as though registering for target device notifications automatically stops any remove requests from completing.

Table 23.2   `IoRegisterPlugPlayNotification` **function**

NTSTATUS IoRegisterPlugPlayNotification	(IRQL==PASSIVE_LEVEL)
Parameter	Description
IN IO_NOTIFICATION_EVENT_CATEGORY EventCategory	EventCategoryDeviceInterfaceChange EventCategoryHardwareProfileChange or EventCategoryTargetDeviceChange
IN ULONG EventCategoryFlags	Optionally PNPNOTIFY_DEVICE_INTERFACE_ INCLUDE_EXISTING_INTERFACES
IN PVOID EventCategoryData	Device GUID, NULL, or file object, respectively
IN PDRIVER_OBJECT DriverObject	The driver object
IN PDRIVER_NOTIFICATION_CALLBACK_ROUTINE CallbackRoutine	Your callback routine name
IN PVOID Context	Context to pass to your callback
OUT PVOID *NotificationEntry	Output value to pass to IoUnregisterPlugPlayNotification

Device Interface Change Notifications

Anyway, device interface change notifications do seem to work, so let's look at the HidKbd device interface change callback, HidKbdDicCallback shown in Listing 23.7. Each callback receives the context pointer and a notification structure pointer. For device interface change events, this is a pointer to a DEVICE_INTERFACE_CHANGE_NOTIFICATION structure.

The notification structure *Event* GUID field says what type of event has occurred. The *SymbolicLinkName* UNICODE_STRING field can be used to open a handle to the device.

When a new device arrives, *Event* contains GUID_DEVICE_INTERFACE_ARRIVAL. For *Device Removal* events, *Event* is GUID_DEVICE_INTERFACE_REMOVAL. HidKbdDicCallback uses the IsEqualGUID macro to detect each of these events. For *Device Arrival* events, the CreateDevice routine is called, and for *Device Removals*, DeleteDevice is called.

Listing 23.7 PnP device interface change notification callback

```
NTSTATUS HidKbdDicCallback(
    IN PVOID NotificationStructure,
    IN PVOID Context)
{
    PDEVICE_INTERFACE_CHANGE_NOTIFICATION dicn =
        (PDEVICE_INTERFACE_CHANGE_NOTIFICATION)NotificationStructure;
    PDRIVER_OBJECT DriverObject = (PDRIVER_OBJECT)Context;

    if( IsEqualGUID( dicn->Event, GUID_DEVICE_INTERFACE_ARRIVAL))
    {
```

Listing 23.7 PnP device interface change notification callback (continued)

```
        DebugPrint("Device arrival: %T", dicn->SymbolicLinkName);
        CreateDevice( DriverObject, dicn->SymbolicLinkName);
    }
    else if( IsEqualGUID( dicn->Event, GUID_DEVICE_INTERFACE_REMOVAL))
    {
        DebugPrint("Device removal: %T", dicn->SymbolicLinkName);
        DeleteDevice(dicn->SymbolicLinkName);
    }
    else
        DebugPrint("Some other device event: %T", dicn->SymbolicLinkName);

    return STATUS_SUCCESS;
}
```

HidKbd Devices

Hold onto your hats, as the HidKbd device handling is a bit complicated.

Remember that the CreateDevice routine is called whenever a HID device is added to the system. HidKbd is just looking for a HID keyboard. However, it must cope if a HID device arrives that is not a keyboard, and if two HID keyboards arrive (it is possible).

HidKbd tries to make things simple by only coping with one keyboard. If a second HID keyboard arrives, it is ignored.

HidKbd creates its own device called \\.\HidKbd for the first HID keyboard that arrives. A user mode program can open a handle to this device and issue read requests. HidKbd handles these by calling the HID device to get an input report. HidKbd does not do anything for Write or IOCTL requests.

Has a HID Keyboard Been Found?

The CreateDevice routine shown in Listing 23.8 starts by checking to see if it has already found a HID keyboard. The HidKbdDo global variable stores a pointer to the HidKbd device object; if this is non-NULL, a suitable keyboard has already been found.

The first job is to open a connection to the HID device and see if it is HID keyboard. While CreateDevice could use ZwCreateFile to open a handle to the HID device, the IoGet-DeviceObjectPointer routine is what is really needed. IoGetDeviceObjectPointer is passed the symbolic link for a device. If the symbolic link is found, IoGetDeviceObjectPointer issues a Create IRP to the device, passing an empty string as the IRP filename parameter[4]. IoGetDeviceObjectPointer returns two pieces of information: the device object pointer and the PFILE_OBJECT pointer.

HidKbd is going to use the HID device object pointer a lot. In addition, it needs a file object pointer when it eventually reads reports from (or writes reports to) the HID class driver. However, in the mean time, CreateDevice closes the file object pointer by calling

---

4. The device to which you are connecting must accept a Create IRP call with an empty filename.

ObDereferenceObject. Why is this done? If a file is open on a device, Windows 2000 will not let a device be removed. The file must be closed to let device removals take place.

CreateDevice now inspects the HID device capabilities using the GetCapabilities routine, which I describe later. If GetCapabilities finds a HID keyboard, *HidKbdUser*, like its user mode equivalent, returns a pointer to the preparsed data and the maximum input and output report lengths.

Creating the HidKbd **Device**

If a HID keyboard is found, CreateDevice can go on to create its own device object. However, it first calls ObReferenceObjectByPointer to reference the HID class driver device object. This ensures that the device object will not disappear from under our feet. When the HidKbd device is deleted, ObDereferenceObject is called to dereference the object. Note that referencing this device object does not stop it from processing removal requests successfully.

The next job is to allocate some memory for a copy of the HID device symbolic link name. This name is stored in a UNICODE_STRING field called *HidSymLinkName* in the new device extension. The HidKbd *Device Removal* event handler only deletes a device if the correct underlying HID device is being removed.

HidKbd now sets up the device name and symbolic link names for the new HidKbd device. These are \Device\HidKbd and \DosDevices\HidKbd respectively, and so the device appears in Win32 as \\.\HidKbd.

HidKbd is finally ready to call IoCreateDevice, with most of the parameters set up as usual. However this time it passes the device type FILE_DEVICE_KEYBOARD, as this seems most appropriate. If the device is created successfully, the global variable, HidKbdDo, stores the device object. Next, CreateDevice sets up the device extension. Finally, CreateDevice calls IoCreateSymbolicLink to create the symbolic link that makes the device visible to Win32 applications.

Note that HidKbd did not call IoAttachDeviceToDeviceStack. If it did make this call, the HidKbd device would be layered over the HID device. Any user mode calls direct to the HID device would arrive at the HidKbd device first, which is not what is wanted.

As a last touch, note that CreateDevice sets up the *StackSize* field of the HidKbdDo device object. If HidKbd had called IoAttachDeviceToDeviceStack, this routine would have set the IRP stack size to be one greater than the HID device stack size. As it did not call IoAttachDeviceToDeviceStack, HidKbd has to do this same job. Later, HidKbd passes IRPs to the HID class driver for processing. Setting the stack size in this way ensures that there will be enough IRP stack locations available.

Listing 23.8 HidKbd CreateDevice **routine**

```
void CreateDevice( IN PDRIVER_OBJECT DriverObject,
    IN PUNICODE_STRING HidSymLinkName)
{
    if( HidKbdDo!=NULL)
    {
        DebugPrintMsg("Already got HidKbdDo");
        return;
    }
```

Listing 23.8 HidKbd CreateDevice **routine (continued)**

```
PFILE_OBJECT HidFileObject = NULL;
PDEVICE_OBJECT HidDevice;
NTSTATUS status = IoGetDeviceObjectPointer( HidSymLinkName,
    FILE_ALL_ACCESS, &HidFileObject, &HidDevice);
if( !NT_SUCCESS(status))
{
    DebugPrintMsg("IoGetDeviceObjectPointer failed");
    return;
}

// Close file object
ObDereferenceObject(HidFileObject);

// Inspect HID capabilities here
PHIDP_PREPARSED_DATA HidPreparsedData = NULL;
USHORT HidInputReportLen, HidOutputReportLen;
if( !GetCapabilities( HidDevice, HidPreparsedData, HidInputReportLen,
    HidOutputReportLen))
{
    DebugPrintMsg("GetCapabilities failed");
    FreeIfAllocated(HidPreparsedData);
    return;
}

// Reference device object
status = ObReferenceObjectByPointer( HidDevice, FILE_ALL_ACCESS, NULL,
    KernelMode);
if( !NT_SUCCESS(status))
{
    DebugPrintMsg("ObReferenceObjectByPointer failed");
    FreeIfAllocated(HidPreparsedData);
    return;
}

// Allocate a buffer for the device ext HidSymLinkName
PWSTR HidSymLinkNameBuffer =
    (PWSTR)ExAllocatePool( NonPagedPool, HidSymLinkName->MaximumLength);
if( HidSymLinkNameBuffer==NULL)
{
    FreeIfAllocated(HidPreparsedData);
    ObDereferenceObject(HidDevice);
    return;
```

Listing 23.8 HidKbd CreateDevice **routine (continued)**

```
    }

#define NT_DEVICE_NAME  L"\\Device\\HidKbd"
#define SYM_LINK_NAME   L"\\DosDevices\\HidKbd"

    // Initialise NT and Symbolic link names
    UNICODE_STRING deviceName, linkName;
    RtlInitUnicodeString( &deviceName, NT_DEVICE_NAME );
    RtlInitUnicodeString( &linkName, SYM_LINK_NAME );

    // Create our device object
    status = IoCreateDevice( DriverObject, sizeof(HIDKBD_DEVICE_EXTENSION),
        &deviceName, FILE_DEVICE_KEYBOARD, 0, FALSE, &HidKbdDo);
    if( !NT_SUCCESS(status))
    {
        HidKbdDo = NULL;
        FreeIfAllocated(HidSymLinkNameBuffer);
        FreeIfAllocated(HidPreparsedData);
        ObDereferenceObject(HidDevice);
        return;
    }

    // Set up our device extension
    PHIDKBD_DEVICE_EXTENSION dx =
        (PHIDKBD_DEVICE_EXTENSION)HidKbdDo->DeviceExtension;
    dx->HidKbdDo = HidKbdDo;
    dx->HidDevice = HidDevice;
    dx->HidPreparsedData = HidPreparsedData;

    dx->HidSymLinkName.Length = 0;
    dx->HidSymLinkName.MaximumLength = HidSymLinkName->MaximumLength;
    dx->HidSymLinkName.Buffer = HidSymLinkNameBuffer;
    RtlCopyUnicodeString( &dx->HidSymLinkName, HidSymLinkName );

    // Create a symbolic link so our device is visible to Win32...
    DebugPrint("Creating symbolic link %T",&linkName);
    status = IoCreateSymbolicLink( &linkName, &deviceName);
    if( !NT_SUCCESS(status))
    {
        DebugPrintMsg("Could not create symbolic link");
        FreeIfAllocated(dx->HidSymLinkName.Buffer);
        IoDeleteDevice(HidKbdDo);
```

Listing 23.8 `HidKbd` `CreateDevice` **routine (continued)**

```
        ObDereferenceObject(HidDevice);
        HidKbdDo = NULL;
        return;
    }


    HidKbdDo->Flags |= DO_BUFFERED_IO;
    HidKbdDo->Flags &= ~DO_DEVICE_INITIALIZING;


    HidKbdDo->StackSize = HidDevice->StackSize+1;


    DebugPrintMsg("Device created OK");
}
```

Deleting the `HidKbd` Device

The `HidKbd` device must be deleted in two circumstances. First, if `HidKbd` is notified that the HID device has been removed. Second, if the `HidKbd` driver is unloaded. The `DeleteDevice` routine shown in Listing 23.9 handles both these cases. When the driver is unloaded the `HidSymLinkName` parameter is NULL. However, if a HID device is being removed, `HidSym-LinkName` contains the symbolic link name of the device.

`DeleteDevice` first checks that a `HidKbd` device has been created. If one has and a device is being removed, `DeleteDevice` calls `RtlCompareUnicodeString` to see if the device name matches the one to which `HidKbd` refers. If it is a different HID device, nothing more is done.

Before the device is deleted, `DeleteDevice` must free any memory that is associated with it (i.e., the preparsed data and the buffer for the copy of the symbolic link name). `DeleteDevice` now remakes the `HidKbd` symbolic link name. The symbolic link name is deleted using `IoDe-leteSymbolicLink`. `ObDereferenceObject` is called to deference the HID device object. Finally, `IoDeleteDevice` deletes the `HidKbd` device.

Listing 23.9 `HidKbd` `DeleteDevice` **routine**

```
void DeleteDevice( IN PUNICODE_STRING HidSymLinkName)
{
    if( HidKbdDo==NULL)
        return;

    PHIDKBD_DEVICE_EXTENSION dx =
        (PHIDKBD_DEVICE_EXTENSION)HidKbdDo->DeviceExtension;
    if( HidSymLinkName!=NULL &&
        RtlCompareUnicodeString( HidSymLinkName, &dx->HidSymLinkName, FALSE)!=0)
    {
        DebugPrintMsg("DeleteDevice: symbolic link does not match our device");
        return;
    }
```

Listing 23.9 `HidKbd DeleteDevice` **routine (continued)**

```
    DebugPrintMsg("Deleting our device");

    FreeIfAllocated(dx->HidPreparsedData);
    FreeIfAllocated(dx->HidSymLinkName.Buffer);

    // Initialise Symbolic link names
    UNICODE_STRING linkName;
    RtlInitUnicodeString( &linkName, SYM_LINK_NAME);

    // Remove symbolic link
    DebugPrint("Deleting symbolic link %T",&linkName);
    IoDeleteSymbolicLink( &linkName);

    ObDereferenceObject(dx->HidDevice);
    IoDeleteDevice(HidKbdDo);

    HidKbdDo = NULL;
}
```

Getting HID capabilities

The `HidKbd GetCapabilities` routine is largely the same as its equivalent in *HidKbdUser*. It returns `true` if the HID device capabilities indicate that it is a HID keyboard.

However, `GetCapabilities` must obtain the device attributes and the preparsed data in a different way from *HidKbdUser*. The `GetPreparsedData` routine shown in Listing 23.10 does this job.

`GetPreparsedData` uses two IOCTLs to obtain the information needed from the HID class driver. `IOCTL_HID_GET_COLLECTION_INFORMATION` returns the attributes and the size of memory buffer required for the preparsed data. Next, `IOCTL_HID_GET_COLLECTION_DESCRIPTOR` is used to get the preparsed data itself. `GetPreparsedData` returns a pointer to the preparsed data memory. This memory must eventually be freed.

The `CallHidIoctl` routine is used to issue the two IOCTLs. This works in a very similar way to the `CallUSBDI` routine shown in Listing 21.2. It calls `IoBuildDeviceIoControlRequest` to build the IOCTL passing the IOCTL code, the output buffer details and a completion event. `CallHidIoctl` calls the HID class driver using `IoCallDriver`, waiting until it completes using the event, if necessary.

Listing 23.10 `HidKbd GetPreparsedData` **routine**

```
bool GetPreparsedData( IN PDEVICE_OBJECT HidDevice,
    OUT PHIDP_PREPARSED_DATA HidPreparsedData)
{
    HID_COLLECTION_INFORMATION HidCi;
    NTSTATUS status = CallHidIoctl( HidDevice,
        IOCTL_HID_GET_COLLECTION_INFORMATION, &HidCi, sizeof(HidCi));
```

Listing 23.10 HidKbd GetPreparsedData **routine (continued)**

```
    if( !NT_SUCCESS(status))
    {
        DebugPrint("IOCTL_HID_GET_COLLECTION_INFORMATION failed %x", status);
        return false;
    }
    DebugPrint("HID attributes: VendorID=%4x, ProductID=%4x,
        VersionNumber=%4x", HidCi.VendorID, HidCi.ProductID,
        HidCi.VersionNumber);

    ULONG PreparsedDatalen = HidCi.DescriptorSize;
    DebugPrint("PreparsedDatalen %d",PreparsedDatalen);
    HidPreparsedData = (PHIDP_PREPARSED_DATA)ExAllocatePool( NonPagedPool,
        PreparsedDatalen);
    if( HidPreparsedData==NULL)
    {
        DebugPrintMsg("No memory");
        return false;
    }

    status = CallHidIoctl( HidDevice, IOCTL_HID_GET_COLLECTION_DESCRIPTOR,
        HidPreparsedData, PreparsedDatalen);
    if( !NT_SUCCESS(status))
    {
        DebugPrint("IOCTL_HID_GET_COLLECTION_DESCRIPTOR failed %x", status);
        return false;
    }

    return true;
}
```

Opening and Closing the HidKbd Device

The DDK documentation for the HID class driver read and write handler says that the IRP file object pointer must be valid. HidKbd obtained a file object using IoGetDeviceObject-Pointer when it first found a HID device. However, this file handle was closed because it stops the HID device from being removed.

When a user mode application opens a handle to a HidKbd device, the Create IRP handler receives another file object pointer. This same file object pointer is passed in subsequent Read, Write, and Close IRPs, etc.

The HidKbd Create IRP handler, HidKbdCreate, therefore, has to tell the HID class driver about this new file object pointer. It does this by passing the Create IRP to the HID class driver. This is actually very easy to do by putting this extra code in the HidKbdCreate routine.

As HidKbd does not need to process the IRP afterwards, there is no need to set a completion routine.

```
// Forward IRP to HID class driver device
IoSkipCurrentIrpStackLocation(Irp);
return IoCallDriver( dx->HidDevice, Irp);
```

The HidKbd Close IRP handler, HidKbdClose, has exactly the same lines in it. This tells the HID class driver that the file handle is being closed.

A side effect of making the HID class driver open a handle for the device is that Windows 2000 will not let the HID device be removed for the duration. This is a perfectly acceptable behavior.

Reading and Writing Data

Our HID kernel mode client is now finally ready to read and write data.

HidKbd currently only supports reading of input reports. The Read IRP expects the provided buffer to be big enough. For a keyboard-input report, the buffer must be at least nine bytes long. The first byte will be 0, with the eight bytes of the input report in the remaining bytes. HidKbd makes no attempt to analyze the data in the same way as *HidKbdUser*. Instead, it simply returns all the information to the user mode application.

The main Read IRP handler, HidKbdRead, eventually calls ReadHidKbdInputReport, shown in Listing 23.11. ReadHidKbdInputReport is passed the precious file object pointer and a pointer to the buffer. It returns a count of the number of bytes transferred.

ReadHidKbdInputReport looks similar to the CallUSBDI and CallHidIoctl routines described before. This time HidKbd must issue a read request to the HID class driver, so it uses IoBuildSynchronousFsdRequest kernel call to build a suitable Read IRP. An event can be used to wait synchronously for the IRP to be completed, so ReadHidKbdInputReport must be called at PASSIVE_LEVEL IRQL.

By default, IoBuildSynchronousFsdRequest does not insert a file object pointer into the IRP. Therefore, HidKbd must do this job by hand. It calls IoGetNextIrpStackLocation to get the stack location that will be seen by the next driver, the HID class driver. ReadHidKbdInputReport then simply stores the PFILE_OBJECT in the stack FileObject field.

Finally, HidKbd runs IoCallDriver to call the HID class driver. If the IRP is still pending when this call returns, ReadHidKbdInputReport waits for the event to become signalled when the IRP does complete.

I have left out one small part of the story. The DDK says that HID class drivers use Direct I/O for their input and output buffers, not Buffered I/O. Luckily, IoBuildSynchronousFsdRequest sorts this out for us. It checks if the called driver uses Direct I/O. If it does, it allocates the required MDL for the passed input or output buffer (and deallocates it on completion).

Listing 23.11 ReadHidKbdInputReport **routine**

```
NTSTATUS ReadHidKbdInputReport( PFILE_OBJECT FileObject, PVOID Buffer,
    ULONG& BytesTxd)
{
    PHIDKBD_DEVICE_EXTENSION dx =
        (PHIDKBD_DEVICE_EXTENSION)HidKbdDo->DeviceExtension;

    BytesTxd = 0;
    if( HidKbdDo==NULL) return STATUS_NO_MEDIA_IN_DEVICE;

    IO_STATUS_BLOCK IoStatus;
    IoStatus.Information = 0;
    KEVENT event;
    LARGE_INTEGER FilePointer;
    FilePointer.QuadPart = 0i64;

    // Initialise IRP completion event
    KeInitializeEvent(&event, NotificationEvent, FALSE);

    PIRP Irp = IoBuildSynchronousFsdRequest(
        IRP_MJ_READ, dx->HidDevice,
        Buffer, dx->HidInputReportLen, &FilePointer,
        &event, &IoStatus);
    if( Irp==NULL) return STATUS_INSUFFICIENT_RESOURCES;

    // Store file object pointer
    PIO_STACK_LOCATION IrpStack = IoGetNextIrpStackLocation(Irp);
    IrpStack->FileObject = FileObject;

    // Call the driver and wait for completion if necessary
    NTSTATUS status = IoCallDriver( dx->HidDevice, Irp);
    if (status == STATUS_PENDING)
    {
        KeWaitForSingleObject( &event, Suspended, KernelMode, FALSE, NULL);
        status = IoStatus.Status;
    }

    // return IRP completion status
```

Listing 23.11 `ReadHidKbdInputReport` **routine (continued)**

```
        DebugPrint("ReadHidKbdInputReport: status %x", status);
        BytesTxd = IoStatus.Information;
        return status;
    }
```

Permanently Allocated IRP

A kernel mode HID client is likely to be reading many input reports. Rather than building up a suitable IRP for each call, it is more efficient to have one at the ready all the time. However, this approach is a bit more complicated to set up. The `HidKbd` driver has this alternative code commented out.

When a `HidKbd` device is created, it must allocate the IRP that will be reused in all subsequent read and write requests. The `SetupHidIrp` routine, shown in Listing 23.12, calls `IoAllocateIrp` to obtain a suitable IRP pointer from the I/O Manager. As IRPs have a variable number of stack locations, `SetupHidIrp` must pass the desired stack size. The second parameter to `IoAllocateIrp` should be `FALSE` for intermediate drivers.

It also makes sense to preallocate a buffer for input and output reports. `SetupHidIrp` works out the size of buffer needed and allocates it from the nonpaged pool. The final preparatory step is to allocate an MDL for this buffer. Remember that the HID class driver uses Direct I/O and so needs an MDL passed in Read and Write IRPs. The call to `IoAllocateMdl` makes a suitable MDL out of the buffer pointer.

Listing 23.12 `SetupHidIrp` **routine**

```
void SetupHidIrp( IN PHIDKBD_DEVICE_EXTENSION dx, IN CCHAR StackSize)
{
    // Work out maximum size of input and output reports
    dx->HidMaxReportLen = dx->HidInputReportLen;
    if( dx->HidOutputReportLen > dx->HidMaxReportLen)
        dx->HidMaxReportLen = dx->HidOutputReportLen;

    DebugPrint("Setting up HidIrp etc %d", dx->HidMaxReportLen);
    if( dx->HidMaxReportLen==0) return;

    dx->HidReport = ExAllocatePool( NonPagedPool, dx->HidMaxReportLen);
    if( dx->HidReport==NULL) return;

    dx->HidIrp = IoAllocateIrp( StackSize, FALSE);
    if( dx->HidIrp==NULL) return;

    dx->HidReportMdl = IoAllocateMdl( dx->HidReport, dx->HidMaxReportLen,
        FALSE, FALSE, NULL);
    if( dx->HidReportMdl==NULL)
    {
```

Listing 23.12 `SetupHidIrp` **routine (continued)**

```
            IoFreeIrp(dx->HidIrp);
            dx->HidIrp = NULL;
    }
}
```

When the `HidKbd` device is removed, the IRP, the buffer memory, and the MDL must be freed. Listing 23.13 shows how the `RemoveHidIrp` routine does this job using the `IoFreeMdl`, `IoFreeIrp`, and `ExFreePool` routines.

Listing 23.13 `RemoveHidIrp` **routine**

```
void RemoveHidIrp( IN PHIDKBD_DEVICE_EXTENSION dx)
{
    DebugPrintMsg("Removing HidIrp etc");
    if( dx->HidReportMdl!=NULL)
    {
        IoFreeMdl(dx->HidReportMdl);
        dx->HidReportMdl = NULL;
    }
    if( dx->HidIrp!=NULL)
    {
        IoFreeIrp(dx->HidIrp);
        dx->HidIrp = NULL;
    }
    if( dx->HidReport!=NULL)
    {
        ExFreePool(dx->HidReport);
        dx->HidReport = NULL;
    }
}
```

I can now discuss how to use this preallocated IRP. Listing 23.14 shows the replacement `ReadHidKbdInputReport` routine. This time, it cannot use `IoBuildSynchronousFsdRequest`, so the IRP and its stack must be built by hand.

The `IoInitializeIrp` call is used to initialize the IRP. `IoInitializeIrp` incorrectly clears the IRP *AllocationFlags* field, so this must be preserved. In W2000, `IoReuseIrp` correctly reinitialises the IRP. `ReadHidKbdInputReport` then stores the MDL for the buffer in the IRP *MdlAddress* field. As before, it calls `IoGetNextIrpStackLocation` to get the next stack location. `ReadHidKbdInputReport` must set up all the stack parameters carefully: the *MajorFunction*, the *Parameters.Read* fields, and the *FileObject*.

Finally, `ReadHidKbdInputReport` needs to set a completion routine so that it knows when the IRP has completed. It passes an event to the completion routine. The completion routine sets the event into the signalled state when it is run). `ReadHidKbdInputReport` waits until the event is set (i.e., when the IRP has been completed by the lower driver. Assuming that the

HID driver has returned data, the final job is to copy the data into the user's buffer, using RtlCopyMemory.

The ReadComplete completion routine returns STATUS_MORE_PROCESSING_REQUIRED. This stops the I/O Manager from deleting the IRP. The IRP will be reused so it must not be deleted.

Listing 23.14 New ReadHidKbdInputReport routine

```
NTSTATUS ReadHidKbdInputReport( PHIDKBD_DEVICE_EXTENSION dx,
    PFILE_OBJECT FileObject, PVOID Buffer, ULONG& BytesTxd)
{
    BytesTxd = 0;
    if( HidKbdDo==NULL || dx->HidIrp==NULL || dx->HidReport==NULL)
    {
        DebugPrintMsg("No HidIrp");
        return STATUS_INSUFFICIENT_RESOURCES;
    }


    RtlZeroMemory( dx->HidReport, dx->HidMaxReportLen);


    // Initialise IRP completion event
    KEVENT event;
    KeInitializeEvent(&event, NotificationEvent, FALSE);


    // Initialise IRP
    UCHAR AllocationFlags = dx->HidIrp->AllocationFlags;
    IoInitializeIrp( dx->HidIrp, IoSizeOfIrp(HidKbdDo->StackSize), HidKbdDo->StackSize);
    dx->HidIrp->AllocationFlags = AllocationFlags;


    dx->HidIrp->MdlAddress = dx->HidReportMdl;


    PIO_STACK_LOCATION IrpStack = IoGetNextIrpStackLocation(dx->HidIrp);
    IrpStack->MajorFunction = IRP_MJ_READ;
    IrpStack->Parameters.Read.Key = 0;
    IrpStack->Parameters.Read.Length = dx->HidInputReportLen;
    IrpStack->Parameters.Read.ByteOffset.QuadPart = 0;
    IrpStack->FileObject = FileObject;


    IoSetCompletionRoutine( dx->HidIrp, (PIO_COMPLETION_ROUTINE)ReadComplete,
        &event, TRUE, TRUE, TRUE);


    NTSTATUS status = IoCallDriver( dx->HidDevice, dx->HidIrp);
    if (status == STATUS_PENDING)
    {
        KeWaitForSingleObject( &event, Suspended, KernelMode, FALSE, NULL);
```

Listing 23.14 New `ReadHidKbdInputReport` routine (continued)

```
        status = dx->HidIrp->IoStatus.Status;
    }


    // return IRP completion status
    DebugPrint("ReadHidKbdInputReport: status %x", status);
    BytesTxd = dx->HidIrp->IoStatus.Information;
    if( BytesTxd>0)
        RtlCopyMemory( Buffer, dx->HidReport, BytesTxd);
    return status;
}


NTSTATUS ReadComplete( IN PDEVICE_OBJECT fdo, IN PIRP Irp, IN PKEVENT Event)
{
    KeSetEvent(Event, 0, FALSE);
    return STATUS_MORE_PROCESSING_REQUIRED;
}
```

Even Better ...

Two problems exist with the permanently allocated IRP solution I have just presented. The first is that the driver will not cope with two "simultaneous" read requests as it uses the same buffer in each call. A quick fix to this problem would be to allow only one read request at a time. The next best solution is dropping the shared buffer; an MDL must then be allocated for the user buffer in each read request.

In fact, the best solution is not to use a permanently allocated IRP, but to reuse the Read IRP. If the HidKbd device uses Direct I/O, the operating system will even do the MDL allocation. In this version, the ReadHidKbdInputReport routine only needs to set up the next IRP stack location appropriately. In fact, calling IoCopyCurrentIrpStackLocationToNext will probably do this job just fine.

The second problem with both the earlier techniques of calling the HID class driver is that they can be inefficient. In both the earlier cases, the call to KeWaitForSingleObject forces the current thread to block waiting for the event to become signalled. As HidKbd may operate in the context of a user thread, this may stop any other overlapped operations from running.[5] The solution to this problem is to modify the completion routine. If the completion routine completes the original Read IRP, there is no need for ReadHidKbdInputReport to wait for the IRP completion event.

This technique should be used wherever possible. The HidKbd Create and Close IRP use this technique as they pass their IRPs to the HID class driver, which completes them in due course. However, it is probably still worth using events in the CallHidIoctl routine for two reasons. The first is that HidKbd needs to know the IRP results. Secondly, my guess is that the

---

5. So far, we have not considered how the kernel calls our driver. However, to be most efficient, we should not block while processing an IRP for very long. Blocking in a Power or PnP IRP handler or a system thread is fine.

HID class driver will be able to complete these IOCTLs straightaway, as it should already have the information at hand.

The CallUSBDI routine in the *UsbKbd* driver is a candidate for this technique, as it is more than likely that the USB class drivers will take some time to process a USBDI request. However, it is usually the case that the USBDI call results are needed. Processing the results in a completion routine is just about possible. However, this will probably lead to code that is very complicated. In the end, it is probably simplest to leave the *UsbKbd* code as it is.

Other HID Class IOCTLs

The DDK header files define several other HID IOCTLs. However, some of these are used by the HID class driver when it talks to a minidriver. It is not clear if any of these are available to HID clients.

Conclusion

This chapter concludes my look at the Human Input Device class driver. It is straightforward to write a user mode application to communicate with a HID device. A Win32 program looks for devices that use the HID device interface. They must then interrogate each device to see if their capabilities are of interest. The program can then read input reports and send output reports.

If need be, you can write a kernel mode HID client. It is best if you use the Plug and Play Notification technique to find any devices that support the HID device interface. Then, you can get the device capabilities and send and receive reports in a broadly similar way to user mode applications.

Both user mode and kernel mode HID clients can make use of the HID parsing routines. These make it much easier to find HID device's capabilities and to generate and understand reports.

Appendix A

Information Resources

Microsoft is the best source of information for most core Windows device driver development issues. Various books and newsgroups are also available to help driver writers. You may well need to check out other sources of information particular to your type of driver. You may need to seek help from vendors, standards bodies, and trade associations, as well as other driver writers.

A Microsoft Developer Network (MSDN) Professional Subscription provides most of the basic information you need to write device drivers. You get all the basic tools, Driver Development Kits (DDKs), and beta test versions of the Microsoft operating systems. However, it is worth keeping an eye on the Microsoft websites, as helpful articles and late-breaking news can often be posted. You will also need a C or C++ compiler, such as Visual Studio. The crucial Microsoft web sites are:

```
http://www.microsoft.com/hwdev/
http://www.microsoft.com/hwdev/driver/
http://msdn.microsoft.com/developer/
```

The DDKs include many example drivers that are very useful. Searching through these examples will often show you how to use a particular function or technique. Alternatively, you can base your entire driver on an existing example. Needless to say, do not use code blindly — make sure you understand what is going on. Finally, the sample driver directories often contain useful documentation in the form of Word or text files.

Table A.1 Information resources

Advanced Configuration and Power Interface Specification, Revision 1.0	http://www.teleport.com/~acpi/
Device Bay Interface Specification, Version 1.0	http://www.device-bay.org
Display Data Channel Standard, Version 3.0 Extended Display Identification Data Standard, Version 2.0	http://www.vesa.org
El Torito—Bootable CD-ROM Format Specification, Version 1.0 Compaq, Intel, Phoenix BIOS Boot Specification, Version 1.01	http://www.ptltd.com/techs/specs.html
Human Input Devices	http://www.usb.org/developers/
IEEE 1394 Information	ftp://ftp.symbios.com/pub/standards/io/
Intel hardware developer site	http://developer.intel.com
Interoperability Specification for ICCs and Personal Computer Systems	http://www.smartcardsys.com
Media Status Notification Support Specification, Version 1.03 Plug and Play specifications	http://www.microsoft.com/hwdev/specs/ Vendor ID registration: pnpid@microsoft.com
MultiProcessor Specification, Version 1.4	Intel part number 242016-002 http://developer.intel.com
PC 99 System design guide	http://www.pcguide.com/ See the following books. PC 99a Addendum published in April 1999.
PCI Local Bus Specification, Revision 2.1 (PCI 2.1) PCI Bus Power Management Interface Specification for PCI to CardBus Bridge, Revision 1.0	PCI Special Interest Group Phone: USA (800) 433-5177 http://www.pcisig.com
Plug and Play Specifications External COM Device Specification, Version 1.0 Industry Standard Architecture (ISA) Specification, Version 1.0a Clarification to Plug and Play ISA Specification, Version 1.0a Parallel Port Device Specification, Version 1.0b Small Computer System Interface Specification, Version 1.0	http://www.microsoft.com/hwdev/specs/
Power Management specifications for device and bus classes Guidelines for audible noise and other OnNow technologies	http://www.microsoft.com/hwdev/onnow.htm

SFF 8070i, SFF 8038i, SFF 8090 (Mt. Fuji specification), and other SFF specifications	SFF Committee publications FaxAccess: USA (408) 741-1600 (fax-back) Fax: USA (408) 867-2115 `ftp://ftp.symbios.com/pub/standards/io/`
SysInternals web site for device driver utilities and information	`http://www.sysinternals.com/`
Smart Battery System Specification Smart Battery Charger Specification, Version 1.0 Smart Battery Selector Specification, Version 1.0	`http://www.sbs-forum.org/`
USB Specification, Version 1.0 USB Class Definition for Communications Devices, Version 0.9 USB Common Class Specification, Version 0.9 USB Device Class Definition for Audio Devices, Version 0.9 USB Device Class Definition for Human Interface Devices, Version 1.0 USB Device Class Definition for Mass Storage Devices, Version 0.9 USB Device Class Definition for Printing Devices, Version 1.0 USB HID Usages Tables, Version 0.9 USB Monitor Control Class Specification, Version 1.0 USB Power Devices Usages Table, Version 0.9	Phone: USA (503) 264-0590 Fax: USA (503) 693-7975 `http://www.usb.org/developers/`
Web-Based Enterprise Management (WBEM)	`http://www.microsoft.com/management/wbem/` `http://www.dmtf.org/`
Windows 98 and Windows 2000 DDKs, including NDIS documentation	MSDN Professional membership `http://msdn.microsoft.com/`
Windows Hardware Instrumentation Implementation Guidelines, Version 1.0 (WHIIG), Microsoft Corporation and Intel Corporation	`http://www.microsoft.com/hwdev/specs`
Wired for Management Baseline Specification, Version 2.0, Intel Corporation.	`http://developer.intel.com/ial/WfM/index.htm`

Newsgroups and Mail Lists

The two primary Newsgroups for DDK information are `comp.os.ms-windows.programmer.nt.kernel-mode` and `comp.os.ms-windows.programmer.vxd.`. The Microsoft news server at `msnews.microsoft.com` also has some useful newsgroups.

The *ntdev* mailing list covers all NT and W2000 development issues, including device drivers. Subscribe by sending an e-mail to `majordomo@atria.com` with the following in the body of the message, "subscribe ntdev your_email_address". A daily digest of this list can be obtained by subscribing to the *ntdev-digest* mailing list instead.

Information on the **DDK-L** mailing list can be found at http://www.chsw.com/DDK-L/. There is a nominal charge after the first 30 days.

Books

PC 99 System Design Guide, Version 1.0. Microsoft Press, 1998. ISBN 0-7356-0518-1.

Baker, Art. *The Windows NT Device Driver Book.* Second edition. Prentice Hall, 1999. ISBN 0-13-0204315.

Dekker, Edward N. and Newcomer, Joseph M. *Developing Windows NT Device Drivers, Vol. 1.* Addison Wesley Longman, 1999. ISBN 0-20169590-1.

Hazzah, Karen. *Writing Windows VxDs and Device Drivers.* Second edition. R&D Books, 1997. ISBN 0-87930-438-3.

Mason, Anthony and Viscarola, Peter G. *Windows NT Device Driver Development.* Macmillan Technical Publishing, 1998. ISBN 1-57870-0582.

Solomon, David A. *Inside Windows NT.* Second edition. Microsoft Press, 1998. ISBN 1-57231-677-2.

Appendix B

PC 99

PC 99 is a hardware and software specification for various types of PCs: Consumer PCs, Office PCs, and Entertainment PCs, including mobile PCs. Windows CE portables are not covered.

Certified PC 99 compliance gets you a nice "Designed for Microsoft Windows" logo on your box.

Drivers

Software plays a large part in achieving PC 99 compliance. Having a driver for both Windows 98 and Windows 2000 is a requirement for all PC types. If a driver is bundled with a PC when it is sold, it must pass the PC 99 compliance tests.

That's already a legacy PC

The PC 99 specification says that a driver must not use INI files, just the registry. INF style installation must be used, if possible. Drivers must be installed in the correct directories. A driver must not use the same filename as a system driver. A driver must be installable without user input. If a driver has any special parameters, a help file for these must be provided.

WDM drivers must fit in with all the bus specifications and device class driver specs.

The Specification

The PC 99 specification is in the MSDN Library CD Books section. It has many useful appendices and checklists. There are required, recommended, and optional features.

The specification is also available in book form. The PC 99 System Design Guide, version 1.0 is available from Microsoft Press (ISBN 0-7356-0518-1).

The general drift of the specification is to move PCs away from the traditional IBM-compatible PC to one dominated by PCI, SCSI, USB, and IEEE 1394 devices. No ISA slots are now allowed!

For each PC type, it defines what standard devices must be available and what are the possible options. For example, a basic PC (a 300Mhz Pentium with MMX with 32Mb memory) must have hard disk, keyboard, mouse, serial, and parallel ports, V.90 modem, a graphics adapter, and two USB ports. If possible, most external devices should use the USB bus.

Standard PC options include wireless operation, DVD drive, network card, floppy disk, audio, MPEG-2, video input and capture, and a TV tuner.

A PC should attempt to reduce the Total Cost of Ownership (TCO) by supporting Plug and Play and OnNow Power Management.

Plug and Play uses the ACPI specification to ease the management of resources as devices are added or removed from the system. Insertions or removals while the power is on (hot-swapping) is supported on appropriate buses.

The OnNow initiative aims to make PCs start up quicker. The PC can be put to sleep, in effect, rather than turned off completely.

It is recommended that the physical design and software support the accessibility initiatives.

IBM-Compatible PCs

The core hardware specification for PCs has hardly changed in years (i.e., it should be hardware compatible with the IBM AT specification).

This original specification defines standard hardware peripherals, either on the system motherboard or on plug-in cards. A PC needed a timer, keyboard controller, interrupt controller, real-time clock, two DMA controllers and page registers, serial ports and parallel ports.

Other peripherals soon became standard (e.g., IDE disk drives, CD-ROM drives, sound cards, and network cards).

These original devices had standard places where they had to live (e.g., they had set I/O port addresses, IRQ levels, and DMA lines). These are listed in Tables B.1, B.2, and B.3.

These devices were originally implemented by a host of different chips on the motherboard. Soon, chipsets appeared that combined all these devices.

Table B.1 Legacy ISA hardware system I/O (PC 99)

I/O Address	Default system function
0000-000F	Slave DMA
0010-0018	System
0001F	System
0020-0021	Master 8259
0040-0043 0048-004B	Programmable interrupt timer (PIT) #1, PIT #2
0050-0052	System
0060	Keyboard/mouse controller
0061	System control port B
0064	Keyboard/mouse status
0070-0071	Nonmaskable Interrupt (NMI) enable/real-time clock
0081-008B	DMA page registers
0090-0091	System
0092	System control port A
0093-009F	System
00A0-00A1	Slave interrupt controller
00C0-00DE	Master DMA controller
00F0-00F1	Coprocessor busy clear/reset
0170-0177	Secondary IDE controller
01F0-01F7	Primary IDE controller
0201	Joystick interface
0220-022F	Sound Blaster
0278-027A	LPT 2 (XT parallel port 3)
02E8-02EF	Alternate COM (4)
02F8-02FF	COM 2
0330-0331	MPU-401
0376	IDE Controller
0378-037A	LPT 1 (XT parallel port 2)
0388-038B	Frequency modulation (FM) synthesis
03B0-03BB	MDA, EGA/video graphics array (VGA)
03BC-03BE	LPT 3 (XT parallel port 1)
03C0-03DF	EGA/VGA

Table B.1 Legacy ISA hardware system I/O (PC 99) (continued)

I/O Address	Default system function
03E0-03E7	PCIC PCMCIA controllers
03E8-03EF	Alternate COM (3)
03F0-03F7	FDC
03F8-03FF	COM 1
0534-0537	Windows Sound System-compatible
0CF8-0CFB	Peripheral Component Interconnect (PCI) ports

Table B.2 Legacy hardware IRQ assignment (PC 99)

Hardware IRQ	Default assignment
IRQ 0	System timer
IRQ 1	Keyboard
IRQ 2	Second programmable interrupt controller (PIC) cascade
IRQ 3	COM 2
IRQ 4	COM 1
IRQ 5	Sometimes LPT 2—not considered fixed
IRQ 6	Standard floppy disk controller (FDC)
IRQ 7	LPT 1
IRQ 8	Real-time clock/CMOS
IRQ 9	—
IRQ 10	Sometimes COM 4—not considered fixed
IRQ 11	Sometimes COM 3—not considered fixed
IRQ 12	PC/2-style mouse
IRQ 13	Coprocessor
IRQ 14	Primary Integrated Device Electronics (IDE) controller
IRQ 15	Secondary IDE controller

Table B.3 Legacy ISA hardware DMA considered fixed (PC 99)

Hardware DMA	System function (default)
DMA 0	ISA expansion
DMA 1	
DMA 2	FDC
DMA 3	extended capabilities port (ECP) parallel port on LPT 1

Table B.3 Legacy ISA hardware DMA considered fixed

Hardware DMA	System function (default)
DMA 4	DMA controller cascading
DMA 5	
DMA 6	
DMA 7	

Over the years, various hardware developments have seen the light of day, such as MCA, EISA, PCI, and SCSI buses, with Plug and Play capability. Not all these developments have won market support.

PC 99 still supports all these "legacy" peripherals, if they are present in the system.

PCs have a Basic I/O System (BIOS) in ROM that runs when the computer is first switched on. This ROM finds the relevant operating system and runs its boot loader to start DOS, Windows 98, Windows 2000, or whatever.

Changing World

Most programs that are used nowadays run using Win32 (or at least Win16) functions. These routines do not allow direct access to I/O ports, etc. For example, when keyboard input is required, W98 or W2000 carries out the necessary operation. In the IBM-compatible environment, it interacts with the keyboard controller using techniques dating from time immemorial.

Less DOS applications are being run. In W98, NT, and W2000, these do not get direct access to the hardware. Instead, a Virtual Device Driver arbitrates access to these peripherals, allowing multiple DOS and Windows applications to run at the same time.

These two developments allow the Windows operating systems to move away from the original IBM-compatible specification. As long as the right keystroke messages get through to user programs, it does not matter where in hardware they come from. For old DOS programs that do try to access the actual hardware, virtual device drivers on faster new machines can emulate the required functionality.

The new wisdom is that the USB and IEEE 1394 (originally called FireWire) devices represent an easier way for users to connect assorted peripherals to the system, including standard system components such as a keyboard and a mouse. The idea is that any devices can be plugged in easily (even when the computer is on) and — within limits — daisy chained together. USB and IEEE 1394 devices cope with being plugged in anywhere. The Windows Plug and Play system sorts out any resource allocations on the fly.

Whether users will take to the new doctrine remains to be seen. Plugging a new USB device in will certainly be easier than dismantling a case, fiddling with jumpers, and inserting a card. The new Device Bay specification will allow easy insertion of cards within the chassis of the PC.

USB's 12Mbs is designed for slower-speed devices, such as keyboards, mice, joysticks, printers, modems, telephones, and monitors. IEEE 1394 runs faster, at 100 Mbs or more, and is more useful for devices such as scanners, storage units, and audio/visual kit, such as digital cameras.

ACPI, OnNow, and Plug and Play

New PCs aim to use less power and switch on more quickly, as well as supporting Plug and Play. Windows and new BIOS chips use the Advanced Configuration and Power Interface (ACPI) specification to help with these tasks.

ACPI-based hardware knows how to turn off or reduce power to some devices. Windows drivers can now help in this process by using the Power Management IRPs and routines, and the power management routines that are relevant for its device class. The aim of the OnNow initiative is that a PC may appear to be off, when in fact it is just "sleeping", waiting for a soft power on button to be pressed. As the operating system and drivers are already loaded into memory, it takes less time to start up the system. Devices simply need to be powered up for the user to get started. A remote control handset or incoming modem call may similarly be used to wake the system up.

Both Windows 98 and Windows 2000 can now take full use of the Plug and Play hardware facilities to allow easy device insertion and removal. Again, drivers need to help in this process, allocating and releasing resources only as instructed by the Plug and Play Manager.

PC 99 specifies that ACPI, OnNow, and Plug and Play must be supported.:

PC 99 Conformance

Microsoft can certify your hardware and software as PC 99 compatible. To do this, contact their Windows Hardware Quality Labs at `http://www.microsoft.com/hwtest/` or e-mail to `whqlinfo@microsoft.com`. When your driver passes these tests, Microsoft gives you a digital signature catalog file that you should include in your driver's INF file.

The Windows DDK includes a simplified version of Microsoft's own tests on their Hardware Compatibility Tests (HCT) CDs. Check out your driver with these tests before submitting your driver for full testing.

Appendix C

Direct Memory Access

This book does not cover Direct Memory Access (DMA) in detail. However, NT 3.51 and NT 4 driver writers will find that they must access the DMA system routines in a new way. All the same routines are there, you just have to call IoGetDmaAdapter first. This returns a pointer to a DMA_ADAPTER object. The *DmaOperations* field in there contains pointers to all the familiar routines. Call the *DmaOperations* PutDmaAdapter callback when you have finished with the DMA_ADAPTER object.

Table C.1 shows the corresponding old and new routines. Just to reiterate, there is no kernel routine directly called PutDmaAdapter. This routine is only available through the DMA_ADAPTER structure, e.g.,

```
(*DmaAdapter->DmaOperations->PutDmaAdapter)(DmaAdapter);
```

Table C.1 Old and new DMA routines

Old	New	New DmaOperations routine	Description
HalGetAdapter	IoGetDmaAdapter		
		PutDmaAdapter	Release DmaAdapter
HalAllocateCommonBuffer		AllocateCommonBuffer	
HalFreeCommonBuffer		FreeCommonBuffer	
IoAllocateAdapterChannel		AllocateAdapterChannel	
IoFlushAdapterBuffers		FlushAdapterBuffers	
IoFreeAdapterChannel		FreeAdapterChannel	
IoFreeMapRegisters		FreeMapRegisters	
IoMapTransfer		MapTransfer	
HalGetDmaAlignmentRequirement		GetDmaAlignment	Get alignment requirements for DMA buffers
HalReadDmaCounter		ReadDmaCounter	Get number of bytes remaining to be transferred
		GetScatterGatherList	
		PutScatterGatherList	

Glossary

Acronyms and Tools

\DosDevices	\??	Different names for the symbolic link directory seen by **WinObj**
AC	Alternating Current	
ACK	Acknowledge	
ACK	USB Acknowledge packet	
ACPI	Advanced Configuration and Power Interface BIOS specification	http://www.teleport.com/~acpi/
AGP	Accelerated Graphics Port	
APC	Asynchronous Procedure Call	Code that operates in the context of a specific user thread
API	Application Programmer Interface	
AML	ACPI Machine Language	
ASL	ACPI Source Language	
ATVEF	Advanced Television Enhancement Forum	HTML-based enhanced television info@atvef.com or, in Europe, dvbinfo@atvef.com
BIOS	Basic Input/Output System	
BSOD	Blue screen of death	NT and W2000 bugcheck screen

build		Command line tool for building drivers
Bulk	A USB data transfer type, for large blocks of data	
CDB	Command Descriptor Block	For SCSI commands
checked build		Nonoptimized debug build of a driver or Windows, with debug symbols
CIM	Common Information Model	WBEM model for WMI
CIMOM	CIM Object Manager	
CMIP	Common Management Information Protocol	
COM	Common Object Model	
Composite device		A USB device with more than one interface
Compound device		A piece of USB hardware with several internal USB devices
Configuration		A set of interfaces in a USB device
Control	A USB data transfer type for small blocks of data	
DATA0, DATA1	USB Data packet IDs	
DC	Direct Current	
DDC	Display Data Channel	Monitor/video channel enumeration
DDK	Driver Development Kit	
DDI	Device Driver Interface	
DebugPrint		PHD driver trace tool
default pipe		A flow of data to USB Endpoint 0
DIP	Dual In-line Package	
DIRQL	Device Interrupt Request Level	
DLL	Dynamic Link Library	
DMA	Direct Memory Access	
DMI	Desktop Management Interface	
DPC	Deferred Procedure Call	Runs at DISPATCH_LEVEL
ECP	Enhanced Centronics Port	IEEE P1284 (ECP)
endpoint		USB logical connection point
Endpoint 0		A control endpoint that is present in all USB devices
EPP	Enhanced Parallel Port	
FDO	Functional Device Object	
frame		A USB frame is the data sent in a 1ms interval
free build		Optimized release retail build of a driver or Windows

GDI	Graphics Device Interface	
GMT	Greenwich Mean Time	Kernel system time is measured in 100ns units since January 1, 1601 in the GMT time zone.
GUID	Globally Unique Identifier	
guidgen		GUID generator tool
HAL	Hardware Abstraction Layer	
HBA	Host Bus Adaptor	
HCT	Hardware Compatibility Tests	
HKLM	HKEY_LOCAL_MACHINE	
HKR	Relative registry key	Used in installation INF files. See Chapter 11 for more details.
HID	Human Input Device	
HMMP	HyperMedia Management Protocol	
hub		A USB device that has downstream ports to let further USB devices be plugged in
IDE	Integrated Development Environment	
IEEE 1394	A high speed serial bus	Originally called FireWire
IFS	Installable File System	
IHV	Independent Hardware Vendor	
I/O	Input and/or Output	
IN		Specifies that a parameter is an input to a kernel routine
IN	USB Input packet	
INF	Installation information file	
interrupt	A USB data transfer type, for small regular data such as user input	
Interface		A set of USB endpoints
IOCTL	I/O Control Code	
IRB	IEEE 1394 Request Blocks	Requests to the IEEE 1394 bus driver
IRP	I/O Request Packet	
IRQL	Interrupt Request Level	
ISA	Industry Standard Architecture PC bus	
ISR	Interrupt Service Routine	Runs at DIRQL
isochronous	A USB data transfer type, for time-critical regular data	
LED	Light Emitting Diode	
MakeDrvr		Batch file used to compile the book software from Visual Studio

makefile		Instructions to *nmake* and *build* to compile and link.
makefile.inc		Additional makefile instructions for driver build process
mof		WMI class file
mofcomp		mof file compiler
MDL	Memory Descriptor List	
MDT	Minidriver Development Tool	Printer mini-drivers
MSDN	Microsoft Developer Network	
NAK	Negative Acknowledge	
NAK	USB Negative Acknowledge packet	
NDIS	Network Driver Interface Specification	
nmake		Command line tool that uses makefiles to issue compile and link commands
nmsym		Utility to generate symbols for NuMega Soft-ICE debugger
NT	New Technology	Windows NT 3.51 and Windows NT 4
NIC	Network Interface Card	
OHCI	USB Open Host Controller Interface	
page size		The size in bytes of a page of memory, i.e., the smallest unit of virtual memory. A page can either be present in physical memory, or temporarily unavailable as it has been swapped out to disk.
OUT		Specifies that a parameter is an output from a kernel routine
OUT	USB Output packet	
PCI	Peripheral Component Interconnect	
PDO	Physical Device Object	
PE	Portable Executable EXE file format	
PHD	PHD Computer Consultants Ltd	
pid	USB Packet ID	
pipe		A flow of data from a USB endpoint
PnP	Plug and Play	
PRE		USB packet ID which high speed devices see for low speed communications

preparsed data		A HID Report Descriptor parsed into a format that is easy to use
rebase		Tool to strip out debug symbols from a driver executable
regedit		Registry editor
rededt32		Registry editor for NT and W2000 only
SCI	System Control Interrupt	ACPI interrupt
SDK	Software Development Kit	
SETUP	USB Setup packet	
SFD	SCSI Filter Driver	
SNMP	Simple Network Management Protocol	
SOF	USB Start Of Frame packet	
SoftICE		Source level debugger from NuMega
SRB	SCSI Request Block	Is translated into a CDB
STALL	USB handshake packet indicating that a USB pipe has stalled	
STI	Still Image Architecture	
TCO	Total Cost of Ownership	
UHCI	USB Host Controller Interface	
UPS	Uninterruptible Power Supply	
URB	USB Request Block	
USB	Universal Serial Bus	
VC++	Visual C++	
VxD	Virtual Device Driver	For Windows 95 and Windows 98 only
W2000	Windows 2000	
W98	Windows 98	
WinDbg		Microsoft Kernel debugger
WinObj		Windows Object viewer tool
WBEM	Web-based Enterprise Management	
WDM	Windows Driver Model	
WHQL	Windows Hardware Quality Labs	Microsoft driver tests
Windows		Used in this book to refer to Windows 98 and Windows 2000
WMI	Windows Management Instrumentation	
WQL	WBEM Query Language	

Index

H

S

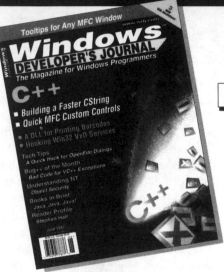

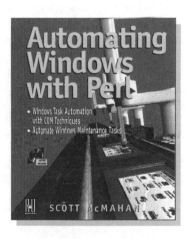

What's on the CD-ROM?

The CD-ROM that accompanies **Writing Windows WDM Device Drivers** contains full source code and executables of tools and drivers that can be employed in your own projects, including:

- DebugPrint — which lets you view the trace debug statements produced by the example drivers,
- PHDIo — a general-purpose driver that can be used to talk to simple interrupt-driven hardware on the ISA bus.

You will find that actual driver code on the CD-ROM often contains more comments than the main text listings.

Installation

You can install drivers and run the test programs direct from the CD-ROM. However, if you want to recompile any of the code you will have to copy all the software to a local hard disk.

It is strongly recommended that you make a directory called `C:\WDMBook` and copy all the files on the CD there. File `WDM Book.dsw` is a Visual Studio 97 workspace that contains projects for all the drivers and test programs in the book software.

Follow all the instructions in the book if you wish to recompile the drivers.

For more information on the CD-ROM's specifications and contents, consult `index.html` and the individual `Read.me` files on the CD.

Late-breaking information about **Writing Windows WDM Device Drivers** is also available at `www.phdcc.com/wdmbook/`